Advanced Structured Materials

Volume 165

Common engineering materials are reaching their limits in many applications, and new developments are required to meet the increasing demands on engineering materials. The performance of materials can be improved by combining different materials to achieve better properties than with a single constituent, or by shaping the material or constituents into a specific structure. The interaction between material and structure can occur at different length scales, such as the micro, meso, or macro scale, and offers potential applications in very different fields.

This book series addresses the fundamental relationships between materials and their structure on overall properties (e.g., mechanical, thermal, chemical, electrical, or magnetic properties, etc.). Experimental data and procedures are presented, as well as methods for modeling structures and materials using numerical and analytical approaches. In addition, the series shows how these materials engineering and design processes are implemented and how new technologies can be used to optimize materials and processes.

Advanced Structured Materials is indexed in Google Scholar and Scopus.

Azman Ismail · Fatin Nur Zulkipli ·
Husna Sarirah Husin · Andreas Öchsner
Editors

Advancements in Materials Science and Technology Led by Women

Springer

Editors
Azman Ismail
Centre for Women Advancement
and Leadership
Universiti Kuala Lumpur Malaysian
Institute of Marine Engineering Technology
Lumut, Perak, Malaysia

Husna Sarirah Husin
Centre for Women Advancement
and Leadership
Universiti Kuala Lumpur Malaysian
Institute of Marine Engineering Technology
Lumut, Perak, Malaysia

Fatin Nur Zulkipli
Faculty of Information Management
Universiti Teknologi MARA
Machang, Kelantan, Malaysia

Andreas Öchsner
Faculty of Mechanical Engineering
Esslingen University of Applied Sciences
Esslingen, Germany

ISSN 1869-8433 ISSN 1869-8441 (electronic)
Advanced Structured Materials
ISBN 978-3-031-21958-0 ISBN 978-3-031-21959-7 (eBook)
https://doi.org/10.1007/978-3-031-21959-7

This Springer imprint is published by the registered company Springer Nature Switzerland AG
The registered company address is: Gewerbestrasse 11, 6330 Cham, Switzerland

Preface

This book is a noteworthy series of works authored by women from diverse research areas and expertise. This book contains research papers from fundamental, experimental, and empirical studies in the fields of mechanical engineering and materials science which are included in this book series. Methods for modelling data, structures and materials using numerical and analytical techniques are described along with experimental data and methodologies.

Lumut, Malaysia	Azman Ismail
Machang, Malaysia	Fatin Nur Zulkipli
Lumut, Malaysia	Husna Sarirah Husin
Esslingen, Germany	Andreas Öchsner

Contents

1 Review: Role of Nanoparticles Growth and Alleviation of Abiotic Stress in Plants ... 1
Shamini Janasekaran, Bassem Farouk Ghorab, Mohamed Al Askari, Marawan Hisham Saad, Habib Ullah Khan, and Abdulla Haroun Idriss
1.1 Introduction ... 2
1.2 Nanoparticles of Metals ... 4
1.2.1 Gold (Au) ... 4
1.2.2 Silver (Ag) ... 5
1.2.3 Copper (Cu) ... 6
1.3 Possible Abiotic Stresses in Plants and Their Treatment with Nanoparticles ... 7
1.3.1 Drought ... 7
1.3.2 Flooding ... 8
1.3.3 Salinity ... 8
1.4 Conclusion ... 9
References ... 10

2 Performance of Fibreglass Mat and Woven for Prosthetic Leg Socket Application ... 13
Wan Nursheila Wan Jusoh, Mohd Yusoff Mohd Haris, Nurhayati Mohd Nur, Khairul Dahri Mohd Aris, Muhammad Amir Ihsan Basri, and Mohd Izhwan Maidu
2.1 Introduction ... 14
2.2 Methodology ... 15
2.2.1 Material ... 15
2.2.2 Fabrication of Composite Split-Mould ... 16
2.2.3 Fabrication of Composite Leg Socket ... 17
2.2.4 Tensile Strength Test ... 18
2.2.5 Static Failure Test ... 18
2.3 Results and Discussion ... 18

2.3.1 Composite Split-Mould ... 18
2.3.2 Composite Leg Socket ... 19
2.3.3 Tensile Strength Test Result ... 20
2.3.4 Static Failure Test Result ... 22
2.4 Conclusion ... 22
References ... 23

3 Investigation of Leakage in Underground Pipelines Using the Radiotracer Technology ... 25
Noraishah Othman, Nolida Yussup, Ahmad Zubair Johan, Iqmal Zulkarnain Abdul Rahman, and Roslanzairi Mostapa
3.1 Introduction ... 26
3.2 Methodology ... 27
3.2.1 Development of Radiotracer Capsule ... 27
3.2.2 Radiotracer Experiments ... 28
3.3 Results and Discussion ... 28
3.3.1 Radiotracer Capsule ... 28
3.3.2 Radiotracer Experiments ... 29
3.4 Conclusion ... 32
References ... 33

4 General Characteristics of Endophytes and Bioprospecting Potential of Endophytic Fungi ... 35
Md Abu Taher, Woei-Yenn Tong, Chean Ring Leong, Syarifah Ab Rashid, and Wen-Nee Tan
4.1 Introduction ... 36
4.2 Characteristics of Endophytes ... 37
4.2.1 Origin ... 37
4.2.2 Distribution ... 37
4.2.3 Host Plant Interaction ... 38
4.3 Endophytic Fungi ... 39
4.4 Natural Products from Endophytic Fungi ... 41
4.4.1 Antimicrobial Compounds ... 42
4.4.2 Anticancer ... 42
4.4.3 Antidiabetic ... 43
4.4.4 Antioxidant ... 44
4.4.5 Immunosuppressive ... 44
4.5 Conclusion ... 46
References ... 46

5 Biochar-Based Graphitic Carbon Nitride Derived from Biomass Waste for Degradation of Pyrene ... 51
Noor Aina Mohd Nazri, Syarifah Nurhusna Qadirah Syed Abd Halim, and Sabrina Karim
5.1 Introduction ... 52
5.2 Methodology ... 53

5.3 Results and Discussion 54
5.4 Conclusion 59
References 61

6 **Response Surface Methodology for the Optimisation of a 96-Well Microtitre Plate Coagulation Activity Assay** 63
Farah Salma Baharudin, Mussarat Saleem, and Robert Thomas Bachmann
6.1 Introduction 64
6.2 Methodology 65
6.2.1 Synthetic Turbid Water Samples 65
6.2.2 Preparation of Coagulant Solutions 65
6.2.3 Development of the 96-Well MTP Coagulation Activity Assay 66
6.2.4 Experimental Design using RSM 67
6.2.5 Coagulation Activity of *M. Oleifera* Coagulant 69
6.2.6 Statistical Analysis 69
6.3 Results and Discussion 69
6.3.1 Statistical Analysis 69
6.3.2 Process Analysis of the MTP Coagulation Activity Assay 72
6.3.3 MTP Coagulation Activity Assay Optimisation and Model Verification 74
6.3.4 MTP Coagulation Activity Assay with Crude *M. Oleifera* Seed Extracts 76
6.4 Conclusion 77
References 77

7 **Nutrient Recovery from Anaerobic Palm Oil Mill Effluent Using Palm Kernel Shell Biochar and Deoiled Spent Bleaching Earth and Their Effect on Oil Palm Growth** 81
Nur Maisarah Ahmad, Nurul Shafinah Foo, Soh Kheang Loh, Khaliesah Abbas, Siew Kooi Ong, Nazatulshima Hassan, Abbas Fadhl Mubarek Al-Karkhi, Mustapha Mohammed Bello, and Robert Thomas Bachmann
7.1 Introduction 82
7.2 Methodology 84
7.2.1 Materials 84
7.2.2 Preparation of Nutrient-Enriched Sorbents 84
7.2.3 Nursery Trial 85
7.2.4 Analytical Techniques 88
7.3 Results and Discussion 89
7.3.1 Nutrient Recovery 89
7.3.2 Effect of Nutrient-Enriched Sorbents on Soil Properties 93

7.3.3 Effect of Nutrient-Enriched Sorbents on Growth of Oil Palm Seedlings ... 96
7.3.4 Comparison of Soil Treatment Performance ... 98
7.3.5 Plant Health by Leaf Colour Data Analysis ... 100
7.4 Conclusion ... 101
References ... 101

8 Red Tilapia By-Product (*Oreochromis Sp.*) Hydrolysates: Bioactivities, Bioprocessing, and Potential Applications ... 105
Emmy Liza Anak Yaji, Nur Suraya Abdul Wahab, Norfahana Abd-Talib, Mohammad Zulkeflee Sabri, Kelly Yong Tau Len, Nadia Razali, and Khairul Faizal Pa'ee
8.1 Introductions to Red Tilapia Production and Its Processing By-Products ... 106
8.2 Bioactive Peptides ... 109
8.2.1 Antioxidant Peptides ... 109
8.2.2 Anticancer Peptides ... 112
8.2.3 Antimicrobial Activity ... 112
8.2.4 Angiotensin I-Converting Enzyme (ACE)-Inhibitory Peptides ... 113
8.3 The Production of ACE-Inhibitory Peptides ... 115
8.3.1 Microbial Fermentation ... 116
8.3.2 Chemical Hydrolysis ... 117
8.3.3 Enzymatic Hydrolysis ... 117
8.4 Purification of Peptides ... 118
8.4.1 Ultrafiltration (UF) ... 120
8.4.2 Chromatographic Approach ... 121
8.5 Red Tilapia By-Product as Protein Precursor for ACE-Inhibitory Peptides ... 123
8.5.1 Amino Acid Compositions of Tilapia By-Products ... 125
8.6 Future Outlook ... 126
8.7 Conclusions ... 128
References ... 128

9 Microhardness and Process Parameter Optimization of Friction Stir Welding on an AA5052 Thin Plate ... 133
Bakhtiar Ariff Baharudin, Mazli Mustapha, Azman Ismail, Fatin Nur Zulkipli, Fauzuddin Ayob, and Azlan Ahmad
9.1 Introduction ... 134
9.1.1 FSW Developments ... 134
9.2 Methodology ... 136
9.2.1 Taguchi's Method ... 136
9.2.2 ANOVA Analysis ... 136
9.2.3 S/N Ratio ... 137
9.2.4 Experimental Procedure ... 137
9.2.5 Microhardness Measurement ... 137

9.3 Results and Discussion 138
9.4 Conclusion 140
References 140

10 Residual Stress Optimization of Friction Stir Welding on an AA5052 Thin Plate 143
Bakhtiar Ariff Baharudin, Mazli Mustapha, Azman Ismail, Fatin Nur Zulkipli, Fauzuddin Ayob, and Azlan Ahmad
10.1 Introduction 144
10.1.1 Development of FSW 145
10.2 Methodology 146
10.2.1 Selection of Orthogonal Arrays 146
10.2.2 ANOVA Analysis 147
10.2.3 Experimental Procedure 147
10.2.4 Residual Stress Measurement 147
10.3 Results and Discussion 148
10.4 Conclusion 150
References 151

11 The Effects of Fly Ash and Aluminium Trihydrate Incorporation on the Tensile and Thermal Properties of Epoxy Resin Mixtures 153
Wan Nursheila Wan Jusoh, Muhd Zulfadhli, Nurafiqah Naqiyah Khalid, Annur Ashran, Muhammad Hafiz, Mohamad Firzani, Rashidi Rahim, Syed Nur Azman Syed Mustaffa, and Norlaila Ramlee
11.1 Introduction 154
11.2 Methodology 155
11.2.1 Preparation of Sample 155
11.2.2 Thermogravimetric Analysis 155
11.2.3 Tensile Strength Test 156
11.2.4 Vertical Flame Test 156
11.3 Results and Discussion 157
11.3.1 Thermogravimetric Analysis 157
11.3.2 Tensile Strength Test 159
11.3.3 Vertical Flame Test 159
11.4 Conclusion 162
References 163

12 Modeling and Simulation of Planar Micro-coils for Invasive Pressure Sensing 165
Norliana Yusof, Syamimi Mohd Norzeli, Siti Nurul Akmal Yusof, Noor Hidayah Mohd Yunus, and Norhayati Soin
12.1 Introduction 166
12.2 Methodology 167
12.3 Results and Discussion 168

12.4 Conclusion 171
References 172

13 Review on Electrochemical and Biosensors and Their Application 173
Norilhamiah Yahya and Nur Afifah Mat Razali
13.1 Introduction 174
13.2 Classification of Sensors 175
13.2.1 Sensor 175
13.2.2 Biosensor 175
13.2.3 Biosensors Based on Bioreceptors 177
13.2.4 Biosensors Based on Transducer 179
13.3 Electrochemical Sensors 180
13.3.1 Potentiometric Biosensors 180
13.3.2 Amperometric Biosensors 185
13.3.3 Conductometric Biosensors 186
13.4 Application of Electrochemical Sensors 187
13.4.1 Electrochemical Biosensors for Glucose Detection 187
13.4.2 Electrochemical Biosensors for Medical Diagnosis 189
13.4.3 Electrochemical Biosensors for Environmental Analysis 191
13.4.4 Electrochemical Biosensors for Food Monitoring 192
13.5 Conclusion 193
References 194

14 Development of a Wireless Solar Power Transmission for Battery Chargers 199
Nur Amirah Liyana Abdul Hadi, Noor Hidayah Mohd Yunus, and Mohd Shahrul Mohd Nadzir
14.1 Introduction 200
14.2 Methodology 201
14.2.1 Project Development 201
14.2.2 Hardware Development 203
14.3 Results and Discussion 205
14.4 Conclusion 206
References 207

15 Short-Term Study on the Potential of Oil Palm Frond Biochar for Acid Sulphate Soil Amelioration 209
Amelia Md Som, Abdul Addahary Abdul Yahya, Padmini Karananidi, and Robert Thomas Bachmann
15.1 Introduction 210
15.2 Methodology 212
15.2.1 Soil Location 212
15.2.2 Oil Palm Frond Collection 212
15.2.3 Preparation of Oil Palm Frond Biochar 212
15.2.4 Preparation of Soil Sample and Incubation 213

15.2.5 pH and Conductivity Measurements ... 213
15.2.6 Particle Size Distribution ... 214
15.2.7 Water-Soluble Sulphate ... 214
15.2.8 Statistical Study ... 214
15.3 Results and Discussion ... 214
15.3.1 Characterization of Soil and Oil Palm Frond Biochar ... 214
15.3.2 Effect on Soil pH ... 215
15.3.3 Effect on Soil Electrical Conductivity ... 216
15.4 Conclusion ... 217
References ... 218

16 The Stress Analysis of the Jack Rod Crankshaft on a Single-Cylinder Engine: A Study Using the Finite Element Method ... 221
Eida Nadirah Roslin, Ikhsan Wafiy Ishak, Mohd Zaki Bahrom, and Hasan Muhamad Abid Hasan
16.1 Introduction ... 221
16.2 Methodology ... 223
16.2.1 Analytical Calculation ... 223
16.2.2 Models of Single-Cylinder Engine Crankshaft ... 226
16.2.3 Material Properties of Crankshaft ... 226
16.2.4 Crankshaft Meshing ... 227
16.2.5 Apply Load and Boundary Conditions ... 228
16.3 Results and Discussion ... 228
16.3.1 Deformation ... 228
16.3.2 Strain ... 229
16.3.3 Stress ... 230
16.4 Conclusion ... 232
References ... 233

17 Antimicrobial Peptides, An Alternative Antimicrobial Agent Against Multi-drug-Resistant Microbes: Source, Application, and Potential ... 235
Nur Husna Hafiza Lyana Ramzah, Tong Woei Yenn, Wing-Hin Lee, Ching-Yee Loo, Wen-Nee Tan, and Leong Chean Ring
17.1 Introduction ... 236
17.2 Source, Mechanism of Action, Application, and Potential of Antimicrobial Peptides ... 237
17.2.1 The Source of AMP and Diversity ... 237
17.2.2 AMP Mechanisms of Action ... 240
17.2.3 The Applications of AMP ... 245
17.3 The Clinical Potential of Antimicrobial Peptides ... 251
17.4 Conclusion ... 252
References ... 253

Chapter 1
Review: Role of Nanoparticles Growth and Alleviation of Abiotic Stress in Plants

Shamini Janasekaran, Bassem Farouk Ghorab, Mohamed Al Askari, Marawan Hisham Saad, Habib Ullah Khan, and Abdulla Haroun Idriss

Abstract The changing environment and advancement in the field of agriculture haves allowed the technological experts to introduce smart and efficient technologies. The experts whether they belong to computer science or agriculture are striving to increase the level of productivity with an intention to balance the uprising demands of world's population. The introduction of nanotechnology has enhanced the production level remarkably by offering technologically advanced equipment. Moreover, agriculture nanotechnology has made it possible to enhance the growth of plants meanwhile lowering the concerns that may arise from biotic and abiotic stresses. Therefore, to identify the available nanoparticles used in the field of agriculture and the alleviation of abiotic stresses through the advantageous properties of nanoparticles, a critical analysis of the role of nanoparticles making contribution in the field of agriculture has been presented.

Keywords Agriculture · Nanotechnology · Nanoparticles · Abiotic stress · Plants growth

---

S. Janasekaran (✉) · B. F. Ghorab · M. Al Askari · M. H. Saad · H. U. Khan · A. H. Idriss
Centre for Advanced Materials and Intelligent Manufacturing, Faculty of Engineering, SEGi University, Built Environment and IT, 47810 Petaling Jaya, Selangor, Malaysia
e-mail: shaminijanasekaran@segi.edu.my

B. F. Ghorab
e-mail: bassem.ghorab@gmail.com

M. Al Askari
e-mail: m7md_askari@hotmail.com

M. H. Saad
e-mail: marawanismailx@gmail.com

H. U. Khan
e-mail: shinay.98@hotmail.com

A. H. Idriss
e-mail: aboody3848@hotmail.com

A. Ismail et al. (eds.), *Advancements in Materials Science and Technology Led by Women*, Advanced Structured Materials 165,
https://doi.org/10.1007/978-3-031-21959-7_1

1.1 Introduction

Agriculture is considered to be an important sector in the world's economy as it feeds living organisms including human beings and animals. The products related to the field of agriculture are comprised of diverse items such as food, textiles, fuel, and furniture (Alston and Pardey 2014). All of the mentioned items are mainly consumed by human beings, whereas the food produced through agricultural means is also consumed by other living organisms. However, factors such as change in the climate, limited space, and emerging plants diseases has challenged the productivity of agricultural sector (Gupta et al. 2018). According to experts, the only possible solution to the agricultural challenges related to the crop's growth rate is the adoption of latest technology (Ku 2019). The development in the agriculture sector can be maintained and further accelerated through newly introduced innovative and smart techniques such as nanotechnology.

The concept of "nanotechnology" was initially introduced by a group of experts including Professor Norio Taniguchi in 1974 (Sandhu 2006). The invention of nanotechnology allowed the experts to introduce an atomic force microscope and scanning tunneling microscope for the first time (Bayda et al. 2019). The additional developments in the field of nanotechnology resulted in advanced arrangements to isolate and demonstrate nanomaterials in a particular and precise way (Thiruvengadam et al. 2018). The remarkable and unique properties of nanoparticles are currently considered by numerous fields such as medical science, engineering, pharmaceuticals, electronic, and agriculture. By definition, any material that is less than 100 nm in size refers to a nanoparticle (NP). According to the European Commission 2018, nanomaterial refers to "a natural, incidental or manufactured material comprising of particles, in an unbound state or as an aggregate or as an agglomerate and where, for 50% or more of the particles in the number size distribution, one to more external dimensions is in the size range 1–100 nm." The basic physical and chemical properties of nanoparticles are dissimilar from those of corresponding substances available in bulk amount (Khan et al. 2019). The studies have provided that experts belonging to biochemical field are actively involved in the synthesis of metal, inorganic, organic, and other hybrid nanoparticles through various biological, physical, and optical activities (Khandel et al. 2018). The methods through which nanoparticles are synthesized or created include two basic approaches. The first approach is the top-down approach, whereas the other refers to bottom-up approach (Kumar and Kumbhat 2016).

The number of commercial nanoparticles available are around 1300 comprising of various applications (Jeevanandam et al. 2018). Additionally, the synthesis of nanoparticles on the basis of diverse chemical and physical properties is considered by numerous fields particularly in agriculture. The nanoparticles used in the field of agriculture are comprised of engineered particles or can be belong to natural sources

(Prasad et al. 2017). The engineered nanoparticles used in agriculture are divided into organic and inorganic and may be of combined materials. Moreover, metals and metal oxides have wide range of applications. The physical characteristics such as shape and size differentiate nanomaterials from each other. In agriculture, experts have reported that engineered nanoparticles are found to be helpful in enhancing the growth of plants (Miralles et al. 2012). The traditional methods such as use of fertilizers and other pesticides are considered to improve the growth in plants and to increase crop yield. However, studies have presented life-threatening adverse effects of traditional methods used for the enhancement of crop yield. Therefore, it has become a need to introduce an innovative and sustainable technology like nanotechnology in the field of agriculture, to eliminate the serious concerns related to the life and to enhance the level of productivity. The nanotechnology has a number of applications that can be used in the field of agriculture for the purpose of enhancing the level of productivity (Pramanik et al. 2020). Adding to this, the applications of nanotechnology include solution to crop protection, identification of toxicity in the soil, genetic manipulation in plants, and diagnosis of diseases in plants (ShangY et al. 2019).

Meanwhile, the Population Institute (Spang et al. 2019) has argued that the increase in population has also increased the demand for food. Conversely, the concept of civilization, industrialization, and pollution has damaged the soil and has adversely affected the agricultural production at global level. The plants growth and crop productivity have also suffered from abiotic factors such as drought, salt, and soil acidity, and extreme temperature. Moreover, plants damaged by abiotic factors are easily attacked by pests and other insects, due to which the quality of outcome has suffered, whereas the excessive use of fertilizers comprising of chemical substances has presented unexpected results in the form of hurting the soil and physiology of plants. However, in the agriculture field and food industry, the applications of nanoparticles are used as nano-fertilizers for various purposes such as for the protection of crop and distribution of nutrients and for the protection of food-related products (Achari and Kowshik 2018). The literature has further supported that the use of nanotechnology is benefiting the agricultural by increasing the crop production. However, the literature has also discussed that due to the changing properties of nanoparticles some of the nanoparticles do not only present beneficial results but can also adversely affect the growth of plants. The studies have shared that nanoparticles can produce oxidative stress and reactive oxygen species upon interaction with cellular mechanisms and other enzymatic systems in plants (Dolez 2015). While reacting to lethal metals, due to oxidative stress, some changes related to biochemical, molecular, and morphophysiological in plants occur. Therefore, in order to surpass the effect of toxic pollutants plants naturally generate protective mechanism. Furthermore, toxic nanoparticles in excessive amount are considered to be harmful to the overall health of plants. Adding to this, the toxicity also damages the structure of plant and adversely affects the photosynthetic mechanism in plants, while on the other hand, a group of researchers have provided that nanoparticles of some metals such as titanium dioxide (TiO_2) improve the efficiency of photosynthetic mechanism in plants (Hasanpour et al. 2015). This shows that nanoparticles on the basis of their physical as well as chemical properties stimulate the growth process in plants.

1.2 Nanoparticles of Metals

The considered nanoparticles are comprised of numerous metals such as gold, silver and copper and metalloids. On the basis of literature, the discussed nanoparticles from various metals are found to be effective in supporting the overall growth in plants.

1.2.1 Gold (Au)

A group of researchers in an experiment considered seeds of the Arabidopsis thaliana plant and presented that the nanoparticles of gold (AuNPs) around 24 nm at a dose of 10 μg/mL provided a three times high productivity in the yield of seed as compared to the controlled germination rate. Furthermore, the application of nanoparticles of gold ranging between 10 and 80 μg/mL allowed the researcher to observe a free radical scavenging movement, by which microRNAs presented a correlation among germination of seed, growth of plant and possible antioxidant potential of plants on the nanoparticles of gold exposure (Kumar et al. 2013).

Moreover, another group of researchers delivered numerous nanoparticles of gold ranging between 30 and 80 nm to the plant of watermelon through the aerosol technique. They observed that the plant of watermelon absorbed the nanoparticles of gold (AuNPs). Basically, the nanoparticles were taken up by the plant through direct diffusion and by stomatal opening and translocating from the leaf of plant to its root by means of phloem transportation mechanism. Additionally, a definite stimulation of the root of watermelon plant while treating with gold nanorods was presented by Wang et al. (2013). Moreover, high concentration of nano-applications of gold presented phytotoxic effects, and an oxidative stress was noticed. This shows a harmful effect of nanoparticles of gold when used in high concentration or beyond normal levels.

Furthermore, a constructive charge of nanoparticles of gold (AuNPs) ranging between 6 and 10 nm was actively absorbed by the roots of various plants including radish, rice, and pumpkin. Additionally, the researcher observed a level of efficiency against the negative charge of nanoparticles of gold during the process of translocation to stems and leaves of plants. The experiment further presented an accumulation of 1.1–2.9 ng/mg of nanoparticles of gold in the shoots of rice plants (Zhu et al. 2012).

The spraying of nanoparticles of gold (AuNPs) over the leaves of Brassica juncea plant presented a positive growth in the physical structure of plant such as gain in the height of plant was observed, including positive change in the diameter of stem, number of leaves and branches also increased per plant, and particularly, an increase in the seed yield was observed. Also, the researchers suggested a dose of 10 ppm of nanoparticles of gold as an effective quantity (Arora et al. 2012). The results of mentioned study were also supported by a group researcher who observed an increase in the length of root, while applying the nanoparticle of AuNPs to the Brassica juncea plant, whereas the applied dose of nanoparticles ranged between 0 and 400 mg/l. Conversely, a minor decrease in the growth of roots and length of shoots of the Oryza sativa plant was observed when a varying dose of 0 mg/l to 2000 mg/l was applied (Ndeh et al. 2017). Adding to this, a decline in the germination of seed, and shortening in the roots and shoots of Artemisia absinthium plant was observed, when a dose of 30 mg/l of AuNPs was provided to the plant during the experiment (Hussain et al. 2017).

1.2.2 Silver (Ag)

The silver nanoparticles (AgNPs) when treated with cucumbers at an absorption level less than 200 mg/l, an enlargement in the roots of cucumber plants has been observed. Meanwhile, the experts have observed that silver nanoparticles when absorbed at a level below 5 mg/l by wheat and radish plants are found to be toxic at the stage of vegetation growth rather than the seed germination stage (Cui et al. 2014). The AgNPs when exposed to the flowers of cucumber plants presented a massive figure for male flower, an increase in the period of flowering, and male node (Golabadi et al. 2015). The authors have further presented an increase in the number of female flowers in cucumber plants found at fifteen-leaf stage. Moreover, the aerosol of silver nanoparticles with 200 ppm at the same stage resulted in a huge number of male flowers. On the contrary, Yan and Chen 2019, argued that the applications of nanoparticles of silver at concentration level of 100 mg/l extended the growth of Napa cabbage plant which is scientifically known as Brassica rapa ssp. Pekinensis (Yan and Chen 2019). Meanwhile, beyond a concentration level of 250 mg/l, phytotoxic effects of silver nanoparticles are noticed (Baskar et al. 2015). Additionally, remarkable figures have been observed in the form of seeds per pod, biological yield, including pods per plant while treating the aerosol of silver nanoparticles ranging between 10 and 100 nm on pea seeding (Mehmood and Murtaza 2017). Adding to this, a group of researchers have observed an enhanced development and crop yield by considering seed soaking and aerosol comprising of biosynthesized nanoparticles of silver ranging between 10 and 100 nm (Rashidi et al. 2016). Furthermore, coating of maize seeds with silver nanoparticles with 10 ppm dose presented an enhanced level of growth, including a positive change in the biochemical and physiological characteristics of maize seedlings (Mahakham et al. 2016).

The low range of silver nanoparticles ranging between 10 and 40 ppm reported a significant increase in rice including improved growth of shoot and root of the seedlings (Gupta et al. 2018). Besides, exposure of seeds of chickpea with silver nanoparticles also reported a significant germination of seed, length of seedling, and improved mechanism of enzymes in Cicer arietinum (Anusuya and Banu 2016). The experts have also observed that the biomass of inherently transmuted hairy root mechanism of Cucumis anguria strongly improved in the presence of AgNPS (Chung et al. 2018).

1.2.3 Copper (Cu)

The application of copper nanoparticles through aerosol presented an improved growth of plants such as maize. Adhikari et al. 2016 observed a significant growth in maize plant caused by copper nanoparticles through spraying. They further noticed in the study that the nanoparticles of copper are able to access to the plant cell because of easy adaption by plants and resulted in a significant growth in the maize plant (Adhikari et al. 2016). The dose of copper microparticles and copper nanoparticles ranging between 0 and 200 mg Cu/kg improved the water content from 6.9 to 12.5% in the plants known as Origanum vulgare, meanwhile, reporting a decrease in the biomass (Du et al. 2018). Adding to this, a group of researchers have claimed an increase in the number of grains and spike dimension while exposing seeds of wheat to 25 ppm copper nanoparticles on the basis of proteomic and biological examination. They further presented that seeds treated with copper nanoparticles (CuNPs) show a significant stress tolerance in wheat seeds mediated by the starch degradation process. The 20 μg/mL dose of biosynthesized nanoparticles of copper is found to increase mitotic index in cell division of Allium cepa species of plants (Nagaonkar et al. 2015).

Similarly, a study presented by López-Vargas et al. 2018 reported that treating tomato plants with copper nanoparticles not only increases the level of production but also improves firmness in fruit. The study further presented that nanoparticles of copper increases the level of vitamin C, meanwhile decreases glutathione peroxidase (López-Vargas et al. 2018). Similarly, it has been found that biogenetic nanoparticles of copper comprising of 20 nm in size present a significant positive change in the height and length of root of pigeon pea seedlings (Shende et al. 2017). Furthermore, a study reported an increases yield in tomato fruit due to the application of copper nanoparticles. The study further claimed an enhancement in the nutraceutical properties of tomato plants and fruits in the form of improved numbers of leaves, including the overall biomass of the roots of tomato plants. Moreover, high content of lycopene and antioxidant capacity is observed, when a dose of copper nanoparticles comprising of 0.02 mg is provided to tomato plants (Hernández et al. 2017).

Additionally, agriculture grafting is considered to be a cultivation method that permits plants to utilize water and other healthy nutrients in an efficient way. The copper nanoparticles with chitosen polyvinyl alcohol hydrogels enhances the overall growth of grated fruits such as watermelon by making changes in micromorphology of leaf. Also, it increases the width of stoma, length of root, and primary stem. This collaboration of copper nanoparticles with mentioned hydrogels increases the growth in plants (Arif et al. 2018).

1.3 Possible Abiotic Stresses in Plants and Their Treatment with Nanoparticles

The abiotic stress over plants mainly includes environmental conditions that adversely influence the growth and overall level of production. Plants are exposed to various abiotic stresses such as drought, flooding, salinity, heat, cold, UV light, soil acidity, and lethal metal pollutants which are considered to be abiotic stresses and are risk factors to plants and ultimately to food. The adverse effects of abiotic stresses could influence the quantity and composition of inferior metabolites in crops and alter the nutritional facts and beneficial characteristics to health (Masarovičová and Kráľová 2019). However, it is considered to be a challenge to protect the plants from abiotic stresses, but the advancement in the technology has made it possible to minimize the harmful effects to abiotic stresses through nanotechnology.

1.3.1 Drought

Drought is considered to be the physiological state of water deficit. Furthermore, in agriculture drought refers to the area where the availability of soil water for irrigation purposes is in inadequate form, due to which the metabolism of plants adversely influences and present harmful effects to the productivity of crops. The use of foliar application of zinc nanoparticles under drought stress caused by ineffective irrigation during the flowering season of wheat plants presents significant impact on drought. The studies have provided beneficial impact of zinc nanoparticles in the form of increased grain yield that get affected by the deficit of water. Adding to this, zinc nanoparticles comprising of 2 g/l dose in flowering period of plants was considered to be an efficient solution to the abiotic stress known as drought (Ghassemi and Farahvash 2018). Furthermore, nanoparticles of cerium dioxide presented a positive impact in improving the leaf carbon adjustment, germination of pollen and finally seed yield in plants particularly in abiotic stresses caused by drought (Djanaguiraman et al. 2018). The combined treatment of applications of nanoparticles of iron and salicylic acid could provide high quality and increased quantity in vitro culture of strawberries that could be considered for the adaption of strawberry plants to drought prior to the transplantation of strawberry plants in the field (Mozafari et al. 2019). Moreover, studies have provided that the foliar treatment of cotton plants with the nanoparticles of titanium oxide and silicon dioxide enhanced the total soluble contents of sugar, phenolics, pigment, proteins, proline, and free amino acids which were actually influenced by the drought stress (Shallan et al. 2016). Moreover, the nanoparticles of mentioned metals and metalloids contributed in improving the yield-related characteristics in cotton plants which were decreased by the drought stress.

1.3.2 Flooding

The nanoparticles of silver were used for the treatment of soybean plants that under the flooding stress slower their growth. Basically, flooding is an abiotic stress that normally occurs due to the run-off of water bodies such as rivers and lakes. The early exposure of soybean plants to flooding was treated with silver nanoparticles comprising of 15 nm and resulted in the growth of soybean plants. Additionally, plants treated in flooding stress with silver nanoparticles presented a small quantity of cytotoxic by-products. The studies have also discussed that the better developmental performance of silver nanoparticles in soybean plants under the situation of flooding stress could be related to the lower deprivation stress of oxygen]48]. Also, fields influenced by flooding stress make it difficult for plants to respire because of the excessive amount of water in the soil, deposition of iron plaque on the roots of plants influences the fate of nanoparticles comprising of metals. Like, treatment with the nanoparticles of copper comprising of 100 mg/l, the presence of iron plaque significantly decreased the copper content in the roots of rice plants and also in the shoots by 89% and 78%, respectively (Peng et al. 2018). Conversely, it has been found that the treatment of soil affected by the waterlogging stress with the nanoparticles of copper oxide and titanium dioxide negatively influenced the microbes, composition, and variety of microbial community. Meanwhile, it has been presented that nanoparticles of copper oxide possess a strong adverse and toxic feature that could harm the soil, whereby the nanoparticles of copper oxide may affect the microbes available in soil by altering the nutrient bioavailability (Xu et al. 2015).

1.3.3 Salinity

The excessive amount of salt in soil degrades its value with respect to crop productivity, and also, it is considered to be harmful to the overall growth of plants. Furthermore, research has provided that the soil salinity is considered to be a main environmental issue in reducing crop productivity at global level, meanwhile threatening the sustainability of agriculture. The crops such as rice, maize, and wheat are found to be prominent crops in feeding the population of world; however, the mentioned crops are exposed to salinity stress (Majeed and Muhammad 2019). The adverse effects of salt include reduction in the soil water level, imbalance related to the ion, and severe toxicity caused by salinity affects seed germination, growth of plants, photosynthesis process, imbalance in nutrients, water relation, crop quality and productivity, and oxidative stress.

In cotton plants, the adverse effects of salinity can be treated with application of zinc nanoparticles comprising of dosage of 200 ppm. Furthermore, studies have suggested diluted seawater for the irrigation of salty fields, although collaboration of phosphorus fertilizer is required to avoid phosphorus and zinc imbalance (Hussein and Abou-Baker 2018). Furthermore, in an experiment, the researchers treated salinity with zinc oxide nanoparticles, which resulted in increased growth of plants such as Lupinus termis. The results of the study presented an enhanced level of photosynthetic pigments, phenols, organic solutes, and ascorbic acid. Adding to this, the nanoparticles of zinc oxide provided best results in the process of salt treatment, whereas the considered level of dosage of nanoparticles of zinc oxide was comprised of 60 mg/l (Abdel Latef et al. 2017). In case of maize, plants grown under the abiotic stress of salinity were applied with zinc oxide nanoparticles and ferric oxide nanoparticles in the form of aerosol, which enhanced the root growth in maize plants. Contrary to this, in an experiment the field experts sprayed wheat plants grown in salinity stress with zinc oxide and ferric oxide nanoparticles in bulk. The results presented plants with enlarged heights due to the involvement of ferric oxide nanoparticles in the experiment. They on the basis of results concluded that the aerosol of nanoparticles may not produce commanding results; therefore, they suggested the use of nanoparticles in bulk form for the purpose of alleviating the issue of soil salinity (Fathi 2017).

Moreover, the use of nanoparticles of titanium dioxide did not present promising impact in seed germination; however, an enlargement in the root of plant and stem was noticed while applying to Zea mays plant.

1.4 Conclusion

The role of agriculture is very crucial in strengthening the world's economy. The agriculture sector is considered to be the only source of providing quality food for the growing world's population. Therefore, challenges are always being there to interrupt the continuous supply of food coming from agriculture. However, innovative and smart technology has made it possible to overcome the challenges concerned to productivity of crop and growth of plants. The introduction of nanotechnology in the field of agriculture has assisted in meeting the demands related to food. However, there is still a lot to discover in the field of nanotechnology as there are numerous controversies on the basis of which nanotechnology is striving to find the answers regarding to the concerned disagreements. But, the role of nanoparticles used in the field of agriculture has allowed the world to overcome alarming issues concerning to the life of human beings.

Acknowledgements This research was supported by SEGi University, grant number: SEGiIRF/2022-Q1/FoEBEIT/001.

References

Abdel Latef AAH, Abu Alhmad MF, Abdelfattah KE (2017) The possible roles of priming with ZnO nanoparticles in mitigation of salinity stress in lupine (Lupinus termis) plants. J Plant Growth Regul 36(1):60–70

Achari GA, Kowshik M (2018) Recent developments on nanotechnology in agriculture: plant mineral nutrition, health, and interactions with soil microflora. J Agric Food Chem 66(33):8647–8661

Adhikari T, Sarkar D, Mashayekhi H, Xing B (2016) Growth and enzymatic activity of maize (Zea mays L.) plant: solution culture test for copper dioxide nano particles. J Plant Nutr 39(1):99–115

Alston JM, Pardey PG (2014) Agriculture in the global economy. J Econ Perspect 28(1):121–146

Anusuya S, Banu KN (2016) Silver-chitosan nanoparticles induced biochemical variations of chickpea (Cicer arietinum L.). Biocatal Agric Biotechnol 8:39–44

Arif N, Yadav V, Singh S, Tripathi DK, Dubey NK, Chauhan DK, Giorgetti, L (2018) Interaction of copper oxide nanoparticles with plants: uptake, accumulation, and toxicity. In Nanomaterials in plants, algae, and microorganisms. Academic Press, pp 297–310

Arora S, Sharma P, Kumar S, Nayan R, Khanna PK, Zaidi MGH (2012) Gold-nanoparticle induced enhancement in growth and seed yield of Brassica Juncea. Plant Growth Regul 66(3):303–310

Baskar V, Venkatesh J, Park SW (2015) Impact of biologically synthesized silver nanoparticles on the growth and physiological responses in Brassica rapa ssp. pekinensis. Environ Sci Pollut Res 22(22):17672–17682

Bayda S, Adeel M, Tuccinardi T, Cordani M, Rizzolio F (2019) The history of nanoscience and nanotechnology: from chemical–physical applications to nanomedicine. Molecules 25(1):112

Chung IM, Rajakumar G, Thiruvengadam M (2018) Effect of silver nanoparticles on phenolic compounds production and biological activities in hairy root cultures of Cucumis anguria. Acta Biol Hung 69(1):97–109

Cui D, Zhang P, Ma YH, He X, Li YY, Zhao YC, Zhang ZY (2014) Phytotoxicity of silver nanoparticles to cucumber (Cucumis sativus) and wheat (Triticum aestivum). J Zhejiang Univ-Sc A 15(8):662–670

Djanaguiraman M, Boyle DL, Welti R, Jagadish SVK, Prasad PVV (2018) Decreased photosynthetic rate under high temperature in wheat is due to lipid desaturation, oxidation, acylation, and damage of organelles. BMC Plant Biol 18(1):1–17

Dolez P (2015) Nanoengineering: global approaches to health and safety issues. Elsevier

Du W, Tan W, Yin Y, Ji R, Peralta-Videa JR, Guo H, Gardea-Torresdey JL (2018) Differential effects of copper nanoparticles/microparticles in agronomic and physiological parameters of oregano (Origanum vulgare). Sci Total Environ 618:306–312

European commission (2018) Definition of a nanomaterial. Retrieved from https://ec.europa.eu/growth/sectors/chemicals/reach/nanomaterials_en

Fathi A (2017) Effect of phosphate solubilization microorganisms and plant growth promoting rhizobacteria on yield and yield components of corn. Sci Agric 18(3):66–69

Ghassemi A, Farahvash F (2018) Effect of nano-zinc foliar application on wheat under drought stress. Fresen Environ Bull 27(7):5022–5026

Golabadi M, Golkar P, Rajabi R (2015) Effects of nano-silver particles on flower expression in greenhouse cucumber (Cucumis sativus L.). Crops Res 49(1–3):48–54

Gupta S, Sharma D, Gupta M (2018) Climate change impact on plant diseases: opinion, trends and mitigation strategies. Microbes for climate resilient agriculture, pp 41–56

Hasanpour H, Maali-Amir R, Zeinali H (2015) Effect of TiO2 nanoparticles on metabolic limitations to photosynthesis under cold in chickpea. Russ J Plant Physiol 62(6):779–787

Hernández HH, Benavides-Mendoza A, Ortega-Ortiz H, Hernández-Fuentes AD, Juárez-Maldonado A (2017) Cu Nanoparticles in chitosan-PVA hydrogels as promoters of growth, productivity and fruit quality in tomato. Emir J Food Agric 29(8):573–580

Hussain M, Raja NI, Mashwani ZUR, Iqbal M, Sabir S, Yasmeen F (2017) In vitro seed germination and biochemical profiling of Artemisia absinthium exposed to various metallic nanoparticles. 3 Biotech, 7(2):1–8

Hussein MM, Abou-Baker NH (2018) The contribution of nano-zinc to alleviate salinity stress on cotton plants. R Soc Open Sci 5(8):171809

Jeevanandam et al (2018) Review on nanoparticles and nanostructured materials: history, sources, toxicity and regulations. Beilstein J Nanotechnol 9:1050–1074

Khan I, Saeed K, Khan I (2019) Nanoparticles: Properties, applications and toxicities. Arab J Chem 12(7):908–931

Khandel et al (2018) Biogenesis of metal nanoparticles and their pharmacological applications: present status and application prospects. J Nanostruct Chem 8:217–254

Ku L (2019) New agriculture technology in modern farming. Plug and Play Tech Center

Kumar V, Guleria P, Kumar V, Yadav SK (2013) Gold nanoparticle exposure induces growth and yield enhancement in Arabidopsis thaliana. Sci Total Environ 461:462–468

Kumar N, Kumbhat S (2016) nanomaterials: general synthetic approaches. Essent Nanosci Nanotechnol 29–74.

López-Vargas ER, Ortega-Ortíz H, Cadenas-Pliego G, de Alba RK, Cabrera de la Fuente M, Benavides-Mendoza A, Juárez-Maldonado A (2018) Foliar application of copper nanoparticles increases the fruit quality and the content of bioactive compounds in tomatoes. Appl Sci 8(7):1020

Mahakham W, Theerakulpisut P, Maensiri S, Phumying S, Sarmah AK (2016) Environmentally benign synthesis of phytochemicals-capped gold nanoparticles as nanopriming agent for promoting maize seed germination. Sci Total Environ 573:1089–1102

Majeed A, Muhammad Z (2019) Salinity: a major agricultural problem—causes, impacts on crop productivity and management strategies. In: Plant abiotic stress tolerance. Springer, Cham, pp 83–99

Masarovičová E, Kráľová K (2019) Applications for contaminated soils. Phytoremed Manage Environ Contam 6:319

Mehmood A, Murtaza G (2017) Application of SNPs to improve yield of Pisum sativum L (pea). IET Nanobiotechnol 11(4):390–394

Miralles P, Johnson E, Church T, Harris AT (2012) Multiwalled carbon nanotubes in alfalfa and wheat: toxicology and uptake. J R Soc Interface 9(77):3514–3527

Mozafari AA, Ghaderi N, Havas F, Dedejani S (2019) Comparative investigation of structural relationships among morpho-physiological and biochemical properties of strawberry (Fragaria× ananassa Duch.) under drought and salinity stresses: a study based on in vitro culture. Sci Horticulturae 256:108601

Nagaonkar D, Shende S, Rai M (2015) Biosynthesis of copper nanoparticles and its effect on actively dividing cells of mitosis in Allium cepa. Biotechnol Prog 31(2):557–565

Ndeh NT, Maensiri S, Maensiri D (2017) The effect of green synthesized gold nanoparticles on rice germination and roots. Adv Nat Sci Nanosci Nanotechnol 8(3):035008

Peng C, Chen S, Shen C, He M, Zhang Y, Ye J, Shi J (2018) Iron plaque: a barrier layer to the uptake and translocation of copper oxide nanoparticles by rice plants. Environ Sci Technol 52(21):12244–12254

Pramanik P, Krishnan P, Maity A, Mridha N, Mukherjee A, Rai V (2020) Application of nanotechnology in agriculture. Environ Nanotechnol 4:317–348

Prasad R, Bhattacharyya A, Nguyen QD (2017) Nanotechnology in sustainable agriculture: recent developments, challenges, and perspectives. Front Microbiol 8:1014

Rashidi R, Abasi S, Kermanchi M, Bastani M, Faraji A (2016) The use of nano-silver particles to increase yield of broad bean. J Pure Appl Microbiol 10(3):1933–1937

Sandhu A (2006) Who invented nano? Nat Nanotechnol 1(2):87–87

Shallan MA, Hassan HMM, Namich AAM, Ibrahim AA (2016) Biochemical and physiological effects of TiO2 and SiO2 nanoparticles on cotton plant under drought stress. Res J Pharm Biol Chem Sci 7(4):1540

ShangY HM, Ahammed GJ, Li M, Yin H, Zhou J (2019) Applications of nanotechnology in plant growth and crop protection: a review. Molecules 24(14):2558

Shende S, Rathod D, Gade A, Rai M (2017) Biogenic copper nanoparticles promote the growth of pigeon pea (Cajanus cajan L. IET Nanobiotechnol 11(7):773–781

Spang ES, Moreno LC, Pace SA, Achmon Y, Donis-Gonzalez I, Gosliner WA, Tomich TP (2019) Food loss and waste: measurement, drivers, and solutions. Annu Rev Environ Resour 44:117–156

Thiruvengadam M, Rajakumar G, Chung IM (2018) Nanotechnology: current uses and future applications in the food industry. 3 Biotech 8(1):1–13

Wang WN, Tarafdar JC, Biswas P (2013) Nanoparticle synthesis and delivery by an aerosol route for watermelon plant foliar uptake. J Nanopart Res 15(1):1–13

Xu C, Peng C, Sun L, Zhang S, Huang H, Chen Y, Shi J (2015) Distinctive effects of TiO2 and CuO nanoparticles on soil microbes and their community structures in flooded paddy soil. Soil Biol Biochem 86:24–33

Yan A, Chen Z (2019) Impacts of silver nanoparticles on plants: a focus on the phytotoxicity and underlying mechanism. Int J Mol Sci 20(5):1003

Zhu ZJ, Wang H, Yan B, Zheng H, Jiang Y, Miranda OR, Vachet RW (2012) Effect of surface charge on the uptake and distribution of gold nanoparticles in four plant species. Environ Sci Technol 46(22):12391–12398

Chapter 2
Performance of Fibreglass Mat and Woven for Prosthetic Leg Socket Application

Wan Nursheila Wan Jusoh, Mohd Yusoff Mohd Haris, Nurhayati Mohd Nur, Khairul Dahri Mohd Aris, Muhammad Amir Ihsan Basri, and Mohd Izhwan Maidu

Abstract In recent years, the public has received the use of polymer composite as a prosthetic leg socket due to its excellent compatibility, ease of customisation, and high strength-to-weight ratio characteristic. The leg socket is the link that connects the prosthetic component to the residual limb. The study aims to fabricate composite split-mould and produce fibreglass leg sockets using fibreglass mat and woven fabric. Tensile test panels are also fabricated to represent the tensile strength of the leg sockets. The compression strength of the leg sockets was evaluated through the static failure test. Results indicated that fibreglass woven has higher tensile strength but lower compression strength than fibreglass mat.

Keywords Prosthetic leg socket · Polymer reinforced composite · Fibreglass

W. N. W. Jusoh (✉) · M. Y. Mohd Haris · N. Mohd Nur · K. D. Mohd Aris · M. A. Ihsan Basri · M. I. Maidu
Aerospace Section, Universiti Kuala Lumpur, Malaysian Institute of Aviation Technology, Kampung Jenderam Hulu, Lot 2891, Jalan Jenderam Hulu, 43900 Sepang, Selangor, Malaysia
e-mail: wannursheila@unikl.edu.my

M. Y. Mohd Haris
e-mail: mohdyusoff@unikl.edu.my

N. Mohd Nur
e-mail: nurhayatimn@unikl.edu.my

K. D. Mohd Aris
e-mail: khdahri@unikl.edu.my

M. A. Ihsan Basri
e-mail: amirihsan62@gmail.com

M. I. Maidu
e-mail: m.izhwan@yahoo.com

A. Ismail et al. (eds.), *Advancements in Materials Science and Technology Led by Women*, Advanced Structured Materials 165,
https://doi.org/10.1007/978-3-031-21959-7_2

2.1 Introduction

In general, lower limb amputation (LLA) removes a part or multiple parts of the lower limbs. There are various reasons for amputation reported, such as war, natural disaster, accidents, or chronic diseases. In Malaysia, according to a study done on "Quality of Life Among Lower Limb Amputees in Malaysia", it is reported that the two major causes of LLA among Malaysians are caused by diabetes (57%) and road traffic accidents (33.3%) (Razak et al. 2016). The study was done on 43 respondents of predominantly Malay Muslim ethnicity. Nurhanisah et al. (2018) and Yusof et al. (2015) reported an increasing trend of diabetes type-2 amputees in Malaysia.

Figure 2.1 shows the combination of the fundamental components for lower limb prosthetic amputation. The dimension and size of each component depend on the amputee's height and the residual limb length. The socket or leg socket is the component connecting the prosthetic leg to the amputee's residual limb. At the same time, the mechanical knee allows the wearer to change to a different movement angle. The mechanical knee and tibia extension or pylon are the key components holding the wearer bodyweight (Amrutsagar et al. 2020; Ramírez et al. 2012; Campbell et al. 2012; Odusote and Kumar 2016). A good prosthetic leg should improve amputee mobility and reduce the dependency on others to carry out regular day-to-day activities. Amputees with LLA need more energy than a normal person to move from one place to another.

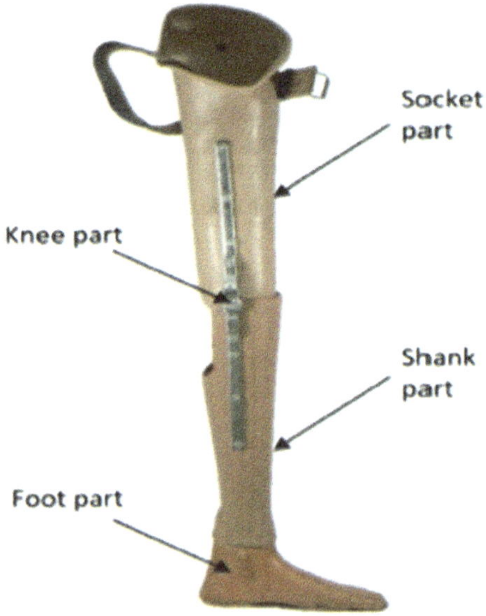

Fig. 2.1 Conventional Jaipur prosthesis leg components Campbell et al. 2012. Reprinted with permission from SpringerNature publishers

The leg socket is typically made from polymeric materials or reinforced polymer composite using synthetic fibres like carbon or glass fibre. These synthetic fibres will increase the leg socket's mechanical properties compared to polymer materials alone. Fibre-reinforced polymers are widely used in various industries, including load-bearing structures, pipelines, and rotor blades, but lately, many medical applications are also moving to reinforced composite. Leg socket is available only as a custom-made product because of the different individual amputee's residual stump, while other prosthesis components like mechanical knee and tibia extension are usually mass-produced in the factory using automation technology (Che Me et al. 2012; Reist et al. 2010). Generally, these components are made from lightweight metallic materials.

Material selection for leg socket fabrication is crucial to ensure the amputee is comfortable using the prosthetic leg. In recent years, many prosthetic leg socket manufacturers choose fibreglass reinforced composite. Fibreglass is lightweight, durable, and cheaper than other synthetic fibres like carbon and kevlar. However, some of the weaknesses found in fibreglass are non-environmental friendly and high hardness (Joshi et al. 2004; Etcheverry and Barbosa 2012). A fibreglass mat is a combination of chopped strand fibreglass in random arrangement and held by mechanical compression pressure or adhesive resin binder.

Woven fabrics are made by numerous fill and warp weaving of a few sets of yarn or roving (Skrifvars et al. 1998). Woven fibre reinforcement enables easier installation for difficult surfaces, lightweight, excellent mechanical performance, and formability (Gherissi 2019). Epoxy or polyester resin alone is not recommended for load-bearing applications because of brittleness and low strength but is widely used for composite matrixes.

Devendra and Rangaswamy 2013 reported that reinforcement structures, matrix types, and additives used in the fabrication process influence the properties of polymer composites produced. Results obtained from this project will be a good input to understand the performance of different fibreglass fabric orientations and the effect of using different resins in their products. The present study is a collaboration with Kedidi Kaki Palsu (dD). This project compares the tensile and static failure strength of fibreglass mat and woven using polyester and epoxy resin as a binder.

2.2 Methodology

2.2.1 Material

The two types of fibreglass, Chopped Strand Mat (CSM) and woven *E*-glass, with 200 g/sm fabric as in Fig. 2.2 were purchased from Chemrex Sdn. Bhd. Epoxy resin and polyester resin were used as the matrix for fabricating the composite leg socket. The epoxy resin set was from Smooth-On Inc., while the polyester resin set was purchased from Multifilla (M) Sdn. Bhd.

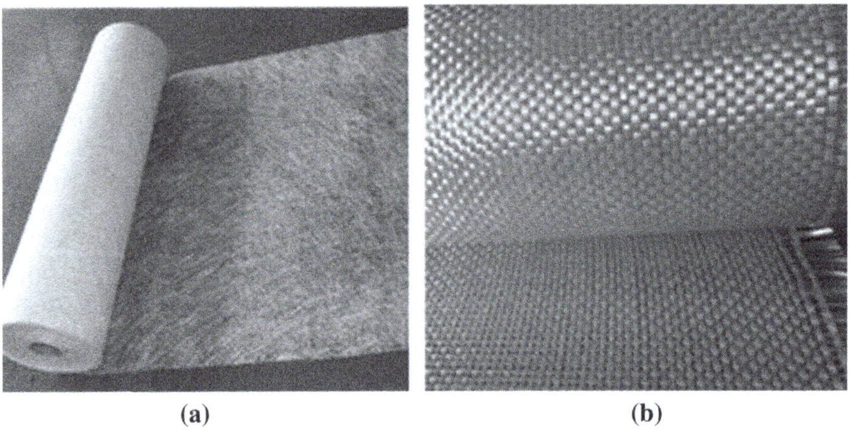

Fig. 2.2 Project uses two types of fibreglass: **a** CSM and **b** woven E-glass

2.2.2 Fabrication of Composite Split-Mould

An actual mould of the leg was from Plaster of Paris (POP) by dD (Fig. 2.3). The POP mould was taken on an actual amputee's residual limb. This mould was used as the female casting mould for the split-mould fabrication. The main reason to fabricate a split-mould is to reuse the same mould to produce four different types of samples. The reuse of the same mould is not possible when using a POP mould since removing the sample from the POP mould will require breaking the POP mould into small pieces.

The fabrication process of split-mould begins by sanding the POP mould with sandpaper grit 120 and 240 to get a smoother surface. A layer of mod podge is later applied to the POP to give better surface finishing. Mod podge is a water-based sealer

Fig. 2.3 POP mould in a wooden frame

or finisher often used by crafters. After the mod podge dried, a split line (half of the mould) was marked. About 30 mm flange was created as a barrier for the mould plug using plasticine. Any gap between the POP mould surface and barrier is sealed using plasticine. Polyvinyl alcohol (PVA) is applied on the exposed side POP surface as a mould release. Later, the tooling gel coat with blue colour ink is poured onto the surface. Once the gel coat becomes tacky, a fibreglass veil is applied using epoxy as a binder. The next consequence layers are the reinforcement layers which are fibreglass with 120, 200, 300, and 450 gsm. The mould was left to dry and cure at room temperature for 24 h. After the fibreglass has been cured, the barrier is removed, and all traces of plasticine are cleaned. Later, the composite mould removes from the POP mould and then trimmed or cut to size. The same steps were repeated on the other side of the POP mould.

2.2.3 Fabrication of Composite Leg Socket

The composite leg socket and test panel were fabricated using the hand layup technique. Each layer had a cross-sectional area of 300 mm × 300 mm for three layers of fibreglass mat and woven. The fibre to resin ratio used was 60:40. The fabrication process begins by applying the release agent three times with 15 min separation time between cycles to prevent the composite from sticking to the mould later. The composite leg socket is prepared by evenly applying a resin layer on the composite mould and then placing a fibreglass fabric layer. A hand brush was used to impregnate the resin into the fibreglass. These steps are repeated until the third layer on both sides of the split-mould and combined properly to the right position. When the composite has properly cured, it is removed from the mould and trimmed for the excess part. Similar steps were conducted for the test panel fabrication, except layup was done on a glass mould and then trimmed according to tensile test panel dimensions. Four samples were prepared using two different resins and fibreglass fabrics, in Table 2.1.

Table 2.1 Sample configuration

S. No.	Sample
1	Fibreglass mat with polyester resin
2	Fibreglass mat with epoxy resin
3	Fibreglass woven with polyester resin
4	Fibreglass woven with epoxy resin

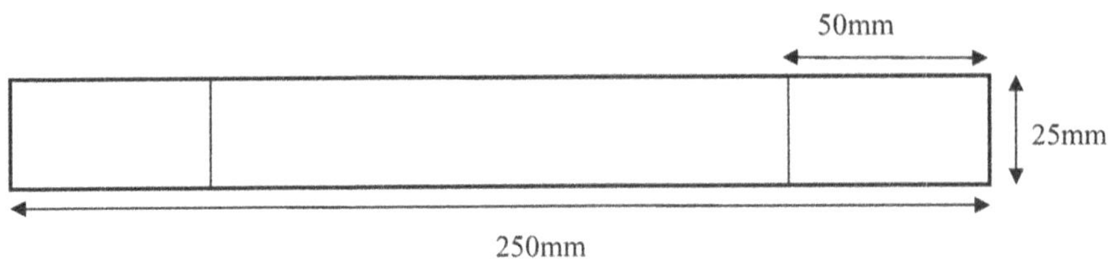

Fig. 2.4 Tensile test specimen dimensions

2.2.4 *Tensile Strength Test*

Five test specimens were prepared and tested according to ASTM D3039 Standard Test Method for Tensile Properties of Polymer Matrix Composite Materials. The specimens were cut into a rectangular cross section with a thickness of 3 mm. The tensile tests were conducted on Shimadzu AG–*X* (50 kN) Universal Testing Machine (UTM). The crosshead speed was set at 2 mm/min with the load applied until failure as per recommendation in the ASTM. The dimensions of the tensile test flat sample are 250 mm × 25 mm × 3 mm with a 50 mm tab (Fig. 2.4).

2.2.5 *Static Failure Test*

Static failure of the composite leg sockets was measured using the same UTM as per the tensile test. The composite leg socket with dimensions of 240 mm × 100 mm was placed inside a metallic box (130 mm × 130 mm × 120 mm) as a holder during the test. According to Gerschutz et al. (2012), there is no fixed testing standard for prosthetic leg testing. Figure 2.5 presents the picture of the testing set-up applied during the static test.

2.3 Results and Discussion

2.3.1 *Composite Split-Mould*

Figure 2.6 is the final product of composite split-mould or duplication of POP mould. The split-mould could replicate the POP mould design pattern with glossy finishing and smooth texture. The final composite split-mould thickness is about 5 mm with about 30 mm extra flange around the perimeter. The extra flange allows proper bonding and assembly between the two sample parts by clipping the moulds together after layup. Few designated points on both moulds were indicated for precise alignment.

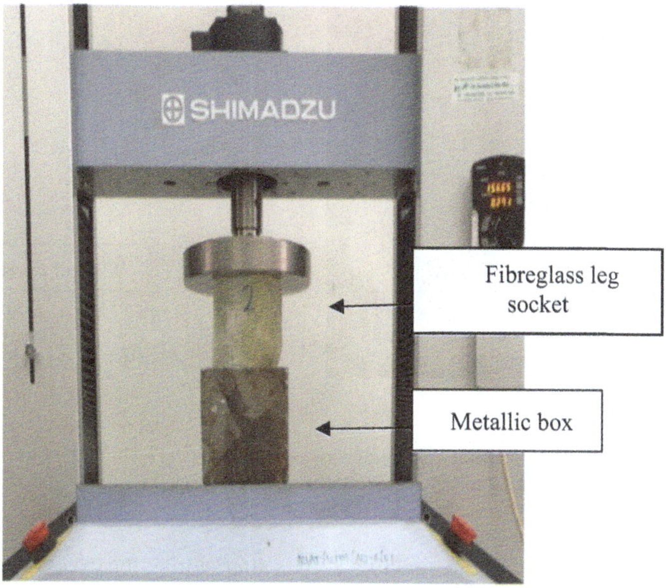

Fig. 2.5 Static failure test set up on the universal testing machine

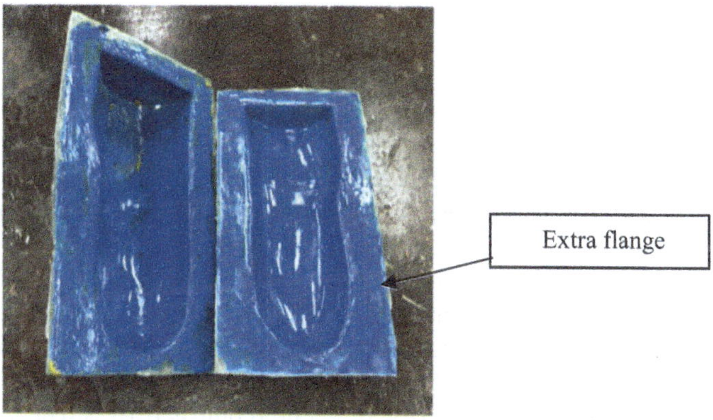

Fig. 2.6 Composite split-mould

2.3.2 Composite Leg Socket

During the leg socket layup process, more resins were required to wet the fibreglass mat properly. This condition is due to the high void content within the fibreglass mat because of the randomly dispersed fibres. As a result, leg sockets using fibreglass mat have a higher thickness of 1 to 1.5 mm than leg sockets made from fibreglass woven. Figure 2.7a is an example of leg sockets using a fibreglass mat with polyester after

(a) (b)

Fig. 2.7 Prosthetic leg socket made from fibreglass mat with polyester resin condition after removed from mould: **a** before trimming to size and **b** after trimming to size

being removed from the mould, and Fig. 2.7b is after cutting the excess fibres. From physical observation, the bonding line between the two parts from the split-moulds is quite good and smooth, without any cracks or voids. Good bonding is achieved by compressing and holding the two parts together (while in the mould) as soon as layup is finished until it is fully cured.

2.3.3 Tensile Strength Test Result

An average of five specimens was calculated for maximum tensile stress, maximum strain and modulus elasticity values. Woven fabrics provide better tensile strength around 34 to 35 N/mm^2 than mat 24 to 25 N/mm^2, as shown in Fig. 2.8a. This result is expected since the non-uniform distribution of fibres in the mat contributes to the lower tensile strength of the socket. The warp and fill fibres support the higher strength in woven fabric tension. Continuous fibres like (0/90)° and (+45)° exhibit better resistance against longitudinal force direction compared to CSM due to a higher percentage of fibres loading in that particular direction designed by the manufacturer hence able to withstand higher load before rupture (Jimit et al. 2017; Banakar et al. 2012; Rajesh et al. 2020; Nayak et al. 2018; Bhaskar and Srinivas 2017). A slightly lower strain value obtained for woven fabrics than mat fabrics, as shown in Fig. 2.8b. The effect of different types of resin on the tensile test results is not significantly witnessed.

Table 2.2 indicates a higher average value of modulus elasticity in tension for the sample with woven fabrics between 4934 and 5227 N/mm^2 compared to the sample with mat fabrics with an average between 3176 and 3767 N/mm^2.

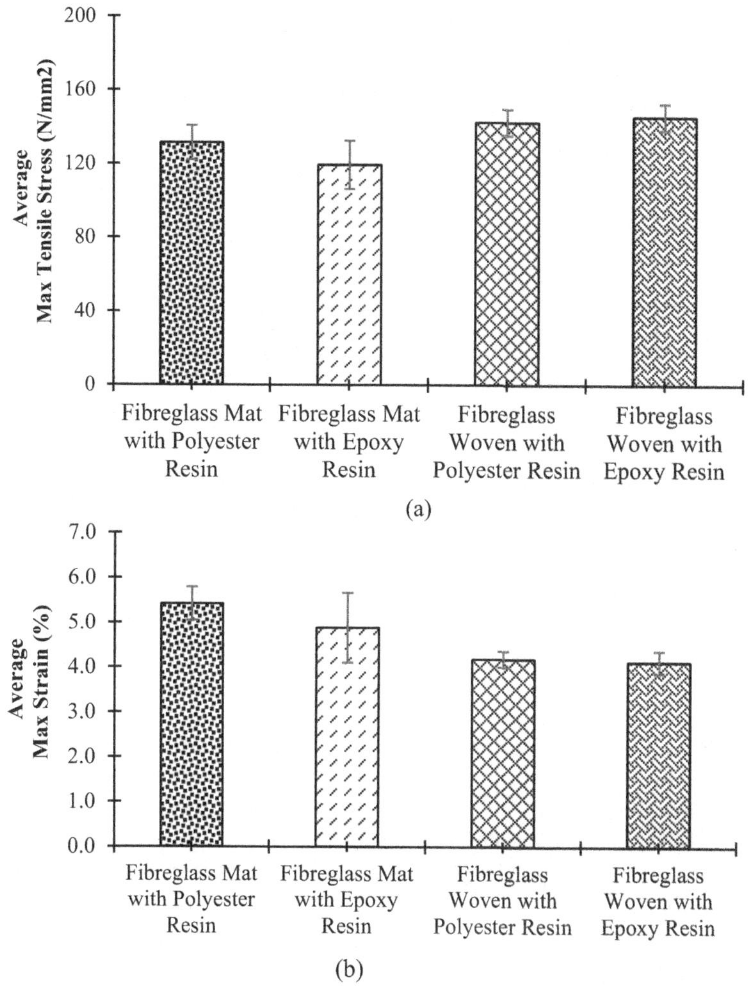

Fig. 2.8 **a** Average maximum for tensile stress and **b** average maximum strain for tensile test

Table 2.2 Average of modulus elasticity in tension

Sample	Average modulus elasticity (N/mm^2)
Fibreglass mat with polyester resin	3176.2
Fibreglass mat with epoxy resin	3767.0
Fibreglass woven with polyester resin	4934.0
Fibreglass woven with epoxy resin	5226.6

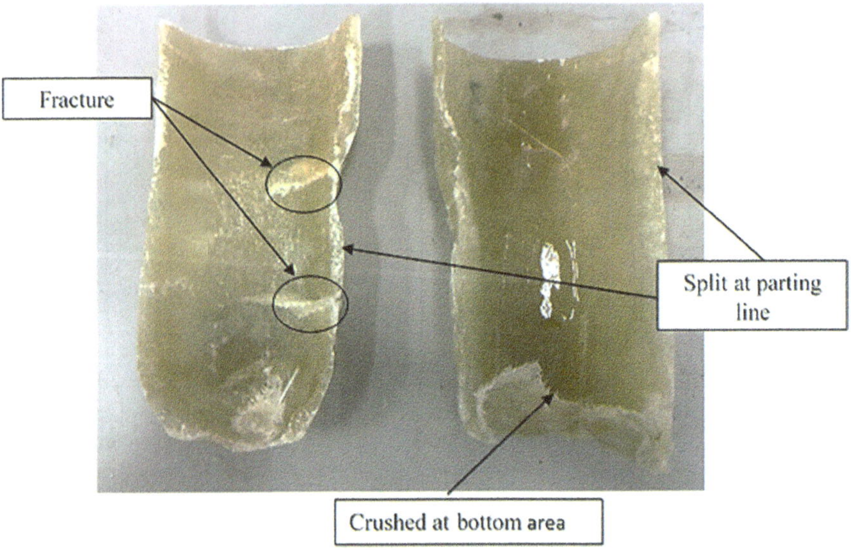

Fig. 2.9 Broken prosthetic leg socket after static failure

2.3.4 Static Failure Test Result

The ability of the prosthetic leg socket to withstand the static load was evaluated under the static failure test. Figure 2.9 shows the prosthetic leg socket split and fractured along the parting line after the static failure test. It is also observed that the leg socket is crushed at the bottom of the leg socket.

The additional resin in fibreglass mat samples increases the load distribution within the fibres. As a result, the leg socket provides better compression or bending resistance, as shown in Fig. 2.10. This observation is consistent with (Varga et al. 2010; Guruprasad and Prakash 2018), who highlighted that interface bonding and load transfer from matrix to fibres play an important role in determining the mechanical properties of a composite. The highest strain was achieved by fibreglass mat epoxy resin at 13%, followed by fibreglass woven epoxy resin at 12.5%, fibreglass mat polyester resin at 12%, and fibreglass woven polyester resin at 11%. The presence of extra resin is believed to improve the ductility properties in samples using fibreglass mat. As shown in the previous tensile test result, the use of the different types of resin does not substantially affect the static strength of the leg socket.

2.4 Conclusion

The split-mould was successfully fabricated and able to replicate the original design from the POP mould. Each leg socket sample has similar dimensions and design. Tensile test results and the modulus elasticity values indicated fibreglass-woven

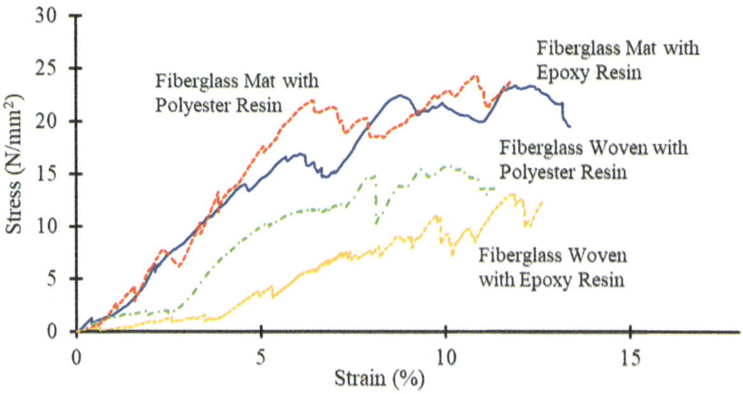

Fig. 2.10 Stress–strain graph for fibreglass leg sockets under static failure test

fabrics would give the leg socket better strength than fibreglass mat. On the contrary, opposite results were obtained from the static failure test. The fibreglass mat shows better compression strength than the fibreglass fabric sample in this test. The presence of extra resin within the fibreglass mat is the main reason for this improvement. The effects of different resins are not significant in either tensile or static failure test results. The data obtained through this study can give some suggestions on the leg socket performance using either fibreglass mat or woven. It is recommended to conduct other mechanical tests such as flexural and impact tests on the sample for future work. Besides improving the mechanical properties of the composite leg socket, the research should also include the capability study of the leg socket in terms of durability test and comfort level of the amputees.

Acknowledgements The authors would like to thank Kedidi Kaki Palsu (dD) for providing the leg socket mould as our reference for mould fabrication.

References

Amrutsagar L, Parit G, Ghyar R, Bhallamudi R (2020) Parametric design and hybrid fabrication of above-knee prosthesis. Indian J Orthop 54(3):381–390

Banakar P, Shivananda HK, Niranjan HB (2012) Influence of fiber orientation and thickness on tensile properties of laminated polymer composites. Int J Pure Appl Sci Technol 9(1):61–68

Bhaskar VV, Srinivas K (2017) Mechanical characterization of glass fiber (woven roving/chopped strand mat E-glass fiber) reinforced polyester composites. AIP Conf Proc 1859:020108

Campbell AI, Sexton S, Schaschke CJ, Kinsman H, McLaughlin B, Boyle M (2012) Prosthetic limb sockets from plant-based composite materials. Prosthet Orthot Int 36(2):181–189

Che Me R, Ibrahim R, Md. Tahir P (2012) Natural based biocomposite material for prosthetic socket fabrication ALAM CIPTA. Int J Sustain Trop Des Res Pract 5(1):27–34

Devendra K, Rangaswamy T (2013) Strength characterization of E-glass fiber reinforced epoxy composites with filler materials. J Miner Mater Charact Eng 01(06):353–357

Etcheverry M, Barbosa SE (2012) Glass fiber reinforced polypropylene mechanical properties enhancement by adhesion improvement. Materials (basel) 5(6):1084–1113
Gerschutz MJ, Haynes ML, Nixon D, Colvin JM (2012) Definitive laminated sockets. J Rehabil Res Dev 49(3):405–426
Gherissi A (2019) Failure study of the woven composite material: 2.5 D carbon fabric/resin epoxy Abderraouf. J Mech Eng Sci 13(3):5390–5406
Guruprasad HL, Prakash JN (2018) Influence on mechanical properties of epoxy polymer matrix composites reinforced with surface treated woven strand mat E-glass fibers. Int J Res Adv Technol 6(8):2191–2196
Jimit R, Zakaria KA, Bapokutty O (2017) Influence of fiber orientation on mechanical properties of fiberglass reinforced composite. Proc Mech Eng Res Day 2017 (May):318–319
Joshi SV, Drzal LT, Mohanty AK, Arora S (2004) Are natural fiber composites environmentally superior to glass fiber reinforced composites? Compos Part A Appl Sci Manuf 35(3):371–376
Nayak SY, Heckadka SS, Thomas LG, Baby A (2018) Tensile and flexural properties of chopped strand E-glass fibre mat reinforced CNSL-epoxy composites. MATEC Web Conf 144:02025
Nurhanisah MH, Hashemi F, Paridah MT, Jawaid M, Naveen J (2018) Mechanical properties of laminated kenaf woven fabric composites for below-knee prosthesis socket application. IOP Conf Ser Mater Sci Eng 368(1)
Odusote J, Kumar V (2016) Mechanical properties of pineapple leaf fibre reinforced polymer composites for application as prosthetic socket. J Eng Technol 6(1):24–32
Rajesh A, Deva Prasad S, Singaravel B, Niranjan T, Shravan Kumar T (2020) Experimental and analytical outcomes of carbon fiber orientation in epoxy resin composite laminate under tensile loading. Adv Unconv Mach Compos 771–783
Ramírez JF, Muñoz EJ, Vélez JA (2012) Algorithm for the prediction of the reactive forces developed in the socket of transfemoral amputees. DYNA 79(173):89–95
Razak MMA, Tauhid MZ, Yasin NF, Hanapiah FA (2016) Quality of life among lower limb amputees in Malaysia. Proc Soc Behav Sci 222:450–457
Reist TA, Andrysek J, Cleghorn WL (2010) Topology optimization of an injection moldable prosthetic knee joint. Comput Aided Des Appl 7(2):247–256
Skrifvars M, Mackin T, Skagerberg B (1998) An Application of experimental design to the development of glass fibre reinforced polyester laminates with enhanced mechanical properties. Polym Test 17(5):345–356
Varga C, Miskolczi N, Bartha L, Lipóczi G (2010) Improving the mechanical properties of glass-fibre-reinforced polyester composites by modification of fibre surface. Mater Des 31(1):185–193
Yusof NM, Ab Rahman J, Zulkifly AH, Che-Ahmad A, Khalid KA, Sulong AF, Vijayasingham N (2015) Predictors of major lower limb amputation among type ii diabetic patients admitted for diabetic foot problems. Singap Med J 56(11):626–631

Chapter 3
Investigation of Leakage in Underground Pipelines Using the Radiotracer Technology

Noraishah Othman, Nolida Yussup, Ahmad Zubair Johan, Iqmal Zulkarnain Abdul Rahman, and Roslanzairi Mostapa

Abstract Leakage in underground pipelines is one of the crucial experiences by industries especially to oil and gas. The seepage of unattended hydrocarbon will easily pollute the groundwater and the environment as a whole. Moreover, Malaysia has multiple pipeline networks which carry hydrocarbon, condensate, oil, gas and water. Hence, early inspection to the respective pipelines should be carried out to avoid the catastrophic incident. Thus, Nuclear Malaysia has established a system which comprises the development of hardware and software to curb the respective problem using the radiotracer technology. Radiotracer is the radioactive material which emits gamma radiation, and the signal is captured by the scintillation detector and saved in the Arduino system that acts as data logger. The electrical components are kept in water sealed capsule, and the system is launched inside the industrial pipeline for leak characterization. However, for this paper, discussion is limited to the proof of concept at which the works are carried out in lab scales.

Keywords Leak detection · Underground pipelines · Radiotracer · Hardware and software development · Proof of concept

---

N. Othman (✉) · N. Yussup · R. Mostapa
Malaysian Nuclear Agency, 43000 Kajang, Selangor, Malaysia
e-mail: noraishah@nm.gov.my

N. Yussup
e-mail: nolida@nm.gov.my

R. Mostapa
e-mail: roslanzairi@nm.gov.my

A. Z. Johan
Jengka Campus Pahang Branch, MARA University of Technology Malaysia, 26400 Bandar Tun Razak, Pahang, Malaysia
e-mail: zubairjohan@gmail.com

I. Z. A. Rahman
MARA University of Technology Malaysia, 40450 Shah Alam, Selangor, Malaysia
e-mail: zulkarnainiqmal605@gmail.com

A. Ismail et al. (eds.), *Advancements in Materials Science and Technology Led by Women*, Advanced Structured Materials 165,
https://doi.org/10.1007/978-3-031-21959-7_3

3.1 Introduction

Leak detection using the radiotracer techniques is one of the most widespread applications of radiotracers in industry. The basic functions of leak detection are the location and size measurement of leaks in sealed systems. Radiotracer techniques are very sensitive, effective and competitive for on-line leak detection (International Atomic Energy Agency 2001, 2008). Radiotracers allow an early detection of small leakages before these develop into major pollution incidents. Radiotracer methods used for on-line leak detection in heat exchangers and underground pipelines can achieve the detection limits up to 0.1% of stream flow (International Atomic Energy Agency 2009). Moreover, Nuclear Malaysia is the sole industrial radiotracer service provider in Malaysia; they are the sole owner of a nuclear research reactor in Malaysia.

The detection of leakage in underground pipeline is a significant challenge to many industries, in particular oil and gas industry. Moreover, Malaysia has multiple pipeline networks which carry hydrocarbon, condensate, oil, gas and water (Verdict Media Limited 2022; PETRONAS 2022). Therefore, Plant Assessment Technology (PAT) team has adopted radiotracer technology (RT) to locate and verify the leakage location. RT is the introduction of radioactive material inside the domain, and the tracing of gamma radiation emitted by this material is conducted externally using a scintillation detector. Any drop of velocity from upstream to downstream fluid flow is susceptible to potential leak. RT is very sensitive, effective and competitive for on-line leak detection (International Atomic Energy Agency 2001). However, the conventional radiotracer technique is very expensive, requires a lot of manpower, is cumbersome and is very time consuming. Thus, a radiotracer capsule is developed to overcome the aforementioned problems (International Atomic Energy Agency 2009).

Insertion of a radiotracer capsule in a pipeline for underground pipe leakage is the state of the art of radiotracer technology. Figure 3.1 shows the principle of the radiotracer system for leak detection in underground pipelines (International Atomic Energy Agency 2009). The liquid radiotracer is injected at the upstream of pipe under investigation as tracer plug. The tracer plug follows the fluid inside the pipe and seeps out from any cracks and stays outside of the pipe and emits gamma rays. The following fluid will wash out the residues of radioactive material inside the pipe. Capsule will be launched afterwards and capture the signals emitted by trapped radiotracer which mimics the leakage. Upon retrieval of the capsule, analysis is carried out by identifying the peak/leak which is the radioactive signal.

The radiotracer capsule is a low-cost prototype system in order to produce a proof of concept in the leak detection of underground pipelines using the radiotracer approach. Therefore, the ratemeter used in the system development is a pocket ratemeter model 9301 manufactured by Minekin, Australia. The crucial ratemeter requirements for the system are small size, adjustable high voltage, detector reading with time stamp and RS232 data communication interface (Minekin 1993). The activities are detected by a NaI(Tl) detector connected to the ratemeter. For standard operation, detector reading is acquired by a computer which acts as terminal via the

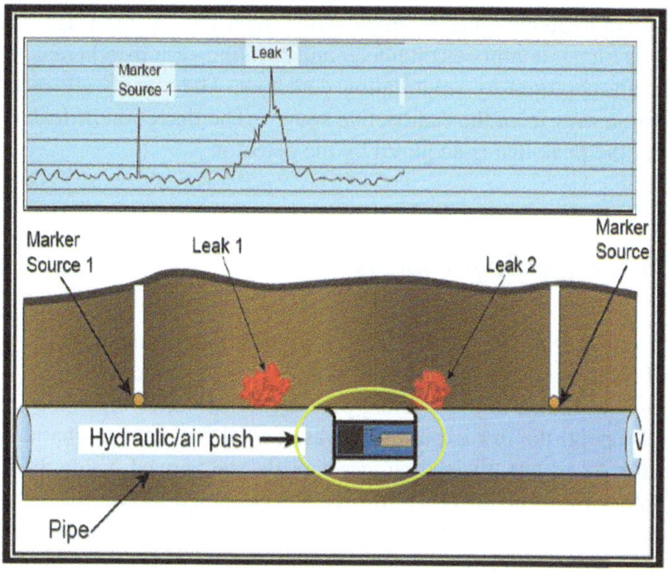

Fig. 3.1 Principle of radiotracer system

RS232 serial interface. However, for the radiotracer system, a small stand-alone data logger is required to acquire and store the data accordingly while the system moves in the pipeline. The RS232 baud rates are selectable from 300 to 38,400 (Minekin 1993). The Arduino microcontroller is used as the main processor for the data logger. The data logger utilizes the simplicity of the Arduino in order to acquire the data from the ratemeter and save the data on the SD card. For the radiotracer system data logger, the selected digital processor for the implementation is the Arduino Uno. Arduino Uno is chosen due to its simplicity and versatility. In this development, the Arduino is programmed to have two functions which are for serial data acquisition and information storage, respectively. Moreover, two shields are utilized for this purpose which are the RS232 shield and the SD card shield.

3.2 Methodology

3.2.1 Development of Radiotracer Capsule

The capsule comprises hardware and software development which is designed to ensure the radioactive data is saved inside the system in a watertight capsule. It consists of radioactive material (radiotracer), scintillation detector, data logger, coaxial cable and battery to power up the system. All these components are incorporated and encapsulated in the waterproofing capsule. Programming is done to the

Arduino system to acquire data from the data logger and save the signals inside the SD card. Data logger comprises ratemeter and Arduino system, whereas the radioactive signals are captured by scintillation detectors. The signals are received from emitted gamma rays from the radioactive source. The detection of leak is obtained by observing the peak that is acquired by the detector.

3.2.2 Radiotracer Experiments

The completed radiotracer system was conducted experimentally as a proof of concept. 10″ of mild steel pipe was used as non-insulated pipe, whereas 2″ thickness of insulation was added to the existing pipe for an insulated pipe study. Since it is a proof of concept, the dry experiment was carried for leakage characterizations. Therefore, for speed test, the radiotracer capsule was moved inside the pipe using the speed controller. The speed controller was designed to control the speed of a DC motor to pull the capsule along the pipeline internally. The study is to optimize the speed of capsule which is capable to pick up the radioactive signals. In this experiment, Caesium-137 (C_s-137) with different activities was used as radioactive sources and installed outside of the pipe to mimic the leakage volume. C_s-137 is emitting gamma rays with half-life of 30 years and an energy of 0.6617 meV and has the ability to penetrate thick and high-density materials. Thus, it is widely used for industrial applications. Moreover, a sodium iodide scintillation (NaI) detector is used to capture the radioactive signal which is stored inside the SD card which is embedded in the Arduino system. Figure 3.2 shows the arrangement of radiotracer experiments, whereas Table 3.1 shows the range of parameters for conducting the works.

3.3 Results and Discussion

3.3.1 Radiotracer Capsule

Figure 3.3 shows the schematic diagram of radiotracer capsule. The ratemeter (1) is attached to detector (2) for data acquisition. However, in order to save the data on the SD card for data storage (5), the Arduino system is developed and acts as data logger. The coaxial cable (4) is connected between the detector (2) and the ratemeter (1). The battery (6) is used to power up the system, and all the components are encapsulated inside the watertight radiotracer casing (3).

Data is acquired from the ratemeter for every one second. Once the Arduino detects the arrival of data, the data is stored on the SD card. The new data is appended in the text file created on the SD card prior to the initialization. The format of the acquired data which is saved on the SD card is shown in Fig. 3.4.

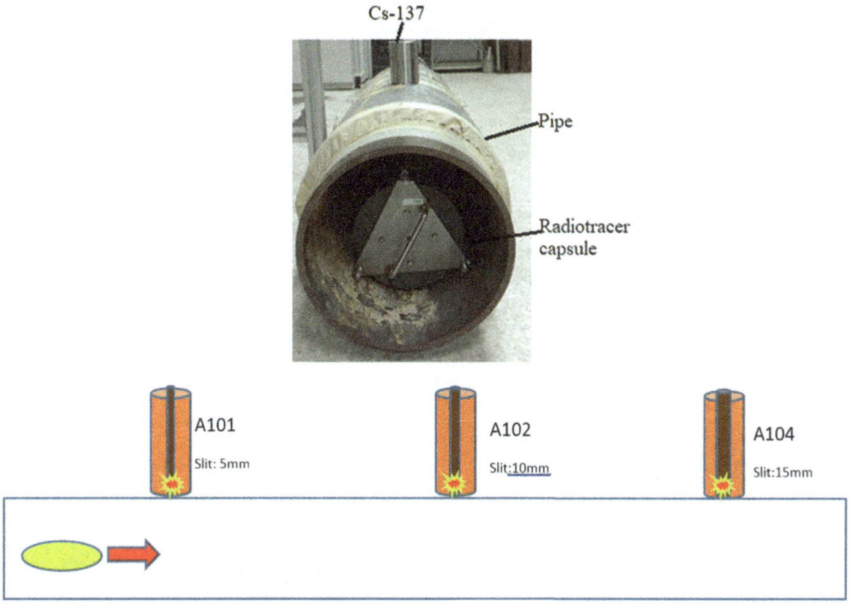

Fig. 3.2 Arrangement of radiotracer experiments

Table 3.1 Parameters of radiotracer experiments

Leakage parameters				
Specimen	Activity (MBq)	Travel speed (m/s)	Type of pipe	Collimator size/slit (mm)
$A101$	9.25	1.1	Non-insulated/insulated	5
$A102$	51.4	1.4	Non-insulated/insulated	10
$A104$	562	1.7	Non-insulated/insulated	15
$A107$	851	1.9	Non-insulated/insulated	–

3.3.2 Radiotracer Experiments

Figure 3.5 shows that the radiotracer capsule is able to capture the signal which is emitted from each different activity of C_s-137. The variation of C_s-137 activities from 9.25, 51.43, 562 and 851 MBq (labelled as $A101$, $A102$, $A104$ and $A107$, respectively) indicates that the system is able to detect the smallest amount of leak which is dissipated outside of the pipe. Activity is the strength or concentration of the said radioactive, and the unit is in MBq. Moreover, Fig. 3.5 also indicates that the system is able to detect the radiation signals from all types of pipes in this case insulated and non-insulated pipe, respectively.

In order to determine the optimum speed of the radiotracer capsule, several range of speed controller manipulation with respect to PWM has been tested. Pulse width

```
M I N E K I N   9 3 0 1   R A T E M E T E R v1.0/1.1
(c)COPYRIGHT 1990-95 MINEKIN AUSTRALIA S/N: 95060001

TIME     COUNTS    EHT

00:10:17 00000003  0575
00:10:18 00000008  0575
00:10:19 00000009  0575
00:10:20 00000008  0575
00:10:21 00000007  0575
00:10:22 00000004  0575
00:10:23 00000008  0575
00:10:24 00000010  0575
00:10:25 00000004  0575
00:10:26 00000008  0575
00:10:27 00000003  0575
```

Fig. 3.3 Format of the logged data

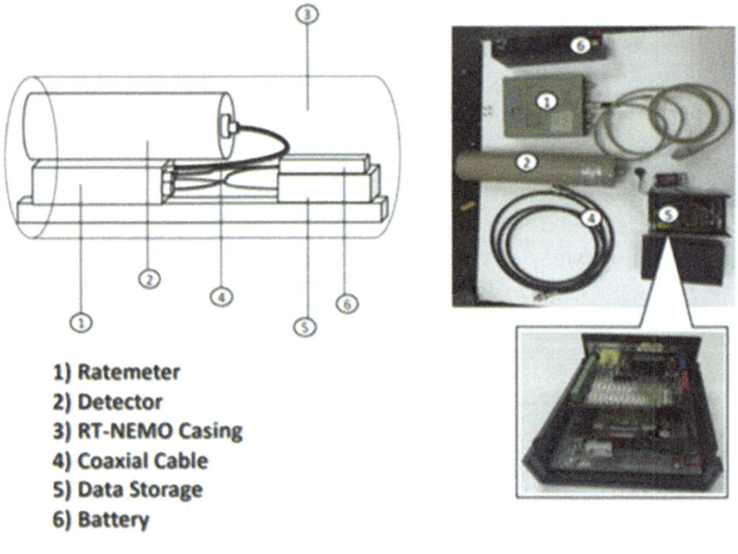

Fig. 3.4 Arrangement of radiotracer experiments

modulation (PWM) is a modulation technique that generates variable-width pulses to represent the amplitude of an analogue input signal. Figure 3.6 shows the wiring diagram of the speed controller that comprises DC motor, time relay delay, PWM controller and power supply that determines the acceleration in accordance with the percentage. The objective is to identify the range of speed of the respective capsule that can be picked up by each C_s-137. As a rule of thumb, the effective speed of industrial pipe inspection gauge (PIG), the device that is moving inside the liquid pipeline, is about 1–5 m/s (Gustavo et al. 2017). Hence, one has to identify the PWM

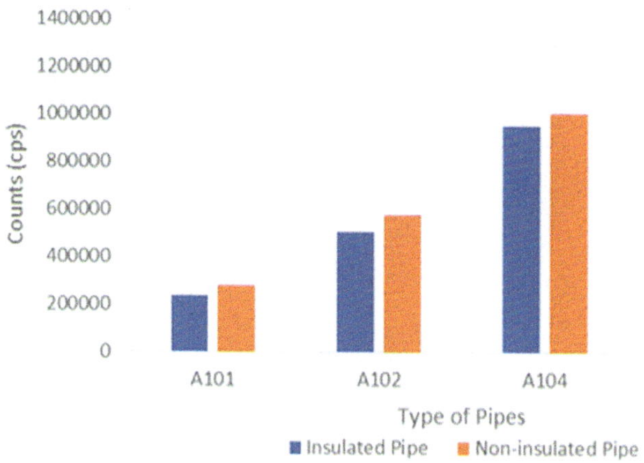

Fig. 3.5 Type of pipe experiments

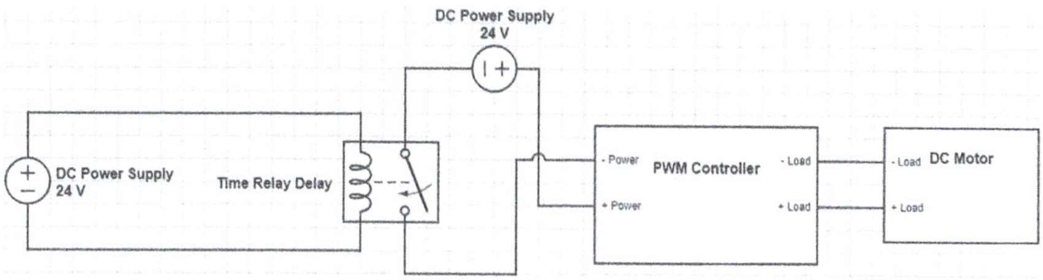

Fig. 3.6 Wiring diagram of the speed controller

(%) in order to obtain the range of speed as shown in Table 3.2. By adopting all these speed arrangements, the radioactive sources, C_s-137 with different activities, are outlined along the outer pipe of line. The overall weight of the radiotracer system is approximately 11 kg.

Moreover, each C_s-137 is having different size of pinhole collimator to mimic the size of leak. The size of pinhole (slit) collimators is 5 mm, 10 mm and 15 mm, respectively. Figure 3.7 shows the ability of capsule to pick up the signal from only 15 mm slit of the collimator which is from the 562 MBq of C_s-137. Thus, it indicates that the radiotracer capsule can only capture radioactive signals from bigger

Table 3.2 Determination of speed parameters

Parameters			
PWM (%)	Timer (s)	Average time (s)	Speed (m/s)
38.00	12	3.88	1.55
40.00	12	3.85	1.56
42.00	11	3.42	1.76
44.00	9	2.88	2.08

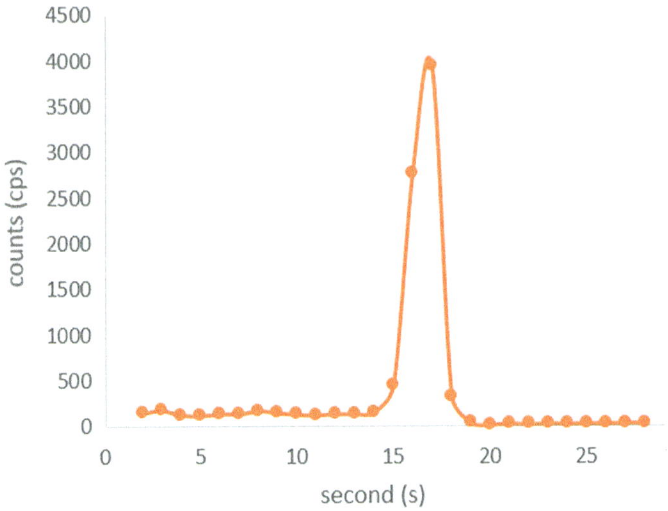

Fig. 3.7 Signal acquisition of 15 mm collimator size (slit) and 562 MBq activity of C_s-137

size of slit (opening) than smaller slit which signifies the size of leaks with higher activity/strength of radioactive source.

3.4 Conclusion

The proof of concept of radiotracer capsule has been successfully carried out. The capsule is capable to detect the radioactive signals which are arranged outside of the pipelines regardless of the type of the pipes (i.e., insulated or non-insulated). Moreover, it is capable to detect the signals within stipulated range of speed, low volume of liquid leaks and last but not least is able to capture signal of 15 mm size of leak. In conclusion, it shows that the capsule is able to be used as a tool of leak identification inside the industrial underground/buried pipelines.

Acknowledgements The authors would like to acknowledge the International Atomic Energy Agency (IAEA) for the financial support under IAEA Research Contract No. 20999/R0 (RC22898) and Plant Assessment Technology team in making this works fruitful.

References

Gustavo FL, Victor CG, Freitas RP, Maitelli AL, Salazar AO (2017) PIG's speed estimated with pressure transducers and hall-effect sensor: An industrial application of sensors to validate a testing laboratory. Sensors. https://doi.org/10.3390/s17092119

International atomic energy agency (2001) Radiotracer technology as applied to industry. IAEA-TECDOC-1262. https://www-pub.iaea.org/MTCD/Publications/PDF/te_1262_prn.pdf. Accessed 6 Jun 2022

International Atomic Energy Agency (2008) Residence time distribution method for industrial and environmental applications. IAEA-TCS-31. https://www-pub.iaea.org/MTCD/Publications/PDF/TCS-31_web.pdf. Accessed 16 Feb 2020

International Atomic Energy Agency (2009) Leak detection in heat exchangers and underground pipelines using radiotracers. IAEA-TCS-38. https://www-pub.iaea.org/MTCD/Publications/PDF/TCS-38_web.pdf. Accessed 2 Feb 2021

Minekin (1993) 9301 Pocket ratemeter instruction manual version 2. Queensland, Australia

PETRONAS (2022) Malaysia's oil and gas landscape: the Malaysian advantage. https://www.petronas.com/mpm/malaysia-oil-gas-landscape/the-malaysian-advantage. Accessed 4 Apr 2022

Verdict media limited (2022) Malaysian super refinery and pipeline project. https://www.hydrocarbons-technology.com/projects/kedah-refinery/. Accessed 4 Apr 2022

Chapter 4
General Characteristics of Endophytes and Bioprospecting Potential of Endophytic Fungi

Md Abu Taher, Woei-Yenn Tong, Chean Ring Leong, Syarifah Ab Rashid, and Wen-Nee Tan

Abstract Endophytes are a group of microorganisms including fungi, actinomycetes, and bacteria that reside within plant tissues without causing apparent disease symptoms in host plants. These underexplored microorganisms are drawing attention as attractive sources of new compounds, due to their genetic diversity leading to structurally diverse secondary metabolites. In this regard, a balanced host–endophyte interaction plays a vital role in facilitating the endophytes to produce bioactive compounds along with mimicry of plant-based metabolites. Endophytic fungi are great resources of naturally derived drugs as they produce various groups of bioactive molecules including alkaloids, flavonoids, peptides, phenolics, quinines, steroids, and terpenoids. These compounds have been found to display broad-spectrum biological activities such as antimicrobial, antidiabetic, anticancer, anti-inflammatory, antioxidant, and immunomodulatory. In this review, endophytes, their interactions with host plants are discussed and their mechanism of non-pathogenicity are also

M. Abu Taher · W.-Y. Tong (✉) · C. R. Leong · S. Ab Rashid
Universiti Kuala Lumpur, Malaysian Institute of Chemical and Bioengineering Technology, Bioengineering Section, Lot 1988 Kawasan Perindustrian Bandar Vendor, Taboh Naning, 78000 Alor Gajah, Melaka, Malaysia
e-mail: wytong@unikl.edu.my

M. Abu Taher
e-mail: md.abu@s.unikl.edu.my

C. R. Leong
e-mail: crleong@unikl.edu.my

S. Ab Rashid
e-mail: syarifahabrashid@unikl.edu.my

W.-Y. Tong
Universiti Kuala Lumpur, Institute of Medical Science Technology, A1-1, Jalan TKS 1, Taman Kajang Sentral, 43000 Kajang, Selangor, Malaysia

W.-N. Tan
Chemistry Section, School of Distance Education, Universiti Sains Malaysia, 11800 Minden, Penang, Malaysia
e-mail: tanwn@usm.my

A. Ismail et al. (eds.), *Advancements in Materials Science and Technology Led by Women*, Advanced Structured Materials 165,
https://doi.org/10.1007/978-3-031-21959-7_4

highlighted. The review also focuses on the therapeutic applications of metabolites derived from plant-associated endophytic fungi. It can be concluded that endophytes are the repository of bioactive metabolites. Therefore, more studies should be conducted on bioprospecting novel compounds from these potential resources.

Keywords Bioactive compound · Bioprospecting · Endophytes · Therapeutic applications

4.1 Introduction

Endophyte is a Greek word where endo means within, and phyton means plant. It was first introduced in 1866 by De Bary and defined as any organism that grows within plant tissues (Yan et al. 2015). The well-accepted concept of endophyte was proposed by Petrini (1991), and the study-defined endophytes are microorganisms that reside within the tissues of almost all living plants, spending a part of their life maintaining mutual interactions with no significant symptom of disease. Some microorganisms can even spend their whole life as endophytes (Wilson 1995). However, many fungi grow as endophytes in the early stage may turn to a pathogenic state in the later stage upon impairment of host immunity followed by stress conditions or down to age (Duan et al. 2019). Endophytes establish asymptomatic colonization within their host, maintaining an equilibrium state of neutral antagonism between the disease-causing or virulence factors and the cellular defensive mechanisms of plants (Wilson 1995).

The endophytes colonize either the intracellular or intercellular regions of several plant tissues, including root, stem, bark, leaf, flower, and seed, without interfering with the average growth rate of the host plant (Dhayanithy et al. 2019). A microbial species living as an endophyte under specific environments may become parasitic or pathogenic in different hosts or physiological conditions. Nevertheless, many endophytes usually associated with appropriate hosts under favorable environmental conditions (Toghueo and Boyom 2019). Usually, plants provide a system that facilitate the growth and reproduction of endophytic microorganisms (Dhayanithy et al. 2019). Endophytes act as the repository of various novel compounds, enabling them to adapt to the environmental or ecological changes and the diverse physiology of their hosts (Segaran and Sathiavelu 2019). More specifically, endophytes produce different types of bioactive metabolites, enzymes, and plant growth hormones (Toghueo and Boyom 2019; Strobel 2003). Endophytes of poaceous grasses, for example, are well-studied due to their economic value, as infected grasses can release alkaloids that cause grazing livestock to get intoxicated. The presence of endophytes increases the host's chances of survival (Strobel 2003). Endophytes that live in unique biotopes can be a useful source of novel bioactive chemicals because new microorganisms have been linked to new bioactive compounds (Strobel 2003). Endophytic microorganism associated with medicinal plants are prominent sources of novel antimicrobial compounds. Hence, they have great potential for industrial applications, including pharmaceutical, agriculture, biotechnology, food, and textile industries (Strobel 2018).

4.2 Characteristics of Endophytes

4.2.1 Origin

Plant hosts facilitate the coexistence of wide ranges of microorganisms. The host and the adopted organisms develop a symbiotic relationship. The stable symbiotic interactions provide benefits to both the parties in terms of growth, survival, biodiversity, and functional roles in the ecosystem (Sasse et al. 2017). In most cases, endophytic fungi originate from the host's surrounding environment, such as the rhizosphere microorganisms or air-borne fungal spores that settle on the epidermal tissues of plants (Sasse et al. 2017; Robertson-Albertyn et al. 2017). A few basic mechanisms explain the establishment of endosymbiosis, leading to colonization of the endophytic fungi. Most of the fungi can enter through the natural openings of roots in the ground or stromata on the plant's aerial parts such as stems, leaves, or flowers, and any other artificial injuries on surfaces. Fungi produce many exoenzymes, including hydrolases, cellulases, pectinases, and xylanases that hydrolyze plant tissues. The fungi with haustoria or appressoria can access the plants through specialized structures (Yan et al. 2015). Various low to high molecular weight compounds, including organic acids, sugars, nucleotides, and polysaccharides exudating from roots, induce fungal penetration into the plant tissues (Busby et al. 2017). However, bacteria and actinomycetes also follow similar mechanisms of plant-entry. In the early stages of adhesion, bacteria are attracted by various organic exudates such as sugars, organic acids, proteins secreted through natural or artificial openings like root hairs or stomata of the plant. These exudates act as signaling molecules inducing the nearby microorganism to accumulate to the openings that allow them to enter the intracellular spaces in the plant tissue (Kawasaki et al. 2016). Endophytic bacteria producing cellulolytic enzymes, including cellulase, pectinase, and endoglucanase, can easily break down the cell wall to enter and spread inside the plant (Naveed et al. 2014). Some bacteria, for example *Gluconacetobacter*, produce exopolysaccharides that mediate biofilm formation and plant colonization (Meneses et al. 2017).

4.2.2 Distribution

Endophytes are very diverse and ubiquitous in the natural environment. They can survive within plants in a wide range of environments ranging from polar regions to temperate to tropical regions, from desert to aquatic environments, rainforests, mangroves to coastal forests (Toghueo and Boyom 2019). They are commonly associated with various types of plants regardless of the taxonomic categories (Chowdhary and Kaushik 2015). Endophytes can be introduced to the new host through vertical or horizontal transmission. In the vertical mode, the endophytes integrated into the seeds are transferred to the offspring during seed germination. On the other side, many fungi transmit via fragmentation of hyphae, dispersion of spores, biotic (herbivores or insects), or abiotic means (wind or rain) from one plant to another plant (Gagic et al. 2018). The route of entry for colonization depends on the type of plants and microorganisms. Many endophytes penetrated in the host tissues remain latent for their entire life or a period until they have the proper environmental conditions (Nogueira-Lopez et al. 2018). However, the endophytes colonizing various plant tissues away from their initial point of colonization are called systemic or true endophytes. Moreover, the transients or non-systemic endophytes remain localized for part of their life (Gagic et al. 2018). In the absence of competition and compromised immunity, the endophytes can distribute randomly and enter the next potential host plants (Higgins et al. 2014).

4.2.3 Host Plant Interaction

The interaction between the host and endophyte follows a symbiotic or mutualistic relationship, and both parties get benefited from such association (Nogueira-Lopez et al. 2018). Once the endophytes successfully evade the host defense systems, both the partners extend cooperation depending on their inherent characteristics and environmental factors. The endophytes generate various metabolites providing chemical signals, and the hosts responses accordingly (Jia et al. 2016). However, the interactions of the endophytes with their hosts may not always be mutualistic during the coexisting periods. Several studies have explained that the symbiotic relationship may extend to the mutuality, neutrality, commensal, synergistic, antagonistic, or parasitic phase depending on the host types, their immunity, nutrient availability, genetics of the microorganisms, and ecological factors (Jia et al. 2016; Rai and Agarkar 2016).

The interactions between the endophytes and the host can be interpreted by a widely accepted hypothesis called balanced antagonism (Higgins et al. 2014). It states that a continuous antagonistic interaction is balanced by the inherent immunity of the host plant and virulence factors of endophyte, leading to the overall asymptomatic colonization. In addition, endophytes develop resistance mechanisms to the plant's toxic metabolites. If the equilibrium is disturbed by any party or environmental factors such as drought, salinity, temperature, and humidity, the balanced antagonism is also disturbed. As a result, the endophytes change their status to pathogens (Schulz et al. 2015). The hypothesis also narrates that the reciprocity between endophytes and pathogens are not universal. For example, endophytic fungus, *Colletotrichum magna,* switches to a pathogenic state due to humidity, leading to anthracnose in the cucurbit plant (Rai and Agarkar 2016). Moreover, the endophyte and host maintain symbiotic or coexisting phases where they impart some of their activities, including induction and execution of metabolism and partially sharing metabolic pathways (Ludwig-Müller 2015). In addition, the specific interactions between the host and endophytes contribute to the quality and quantity of bioactive constituents of the plant.

Plants and endophytes interactions are mutualistic (Aly et al. 2011). The host provides a shelter, facilitates nutrients and ecological niches appropriate for the growth and survival, and protects the endophytes from their physical, chemical, and biological enemies (Yan et al. 2015; Sasse et al. 2017). On the other hand, endophytes are the synthesizers of bioactive molecules, many of which are utilized by the hosts (Tidke et al. 2017). They produce metabolites supplementing nutrients, fixing nitrogen, solubilizing minerals, essential microelements, and cofactors of enzymes. They also release growth hormones that enhance biomass and plants' survival (Aly et al. 2011). As part of metabolisms, they also synthesize bioactive compounds conferring host defense to the pathogens (Singh et al. 2017). Endophytes enable the host plants to manage the herbivores in several ways, including reducing insect growth rate, compelling feeding deterrence, and reducing survival and oviposition (Tidke et al. 2017). Several investigations reported that endophytes secrete cell wall degrading extracellular enzymes to break complex substrates into simple molecules that can be easily up taken by the host plant (Zhejian et al. 2015).

4.3 Endophytic Fungi

Endophytic fungi are a group of fungal microorganisms, inter or intra-cellularly living the whole or part of their life cycle inside the plant tissues without causing any apparent illness to the host (Yan et al. 2015). Endophytic fungi have been widely described to produce various bioactive molecules comprising alkaloids, flavonoids, peptides, phenolics, quinines, steroids, and terpenoids, which play a vital role as antimicrobial, antiviral, anti-insecticidal, anticancer, and antidiabetic agents (Strobel 2003). In addition, endophytic fungi can survive in hostile environmental conditions such as extreme temperature, pressure, and pH. These adverse conditions trigger them to produce secondary metabolites as the survival strategies, which, upon recovery, may possess various bioactive properties (Toghueo 2020).

In this context, endophytic fungi act as sustainable natural reservoirs of compounds and have been drawing the attention of the researchers. The bioactive compounds derived from biological sources are useful and relatively less toxic, with a negligible effect on the environment (Pham et al. 2019). This group of underexplored endophytic fungi exhibits diverse genetic and metabolic capabilities, and these characteristics enable them to produce a high abundance of bioactive compounds and adapt well to environmental stresses. As a part of mutual relationships, endophytes generate many biologically active compounds protecting themselves and their hosts from the enemies (Yan et al. 2015; Chen et al. 2016). However, selecting a suitable host plant is the prerequisite to successful bioprospecting of novel bioactive compounds from the endophytic fungi. From the beginning, ethnobotanical knowledge has helped the discovery of novel bioactive compounds (Alvin et al. 2014). Usually, the plants having the ethnobotanical history of applications by the local and indigenous people are good sources of potential endophytes (Strobel and Daisy 2003). In addition, plants surrounded by the pathogen-infected plants are potential candidates that harbor antimicrobial compounds producing endophytes.

Endophytic fungi are heterogeneous groups of microorganisms and present in almost every plant, and till now, no study has found the absence of endophyte in plants (Nisa et al. 2015). Although endophytic fungi are ubiquitous and morphologically and functionally diverse microorganisms, a significant part of them is still unexplored. They are usually associated with various tissues, including roots, stems, and leaves of host plants (Tidke et al. 2017). Most of the endophytic fungi are non-specific to the host plants, while some are considered genus or family specific. Besides, several factors, including geographic location, seasonal conditions, initial endophyte colonization, the type of tissue, and physicochemical states of plants, regulate the endophytic establishment (Toghueo and Boyom 2019). The fungal genera dominating as endophytes are *Alternaria, Aspergillus, Botryosharia, Chaetomium, Cladosporium, Colletotrichum, Diaporthe, Fusarium, Nigrospora, Pestalotiopsis, Phoma, Taxomyces*, and *Xylaria* (Rashmi et al. 2019). *Colletotrichum, Phoma, Phomopsis*, and *Phyllosticta* are the most common fungal genera distributed in a broad range of host plants (Segaran and Sathiavelu 2019).

The endophytic fungi obtain growth and survival facilities from the host and return some benefits such as protection from biotic agents like pathogens and herbivores, and abiotic stresses (Yao et al. 2017). Many studies have reported that *Aspergillus, Penicillium*, and *Phoma* can produce compounds acting against common bacterial and fungal pathogens (Shikano et al. 2017). For example, the host plants excrete excess salicylic acid along with some defensive proteins and enzymes, providing immunity to downy mildew pathogen, *Sclerospora graminicola* in pearl millet when *Trichoderma hamatun* colonizes as an endophytic fungus (Siddaiah et al. 2017).

Fungal endophytes are broadly classified into two main groups. These are clavicipitaceous or balansiaceous endophytes, and non-clavicipitaceous or non-balansiaceous endophytic fungi (Strobel and Daisy 2003). Moreover, all endophytic fungi are aligned into four classes based on taxonomy or phylogeny, ecological roles, modes of transmission, life-history traits, host plant range, and fitness benefits (Santangelo et al. 2015). Class 1 belongs to the Clavicipitaceous group, and the rest of the classes (classes 2, 3, and 4) are included in the non-clavicipitaceous group (Tidke et al. 2017). Besides, endophytic fungi can be systemic and transient endophytes based on their genetics and transmission modes. Systemic endophytes complete their whole life cycle within the plant, and they are vertically transmitted from the parents to offspring plants. On the other hand, non-systemic fungal endophytes horizontally colonize, and they are transient inhabitants of host plants (Wani et al. 2015).

4.4 Natural Products from Endophytic Fungi

The term bioactive (the Greek, bio-life, and active from the Latin activus-dynamic, or full of energy or involves to activity) or biologically active compound means any substance with an effect, triggers a reaction, or induces a response to the living organism or tissue. The effects may either be positive or negative, depending on the type of compounds, dose, and bioavailability (Guaadaoui et al. 2014). Chemicals with medicinal benefits are called active ingredients of natural medicines (Zhang et al. 2018). The compounds have a wide range of functional spectra, including antiallergens, anti-cancers, antidiabetics, anti-inflammatory, antimicrobials, antioxidants, antispasmodics, anti-ulcers, auxins, herbicides, natural pesticides, neuro-active, and toxins (Strobel 2003).

Bioprospecting is the process of exploring and commercializing of potential new products from biological resources. It should involve a systematic search for, and development of new products from compounds, genes, micro- or macroorganisms in the nature. Therefore, bioprospecting indicates the scientific ways of exploring natural products from the biological sources (Tidke et al. 2017). Searching for novel compounds has been practicing for centuries from biological sources, especially microorganisms, plants, or animals. This interest has recently been significantly increased, focusing additional attention on endophytes for several reasons such as treatment of emerging or recurring infections, antimicrobial resistance to existing antibiotics, managing wounds, and transplant infections (Guaadaoui et al. 2014). The microorganisms being renewable sources may act as a good source of bioactive compounds. Simultaneously, this also bypasses the dependency on the synthetic compounds that are sometimes xenobiotic, difficult to metabolize in the body, and selectively detrimental to the environment due to non-degradability (Aly et al. 2011).

The discovery of new therapeutic alternatives is a continuous process of drug development study. The common ways of developing the medically important compounds are logical drug design to specific targets, synthesis of molecules using combinatorial chemistry, and natural products discover from biological sources (Alvin et al. 2014). Although the first two methods use advanced technological tools, they are very costly, having long-term effects on health and the environment (Strobel and Daisy 2003). Therefore, continuous studies are conducted to explore industrially important compounds from renewable biological sources like endophytic fungi. The critical issues of medicinal compounds are efficacy and cytotoxicity. Since endophytes' products are naturally derived or analogous to plant compounds, they are relatively less cytotoxic than compounds produced by the hosts (Strobel 2003; Strobel and Daisy 2003).

Table 4.1 Antimicrobial compounds produced by endophytic fungi

Product	Endophyte	Host	References
Berberine	*Alternaria* sp.	*Phellodendron amurense*	Strobel (2003)
Camptothecin	*Fusarium solani*	*Camptotheca acuminate*	Abdalla and McGaw (2018)
Cinchonidine	*Diaporthe* sp.	*Cinchona ledgeriana*	Maehara et al. (2012)
Pestalotiopamide E	*Aureobasidium pullulans*	*A. vera*	El-Amrani et al. (2016)
Thiodiketopiperazines	*Phoma* sp.	*Glycyrrhiza glabra*	Arora et al. (2016)
Fumigaclavine C	*Aspergillus* sp.	*Bauhinia guianensis*	Ma et al. (2006)
Isoquinolin	*Penicillium* sp.	*Nerium indicum*	Ma et al. (2017)
Chaetomugilin A, D	*Chaetomium globosum*	*Ginkgo biloba*	Qin et al. (2009)
Griseofulvin	*Xylaria* sp.	*Abies holophylla*	Park et al. (2005)
Gavodermside	*Phomopsis* sp.	*Aconitum carmichaeli*	Wu et al. (2013)
Piperine	*C. gloeosporioides*	*Piper nigrum*	Chithra et al. (2014)
Quinine	*Diaporthe* sp.	*Cinchona ledgeriana*	Maehara et al. (2012)

4.4.1 Antimicrobial Compounds

Endophytic fungi are the common producers of various bioactive compounds, which are able to inhibit the growth of microorganisms including bacteria, fungi, viruses, and protozoa (Strobel 2003). Penicillin is the first antimicrobial compound discovered by Sir Alexander Fleming in 1929 from a fungus later identified as *Penicillium notatum*. Since then, numerous studies have been conducted to search for new antibiotics from the endophytic fungi (Toghueo and Boyom 2019). Table 4.1 presents a list of antimicrobial compounds produced by endophytic fungi.

4.4.2 Anticancer

Endophytic fungi are the most promising sources of compounds for treating various cancers due to the unregulated multiplication of abnormal or functionally deviated cells. The available chemotherapeutic agents have proven to have many non-specific cyto-toxicities leading to severe complications of patients (Strobel 2003). Taxol, the world's first billion-dollar anticancer medication, was discovered from *Taxomyces*

Table 4.2 Anticancer compounds produced by endophytic fungi

Product	Endophyte	Host	References
Altertoxin	*Alternaria* sp.	*Broussonetia papyrifera*	Zhang et al. (2018)
Arugosin J, Xylarugosin	*Xylaria* sp.	*Curcuma xanthorrhiza*	Hammerschmidt et al. (2015)
Camptothecin	*Fusarium solani*	*Camptotheca acuminate*	Abdalla and McGaw (2018)
5-hydroxyramulosin	*Phoma* sp.	*Cinnamomum mollisimum*	Santiago et al. (2012)
Paclitaxel	*Taxomyces andreanae*	*Taxus brevifolia*	Stierle et al. (1993)
Podophyllotoxin	*Aspergillus fumigatus*	*Juniperus communis*	Kusari et al. (2012)
Vinblastine, Vincristine	*F. oxysporum*	*Catharanthus roseus*	Kumar et al. (2013)

andreanae, an endophytic fungus of the pacific yew *Taxus brevifolia* (Stierle et al. 1993). Because only the inner bark of matured pacific yew can produce taxol, the drug's price is high, and the yield is minimal. Because the same molecule can be synthesized by its endophyte, the drug's manufacturing costs can be reduced because industrial fermentation can be used to harvest the compound. After discovering the renowned anticancer drug paclitaxel produced by *T. andreanae*, an endophyte of *T. brevifolia*, several investigations have found endophyte-derived potent anticancer agents (Stierle et al. 1993).

Camptothecin, a pentacyclic quinoline alkaloid with anti-neoplastic action by blocking intranuclear topoisomerase I, is found in the wood of *Camptotheca acuminota* (Puri et al. 2018). Because of the increasing demand for the medicine, massive cropping of the plants occurred in China and India, necessitating the search for a microbiological source of the cancer chemotherapeutic treatment. The first report of camptothecin from fungal endophyte was from *Entrophospora infrequens* residing in *Nothapodytes foetida* (Puri et al. 2018). Table 4.2 summarizes the related studies conducted on anticancer molecules identified from the endophytic fungi. The toxic compounds are released as a part of the defense strategies and spread at the site of colonization of the endophytes (Rajamanikyam et al. 2017). Consequently, the secreted compounds also protect their natural habitat for the host.

4.4.3 Antidiabetic

Endophytic fungi are potential resources to explore the effective and safe therapeutic alternative for the treatment of diabetes. Several compounds derived from endophytes have been tested for antidiabetic and hypolipidemic activity. A study was performed to investigate the effects of extracts of *Aspergillus* sp. on Wistar albino rats with glucose and alloxan induced diabetes (Dhankhar et al. 2013). They concluded that the fungal species produce antidiabetic compounds, significantly lowering blood glucose levels. *Alternaria* sp. and *Xylaria* sp. produce N-acetylgalactosamine and 8-hydroxy-6,7-dimethoxy-3-methyl isocoumarin, respectively, which act as inhibitors of α-glucosidase (Govindappa et al. 2015). The enzyme hydrolyzes dietary carbohydrates to glucose, of which increased level in blood is an indicator of diabetic condition. Table 4.3 refers to some products with antidiabetic activity of endophytic fungi.

4.4.4 Antioxidant

The study of antioxidants is becoming the topic of interest due to the degenerative diseases caused by various free radicals generated through physiological processes. Endophytic fungi generate metabolites with antioxidant activity to capture the oxygen-derived toxic free radicals based on radical scavenging assays. The methanolic extract of *Fusarium lateritium*, an endophyte of *Rhizophora mucronata*, has free radical scavenging property (Hamzah et al. 2018). An effective antioxidant, L-ascorbic acid, was isolated from endophytic *Penicillium citrinum*, residing in *Tragia involucrate* (Danagoudar et al. 2018).

4.4.5 Immunosuppressive

Immunosuppressive compounds are essential mediators to deal with immunological disorders, especially in graft rejection and autoimmune diseases (Rajamanikyam et al. 2017). Subsequently, several fungal metabolites with immunomodulatory functions have been isolated. Two well-known immunosuppressive agents, subglutinol A and subglutinol B, were isolated from *Fusarium subglutinans*, an endophyte of fungus found in *Tripterygium wilfordii* (Lee et al. 1995). The drugs relieve the patients undergoing organ transplant, hence reduce the chance of allograft rejection in the treatment of autoimmune diseases (Padhi et al. 2013). The intense immunomodulatory activity was shown with the crude extracts of *Penicillium* species. Delayed-type hypersensitivity responses were noticed with the chloroform and methanolic extracts (Pur et al. 2007). Recently, the immunosuppressive cytochalasins produced by the endophytic *Xylaria* sp. can resist concanavalin-A-induced T-lymphocyte and lipopolysaccharide-induced B-lymphocyte proliferation (Wang et al. 2019).

Table 4.3 Antidiabetic, antioxidant, and immunosuppressive compounds produced by endophytic fungi

Product	Endophyte	Host	References
Antidiabetic			
N-acetylgalactosamine	*Alternaria* sp.	*Viscum album*	Govindappa et al. (2015)
8-hydroxy-6,7-dimethoxy-3-methylisocoumarine	*Xylaria* sp.	*Quercus gilva*	Indrianingsih and Tachibana (2017)
2,6-di-tertbutyl-p-cresol	*Aspergillus* sp. JPY1	*Salvadora oleoides*	Dhankhar et al. (2013)
Antioxidant			
L-ascorbic acid	*Penicillium citrinum*	*Tragia involucrate*	Danagoudar et al. (2018)
Cajaninstilbene acid	*Fusarium* sp.	*Cajanus cajan*	Zhao et al. (2012)
Gallic acid	*Aspergillus* sp.	*Calotropis procera*	Khiralla et al. (2015)
Graphislactone A	*Cephalosporium* sp.	*Trachelospermum jasminoides*	Song et al. (2005)
Salidrosides, p-tyrosol	*Rhodiola* sp.	*Alpine plants*	Cui et al. (2016)
Immunosuppressive			
Colutellin A	*Colletotrichum* sp.	*Pteromischum sp.*	Ren et al. (2008)
Rigidiusculamide E, Oxysporidinone	*Fusarium tricinctum*	*Panax notoginseng*	Sun et al. (2018)

4.5 Conclusion

Endophytes are rich and a reliable source of various compounds with various medicinal and agricultural applications. These poorly investigated microorganisms are proven to be the reservoir of diverse bioactive molecules. These compounds having unique chemical structures exhibit a wide range of biological activities such as antimicrobial, antidiabetic, anti-inflammatory, antioxidant, anticancer, and immunosuppression. However, further studies should be conducted at the molecular level in this field for a clear understanding of the interactions between host and endophyte which may expedite the discovery of novel compounds with diverse bioactivities.

Acknowledgements The authors are thankful to Universiti Kuala Lumpur.

References

Abdalla MA, McGaw LJ (2018) Bioprospecting of South African plants as a unique resource for bioactive endophytic microbes. Front Pharmacol 9:1–18

Alvin A, Miller KI, Neilan BA (2014) Exploring the potential of endophytes from medicinal plants as sources of antimycobacterial compounds. Microbiol Res 169(7–8):483–495

Aly AH, Debbab A, Proksch P (2011) Fungal endophytes: unique plant inhabitants with great promises. Appl Microbiol Biotechnol 90(6):1829–1845

Arora P, Wani ZA, Nalli Y et al (2016) Antimicrobial potential of Thiodiketopiperazine derivatives produced by *Phoma* sp., an endophyte of *Glycyrrhiza glabra* Linn. Microb ecol 72(4):802–812

Busby PE, Soman C, Wagner MR et al (2017) Research priorities for harnessing plant microbiomes in sustainable agriculture. Plos One Biol 15(3):1–14

Chen L, Zhang QY, Jia M et al (2016) Endophytic fungi with antitumor activities: their occurrence and anticancer compounds. Crit Rev Microbiol 42(3):454–473

Chithra S, Jasim B, Sachidanandan P et al (2014) Piperine production by endophytic fungus *Colletotrichum gloeosporioid*es isolated from *Piper nigrum*. Phytomedicine 21(4):534–540

Chowdhary K, Kaushik N (2015) Fungal endophyte diversity and bioactivity in the Indian medicinal plant *Ocimumsanctum* Linn. PLoS ONE 10(11):1–25

Cui J, Guo T, Chao J et al (2016) Potential of the endophytic fungus *Phialocephala fortinii* Rac56 found in Rhodiola plants to produce salidroside and p-tyrosol. Mol 21(4):502

Danagoudar A, Joshi CG, Ravi SK et al (2018) Antioxidant and cytotoxic potential of endophytic fungi isolated from medicinal plant *Tragia involucrata* L. Pharmacognosy Res 10(2):188–194

Dhankhar S, Dhankhar S, Yadav JP (2013) Investigations towards new antidiabetic drugs from fungal endophytes associated with *Salvadora oleoides* Decne. Med Chem 9(4):624–632

Dhayanithy G, Subban K, Chelliah J (2019) Diversity and biological activities of endophytic fungi associated with *Catharanthus roseus*. BMC Microbiol 19:1–14

Duan X, Xu F, Qin D et al (2019) Diversity and bioactivities of fungal endophytes from *Distylium chinense*, a rare waterlogging tolerant plant endemic to the Three Gorges Reservoir. BMC Microbiol 19:1–14

El-Amrani M, Ebada SS, Gad HA et al (2016) Pestalotiopamide E and pestalotiopin B from an endophytic fungus *Aureobasidium pullulans* isolated from *Aloe vera* leaves. Phytochem Lett 18:95–98

Gagic M, Faville MJ, Zhang W et al (2018) Seed transmission of *Epichloë* endophytes in *Lolium perenne* is heavily influenced by host genetics. Front Plant Sci 9:1–16

Govindappa M, Sadananda TS, Ramachandra YL et al (2015) In vitro and in vivo antidiabetic activity of lectin (N-acetylgalactosamine, 64 kDa) isolated from endophytic fungi, *Alternaria* species from *Viscum album* on alloxan induced diabetic rats. Intergr Obes Diabetes 1(1):11–19

Guaadaoui A, Benaicha S, Elmajdoub N et al (2014) What is a bioactive compound? A combined definition fora preliminary consensus. Int J Nutr Food Sci 3(3):74–179

Hammerschmidt L, Ola A, Mueller WE et al (2015) Two new metabolites from the endophytic fungus *Xylaria* sp. isolated from the medicinal plant *Curcuma xanthorrhiza*. Tetrahedron Lett 56(10):1193–1197

Hamzah TNT, Lee SY, Hidayat A et al (2018) Diversity and characterization of endophytic fungi isolated from the tropical mangrove species, *Rhizophora mucronata*, and identification of potential antagonists against the soil-borne fungus, *Fusarium solani*. Front Microbiol 9:1–17

Higgins KL, Arnold AE, Coley P et al (2014) Communities of fungal endophyte in tropical forest grasses: highly diverse host and habitat generalists characterized by strong spatial structure. Fungal Ecol 8(1):1–11

Indrianingsih AW, Tachibana S (2017) α-Glucosidase inhibitor produced by an endophytic fungus, *Xylariaceae* sp. QGS 01 from *Quercus gilva* Blume. Food Sci Hum Wellness 6(2):88–95

Jia M, Chen L, Xin H et al (2016) A friendly relationship between endophytic fungi and medicinal plants: a systematic review. Front Microbiol 7:1–14

Kawasaki A, Donn S, Ryan PR et al (2016) Microbiome and exudates of the root and rhizosphere of *Brachypodium distachyon*, a model for wheat. PLoS ONE 11(10):1–25

Khiralla A, Mohamed I, Thomas J et al (2015) A pilot study of antioxidant potential of endophytic fungi from some Sudanese medicinal plants. Asian Pac J Trop Med 8(9):701–704

Kumar A, Patil D, Rajamohanan PR et al (2013) Isolation, purification and characterization of vinblastine and vincristine from endophytic fungus *Fusarium oxysporum* isolated from *Catharanthus roseus*. PLoS ONE 8(9):e71805

Kusari S, Hertweck C, Spiteller M (2012) Chemical ecology of endophytic fungi: origins of secondary metabolites. Chem Biol 19(7):792–798

Lee JC, Lobkovsky E, Pliam NB et al (1995) Subglutinol A and B: immunosuppressive compounds from the endophytic fungus *Fusarium subglutinans*. J Org Chem 60(22):7076–7077

Ludwig-Müller J (2015) Plants and endophytes: equal partners in secondary metabolite production? Biotechnol Lett 37(7):1325–1334

Ma HY, Song YC, Mao YY et al (2006) Endophytic fungal metabolite fumigaclavine C causes relaxation of isolated rat aortic rings. Planta Med 72(05):387–392

Ma YM, Qiao K, Kong Y et al (2017) A new isoquinolone alkaloid from an endophytic fungus R22 of *Nerium indicum*. Nat Prod Res 31(8):951–958

Maehara S, Simanjuntak P, Kitamura C et al (2012) Bioproduction of Cinchona alkaloids by the endophytic fungus *Diaporthe* sp. associated with *Cinchona ledgeriana*. Chem Pharm Bull 60(10):1301–1304

Meneses C, Gonçalves T, Alquéres S et al (2017) *Gluconacetobacter diazotrophicus* exopolysaccharide protects bacterial cells against oxidative stress in vitro and during rice plant colonization. Plant Soil 416:133–147

Naveed M, Mitter B, Yousaf S et al (2014) The endophyte *Enterobacter* sp. FD17: a maize growth enhancer selected based on rigorous testing of plant beneficial traits and colonization characteristics. Biol Fertil Soils 50(2):249–262

Nisa H, Kamili AN, Nawchoo IA et al (2015) Fungal endophytes as prolific source of phytochemicals and other bioactive natural products: a review. Microb Pathog 82:50–59

Nogueira-Lopez G, Greenwood DR, Middleditch M et al (2018) The apoplastic secretome of *Trichoderma virens* during interaction with maize roots shows an inhibition of plant defence and scavenging oxidative stress secreted proteins. Front Plant Sci 9:1–23

Padhi L, Mohanta YK, Panda SK (2013) Endophytic fungi with great promises: a review. J Adv Pharm Edu Res 3(3):152–170

Park JH, Choi GJ, Lee SW et al (2005) Griseofulvin from *Xylaria* sp. strain F0010, an endophytic fungus of *Abies holophylla* and its antifungal activity against plant pathogenic fungi. J Microbiol Biotechnol 15(1):112–117

Petrini O (1991) Fungal endophytes of tree leaves. In: Andrews JH, Hirano SS (eds) Microbial ecology of the leaves, 1st edn. Springer, New York

Pham JV, Yilma MA, Feliz A et al (2019) A review of the microbial production of bioactive natural products and biologics. Front Microbiol 10:1–27

Pur S, Amna T, Khajuria A et al (2007) Immunomodulatory activity of an extract of the novel fungal endophyte *Entrophospora infrequens* isolated from *Nothapodytes foetida* (Wight) Sleumer. Acta Microbiol Imm H 54(3):237–260

Puri SK, Habbu PV, Kulkarni PV et al (2018) Nitrogen containing secondary metabolites from endophytes of medicinal plants and their biological/pharmacological activities—a review. Syst Rev Pharm 9(1):22–30

Qin JC, Zhang YM, Gao JM et al (2009) Bioactive metabolites produced by *Chaetomium globosum*, an endophytic fungus isolated from *Ginkgo biloba*. Bioorganic Med Chem Lett 19(6):1572–1574

Rai M, Agarkar G (2016) Plant-fungal interactions: what triggers the fungi to switch among lifestyles? Crit Rev Microbiol 42(3):428–438

Rajamanikyam M, Vadlapudi V, Amanchy R et al (2017) Endophytic fungi as novel resources of natural therapeutics. Braz Arch Biol Technol 60:1–26

Rashmi M, Kushveer JS, Sarma VV (2019) A worldwide list of endophytic fungi with notes on ecology and diversity. Mycosphere 10(1):798–1079

Ren Y, Strobel GA, Graff JC et al (2008) Colutellin A, an immunosuppressive peptide from *Colletotrichum dematium*. Microbiology 154(7):1973–1979

Robertson-Albertyn S, Terrazas RA, Balbirnie K et al (2017) Root hair mutations displace the barley rhizosphere microbiota. Front Plant Sci 8:1–15

Santangelo JS, Turley NE, Johnson MT (2015) Fungal endophytes of *Festuca rubra* increase in frequency following longterm exclusion of rabbits. Botany 93(4):233–241

Santiago C, Fitchett C, Munro MH et al (2012) Cytotoxic and antifungal activities of 5-hydroxyramulosin, a compound produced by an endophytic fungus isolated from *Cinnamomum mollisimum*. Evid Based Complement Alternat Med 2012

Sasse J, Martinoia E, Northen T (2017) Feed your friends: doplant exudates shape the root microbiome? Trends Plant Sci 23(1):25–41

Schulz B, Haas S, Junker C et al (2015) Fungal endophytes are involved in multiple balanced antagonisms. Curr Sci 109(1):39–45

Segaran G, Sathiavelu M (2019) Fungal endophytes: a potent biocontrol agent and a bioactive metabolites reservoir. Biocatal Agric Biotechnol 21:1–17

Shikano I, Rosa C, Tan CW et al (2017) Tritrophic interactions: microbe-mediated plant effects on insect herbivores. Annu Rev Phytopathol 55:313–331

Siddaiah CN, Satyanarayana NR, Mudili V et al (2017) Elicitation of resistance and associated defense responses in *Trichoderma hamatum* induced protection against pearl millet downy mildew pathogen. Sci Rep 7:1–18

Singh M, Kumar A, Singh R et al (2017) Endophytic bacteria: a new source of bioactive compounds. 3 Biotech 7(5):1–14

Song YC, Huang WY, Sun C et al (2005) Characterization of graphislactone A as the antioxidant and free radical-scavenging substance from the culture of *Cephalosporium* sp. IFB-E001, an endophytic fungus in *Trachelospermum jasminoides*. Biol Pharm Bull 28(3):506–509

Stierle A, Strobel G, Stierle D (1993) Taxol and taxane production by *Taxomyces andreanae*, an endophytic fungus of Pacific yew. Science 260:214–216

Strobel GA (2003) Endophytes as sources of bioactive products. Microbes Infect 5(6):535–544

Strobel GA (2018) The emergence of endophytic microbes and their biological promise. J Fungi 4:1–19

Strobel GA, Daisy B (2003) Bioprospecting for microbial endophytes and their natural products. Microbiol Mol Biol Rev 67(4):491–502

Sun WJ, Zhu HT, Zhang TY et al (2018) Two new alkaloids from *Fusarium tricinctum* SYPF 7082, an endophyte from the root of *Panax notoginseng*. Nat Prod Bioprospect 8(5):391–396
Tidke SA, Kumar AKL, Ramakrishna D et al (2017) Current understanding of endophytes: their relevance, importance, and industrial potentials. J Biotechnol Biochem 3(3):43–59
Toghueo RMK (2020) Bioprospecting endophytic fungi from *Fusarium* genus as sources of bioactive metabolites. Mycology 11(1):1–21
Toghueo RMK, Boyom FF (2019) Endophytic fungi from *Terminalia* species: a comprehensive review. J Fungi 5(2):1–20
Wang WX, Lei X, Ai HL et al (2019) Cytochalasans from the endophytic fungus *Xylaria* cf. curta with resistance reversal activity against fluconazole-resistant *Candida albicans*. Org Lett 21(4):1108–1111
Wani ZA, Ashraf N, Mohiuddin T et al (2015) Plant-endophyte symbiosis, an ecological perspective. Appl Microbiol Biotechnol 99(7):2955–2965
Wilson D (1995) Endophyte: the evolution of a term, and clarification of its use and definition. Oikos 73(2):274–276
Wu SH, Huang R, Miao CP et al (2013) Two new steroids from an endophytic fungus *Phomopsis* sp. Chem Biodivers 10(7):1276–1283
Yan JF, Broughton SJ, Yang SL et al (2015) Do endophytic fungi grow through their hosts systemically? Fungal Ecol 13:53–59
Yao YQ, Lan F, Qiao YM et al (2017) Endophytic fungi harbored in the root of *Sophora tonkinensis* Gapnep: diversity and biocontrol potential against phytopathogens. Microbiol Open 6(3):1–17
Zhang QW, Lin LG, Ye WC (2018) Techniques for extraction and isolation of natural products: a comprehensive review. Chin Med 13:1–26
Zhao J, Fu Y, Luo M et al (2012) Endophytic fungi from pigeon pea [*Cajanus cajan* (L.) Millsp.] produce antioxidant cajaninstilbene acid. J Agricul Food Chem 60(17):4314–4319
Zhejian W, Zhao M, Lili W et al (2015) Active anti-acetylcholinesterase component of secondary metabolites produced by the endophytic fungi of *Huperzia serrata*. Electron J Biotechnol 18(6):399–405

Chapter 5
Biochar-Based Graphitic Carbon Nitride Derived from Biomass Waste for Degradation of Pyrene

Noor Aina Mohd Nazri, Syarifah Nurhusna Qadirah Syed Abd Halim, and Sabrina Karim

Abstract A series of biochar (BC)/graphitic carbon nitride (g-C_3N_4) derived from different biomasses were synthesized via the thermal polycondensation method. The photocatalyst was characterized by FTIR, N_2 sorption isotherm, Tauc plot method and XRD. The performance of the BC/g-C_3N_4 photocatalyst was evaluated by assessing the efficiency of the photocatalyst for the degradation of pyrene as model pollutants. The types of biomasses used and surface area generally influenced the photocatalytic performance. Experiment results revealed that the BC/g-C_3N_4 demonstrated a higher band gap (~ 3.27) as compared to pristine g-C_3N_4. Among the photocatalysts tested, g-C_3N_4 derived from sugarcane bagasse exhibited the highest degradation of pyrene. Several factors such as photocatalyst dosage and initial concentration of pyrene during photocatalytic experiment influenced the performance of BC/g-C_3N_4. The experimental results demonstrated the potential of utilization of abundance biomass waste as low-cost material to produce a biochar-based photocatalyst which is part of the effort to promote a green and sustainable solution for water reclamation.

Keywords Biochar · Graphitic carbon nitride · Pyrene · Photocatalytic

N. A. M. Nazri (✉) · S. N. Q. S. A. Halim
Universiti Kuala Lumpur, Malaysian Institute of Chemical and Bioengineering Technology, Lot 1988, Kawasan Perindustrian Bandar Vendor, Simpang Ampat 78000. Alor Gajah, Melaka, Malaysia
e-mail: nooraina@unikl.edu.my

S. N. Q. S. A. Halim
e-mail: syarifah.halim19@s.unikl.edu.my

S. Karim
Universiti Kuala Lumpur, Institute of Medical Science Technology, A1, 1, Jalan TKS 1, Taman Kajang Sentral, 43000 Kajang, Selangor, Malaysia
e-mail: sabrinakarim@unikl.edu.my

A. Ismail et al. (eds.), *Advancements in Materials Science and Technology Led by Women*, Advanced Structured Materials 165,
https://doi.org/10.1007/978-3-031-21959-7_5

5.1 Introduction

The photocatalysis technology has attracted significant attention by researchers due to its green, high efficiency and low energy consumption. Recently, the technology has emerged as one of the promising alternatives for degradation of various pollutants (i.e., pesticide, pharmaceutical compounds, textile wastewater, etc.) (Koe et al. 2020). However, in spite of the broad application of the photocatalysis process, several issues such as low efficiency under visible light and fast recombination rate of the photoinduced electron–hole which could decrease the efficiency of photocatalyst could hinder its wider application particularly for industrial applications (Cui et al. 2020; Hasija et al. 2021; Karpuraranjith et al. 2021). For example, g-C_3N_4 has received immense attention for various applications due to its high chemical stability, non-toxicity and low band gap of around 2.7 eV which makes it as an attractive visible light active. However, its relatively low visible light absorption, quick recombination of excited charges and low surface area have become main concern by researchers (Qi et al. 2020).

In this regard, various strategies such as doping with foreign elements, introduction of co-catalysts and integration with noble metals have been proposed to further enhance the properties and performance of the g-C_3N_4 (Hasija et al. 2021; Karpuraranjith et al. 2021; Zhu et al. 2018; Fronczak et al. 2018). The strategies also include the utilization of low-cost photocatalysts derived from biomass waste. Biomass waste with high carbon content that can be converted into a value-added material can be used to produce biochar-based photocatalysts with the addition of a precursor (i.e., melamine, urea). The carbon-modified (biochar-based) photocatalyst can be considered as a potential new class of green photocatalyst due to its promising visible-light-active property and its versatile characteristics such as easy to synthesize, low-cost starting compounds and considerable physicochemical stability (Zhu et al. 2018; Ahmed 2022; Lu et al. 2022; Tseng et al. 2022). Waste biomass derived from agriculture that consists of cellulose, lignin and hemicellulose can be converted to value-added carbon material via the pyrolysis method. This approach provides an attractive route for environmentally friendly solutions (Wang and Wang 2022; Shi et al. 2020).

Recently, different types of biomasses such as hazelnut shell, olive residue, withered magnolia blossoms, bamboo charcoal, sugarcane bagasse, urea and chestnut leaves have been used to synthesize the BC/g-C_3N_4 photocatalyst (Zhu et al. 2018; Lu et al. 2022; Tseng et al. 2022; Jeon et al. 2017; Pi et al. 2015; Sliem et al. 2018). Recent studies highlighted that BC/g-C_3N_4 derived from biomasses (i.e., leaf, red mud) demonstrated a long-term stability, better separation efficiency for the degradation of various pollutants such as 2-mercaptobenzothiazole, antibiotics and dyes under visible light due to enhanced properties of BC/g-C_3N_4 with the introduction of biomass (Zhu et al. 2018; Shi et al. 2020; Pi et al. 2015).

In this study, BC/g-C_3N_4 were synthesized from different types of biomasses (i.e., sugarcane bagasse (SB), palm kernel shell (PKS) and rice husk (RH)) via a one-step thermal polycondensation method using melamine as precursor. The photocatalytic degradation performance of the BC/g-C_3N_4 was evaluated with pyrene, a polycyclic aromatic hydrocarbons (PAHs) as model pollutants. The performance of the BC/g-C_3N_4 photocatalyst synthesized at different ratio of biomass to precursor was also evaluated at various parameters such as initial concentration of pyrene and photocatalyst dosage. Up to date, limited studies have been performed to understand the performance of BC/g-C_3N_4 derived from different types of biomasses. Thus, this study could provide a better understanding on the selection of biomass to synthesize photocatalyst as well as better insight into the performance of BC/g-C_3N_4 for degradation of pollutants from various wastewater. This study also provides an attractive sustainable approach for water treatment by the utilization of biomass waste as low-cost material.

5.2 Methodology

BC/g-C_3N_4 were synthesized via the thermal polycondensation method. Prior to synthesis, all the biomass wastes were firstly pre-treated by washing and drying at 100 °C for 8 h. This was followed by crushing the pre-treated biomasses with grinder (12,000 rpm) and sieving through a 30 mesh sieve (0.6 mm). In the first part of the experiment, BC/g-C_3N_4 at ratio of melamine to biomass of 75:25 were selected to understand the effect of temperature and time during synthesis of the photocatalyst. The samples were then heated in a muffle furnace at various synthesis temperature (450–600 °C) and time (2, 3, 4 h) at constant rate of 5 °C/min. The optimum temperature and time which were identified based on the first part of the experiment were used to synthesize BC/g-C_3N_4 at different ratios (second part).

For the second part, in a typical experiment, melamine which was used as a precursor was mixed with the biomass waste (i.e., SB, PKS, RH) according to the specified ratio as shown in Table 5.1. The prepared samples were then collected and used as a photocatalyst in the photocatalytic degradation experiments.

The Fourier transform infrared (FTIR) spectra were collected from Thermo Scientific, US in the range of 400–4500 cm^{-1} with a resolution of 4 cm^{-1} and 8 accumulation modes. The specific surface area of the prepared samples was analyzed via N_2 sorption isotherm by the Brunauer–Emmett–Teller (BET) method. The UV–vis diffuse reflection spectra (UV-DRS) were obtained from a Scan UV–vis spectrophotometer (Perkin Elmer Lambda 35) equipped with a 60 mm integrating sphere assembly and BaSO4 as the reflectance sample. The optical band gap of the synthesized photocatalyst was determined from Tauc plot methods. The X-ray diffraction (XRD) patterns were recorded on a X´pert, PANalytical Advance diffractometer with Cu K radiation (1.54050 Å).

Table 5.1 Ratio of melamine to biomass for the prepared samples

Sample name	Melamine (Wt. %)	Biomass (Wt. %)
BC(SB)	0	100
SB/g-C_3N_4-M75	75	25
SB/g-C_3N_4-M50	50	50
SB/g-C_3N_4-M25	25	75
BC (PKS)	0	100
PKS/g-C_3N_4-M75	75	25
PKS/g-C_3N_4-M50	50	50
PKS/g-C_3N_4-M25	25	75
BC (RH)	0	100
RH/g-C_3N_4-M75	75	25
RH/g-C_3N_4-M50	50	50
RH/g-C_3N_4-M25	25	75

SB sugarcane bagasse, *PKS* palm kernel shell and *RH* rice husk

5.3 Results and Discussion

Effect of Operating Conditions During Synthesis of BC/g-C_3N_4 via Thermal Polycondensation

In this study, BC/g-C_3N_4 derived from sugarcane bagasse (SB), palm kernel shell (PKS) and rice husk (RH) were synthesized via a one-step thermal polycondensation method. The performance of BC/g-C_3N_4 was evaluated by varying the synthesis temperature (450–600 °C) and synthesis time (2–4 h). Figure 5.1 implies that the highest degradation efficiency can be observed from BC(SB)/g-C_3N_4-M75 synthesized at 600 °C with synthesis time of 3 h. Previous studies highlighted that the synthesis temperature of 500–650 °C can be used to synthesize g-C_3N_4 (Alwin et al. 2020). Generally, synthesis temperature plays an importance role which could affect the surface area and pore volume of the synthesized g-C_3N_4 and consequently the efficiency of the photocatalyst (Alwin et al. 2020). It was also reported that higher degree of condensation and lower interplanar spacing was observed at g-C_3N_4 prepared from melamine as precursor at 600 °C (Alwin et al. 2020). Thus, it can be concluded that the highest degradation efficiency could be obtained from BC(SB)/g-C_3N_4 synthesized at 600 °C which is probably due to the increase in surface area/pore volume and complete condensation (fully condensed g-C_3N_4).

The photocatalytic performance of BC/g-C_3N_4 indicated that BC/g-C_3N_4 produced at thermal polycondensation time more than 2 h demonstrated higher degradation efficiency for pyrene. This condition might be due to the increase in pore volume of the photocatalyst which is helpful during adsorption-photocatalytic degradation of pyrene (Alwin et al. 2020). However, further increase of synthesis time more than 3 h could result in lower degradation efficiency. Thus, based on the results for this part, temperature of 600 °C and time of 3 h were used further to synthesize BC/g-C_3N_4.

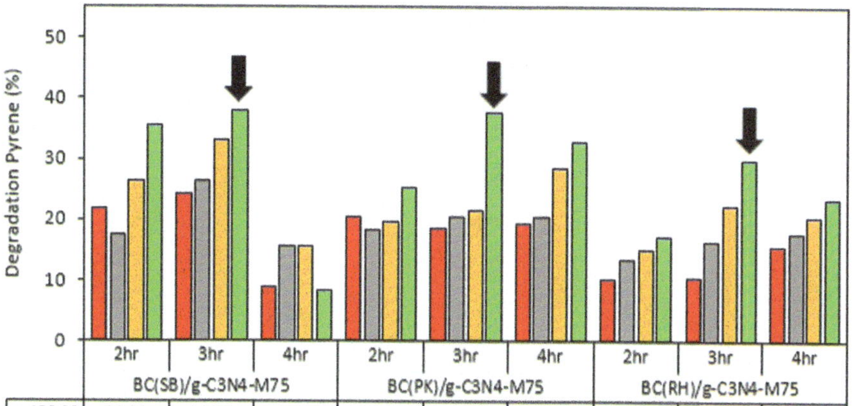

	BC(SB)/g-C3N4-M75 2hr	BC(SB)/g-C3N4-M75 3hr	BC(SB)/g-C3N4-M75 4hr	BC(PK)/g-C3N4-M75 2hr	BC(PK)/g-C3N4-M75 3hr	BC(PK)/g-C3N4-M75 4hr	BC(RH)/g-C3N4-M75 2hr	BC(RH)/g-C3N4-M75 3hr	BC(RH)/g-C3N4-M75 4hr
450 °C	21.83	24.37	9.10	20.63	18.49	19.35	10.39	10.66	15.80
500 °C	17.47	26.38	15.61	18.46	20.49	20.58	13.55	16.49	17.85
550 °C	26.36	33.23	15.58	19.61	21.46	28.46	15.25	22.50	20.55
600 °C	35.55	38.00	8.37	25.30	37.63	32.96	17.38	29.87	23.36

Fig. 5.1 Photocatalytic degradation of pyrene by BC/g-C_3N_4-M75 synthesized at different temperature and synthesis duration

Characterization of BC/g-C_3N_4

In order to investigate the structural properties of BC/g-C_3N_4, FTIR analysis was performed in the range of 4000–500 cm^{-1}. Figure 5.2 illustrates the FTIR spectra of BC(PKS)/g-C_3N_4 which is a typical FTIR spectra for BC/g-C_3N_4 for other photocatalyst derived from other biomasses (i.e., SB and RH). The peak observed at 3100–3400 cm^{-1} indicated the stretching mode of hydroxyl radicals and/or water. The band at around 2900 cm^{-1} was attributed to the interaction of C–H with the carbon. For the g-C_3N_4 and BC(SB)/g-C_3N_4, the band observed at 3200–3600 cm^{-1} was correlated to the vibration mode of N–H and O–H (Lu et al. 2022). Additionally, series of bands at around 1200–1650 cm^{-1} indicated the stretching vibration modes of C = C, CN heterocycles and N–H deformation (Lu et al. 2022). As compared to the other samples, pure g-C_3N_4 demonstrated a strong bending vibration mode of the tri-s-triazine unit at around 800 cm^{-1} (Lu et al. 2022; Pi et al. 2015). It was also noted that the relative intensity of peak at around 1600 cm^{-1} (CN heterocycles) to 1400–1500 (N–H deformation) for biochar-based g-C_3N_4 is higher as compared to pristine g-C_3N_4 (Lu et al. 2022; Nitayaphat et al. 2009). This verified the presence of biochar in the photocatalyst.

N_2 sorption isotherm by the BET method was used to evaluate the specific surface area S_{BET} (m^2 g^{-1}) of the prepared samples, as shown in Table 5.2. Overall, it can be observed that photocatalyst derived from SB demonstrated higher S_{BET} as compared to photocatalyst derived from PKS and RH, suggesting that the higher specific surface area could enhance the adsorption of pollutants onto the surface of the photocatalyst. It was also noted that the surface area of BC/g-C_3N_4 is significantly

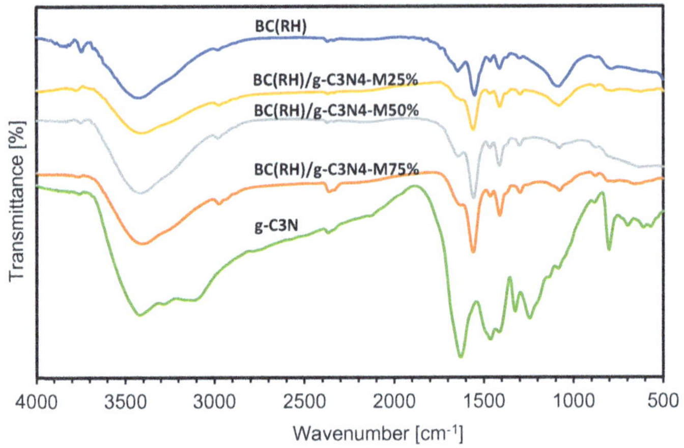

Fig. 5.2 FTIR spectra of the synthesized BC(RH)/g-C_3N_4

higher as compared to pristine g-C_3N_4, indicating that the introduction of biochar promotes better surface area. In addition, the optical band gap of the synthesized photocatalyst which was determined from Tauc plot methods is also shown in Table 5.2. It was noted that the pristine g-C_3N_4 prepared from melamine alone demonstrated the lowest band gap of ~ 2.6 eV which was in accordance with previous studies (Hasija et al. 2021; Karpuraranjith et al. 2021). From Table 5.2, it was noted that the band gap of g-C_3N_4 derived from biomass waste was around 3.27–3.28 eV. The results were also contradicted with previous literature that stated that introduction of BC would narrow the band gap of photocatalyst which could promote higher photocatalytic activity (Cui et al. 2020; Zhu et al. 2020). It was believed that the types of biomasses might have different influence to the values of the band gap. Up to date, limited studies have been conducted to understand the performance of BC/g-C_3N_4 derived from SB, PKS and RH.

Table 5.2 Specific surface area (S_{BET}, $m^2\ g^{-1}$) and bandgap (eV) of the prepared samples

Samples	S_{BET} ($m^2\ g^{-1}$)	Bandgap (eV)
g-C3N4	14.7	2.60
BC(SB)/g-C3N4-M75	188.3415	3.27
BC(SB)/g-C3N4-M50	327.4769	3.28
BC(SB)/g-C3N4-M25	360.5751	3.28
BC(SB)	425.0549	3.28
BC(PKS)/g-C3N4-M75	53.4258	3.28
BC(PKS)/g-C3N4-M50	140.1428	3.28
BC(PKS)/g-C3N4-M25	233.6312	3.28
BC(PKS)	236.7198	3.30

(continued)

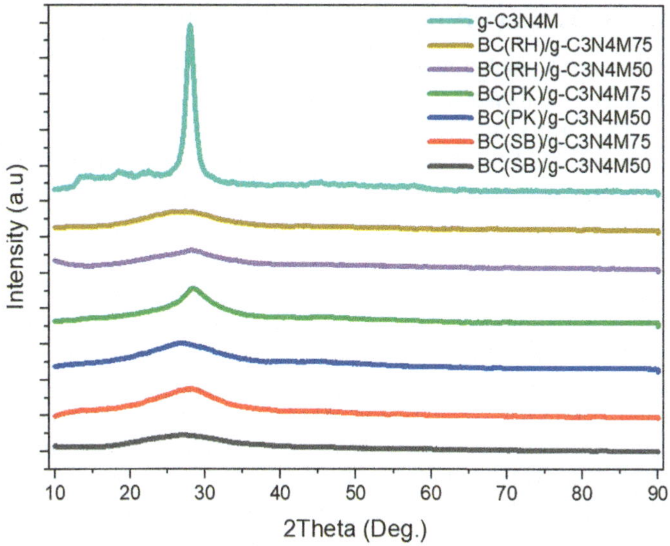

Fig. 5.3 XRD spectra of the photocatalyst prepared with different precursor/biochar ratio

(continued)

Samples	S_{BET} (m^2 g^{-1})	Bandgap (eV)
BC(RH)/g-C3N4-M75	51.9355	3.30
BC(RH)/g-C3N4-M50	124.5723	3.28
BC(RH)/g-C3N4-M25	137.8122	3.28
BC(RH)	167.7793	3.28

XRD analysis was helpful to assess the crystallinity of the prepared photocatalyst. It was reported that typical graphitic carbon nitride has in-plane s-triazines unit with crystallinity at 13° and a conjugated graphitic stacking is normally observed at 27.5° (Jeon et al. 2017). In addition, biochar which is an amorphous carbon normally did not show any characteristic peak for XRD spectra. Figure 5.3 shows the XRD spectra of the photocatalyst prepared with different precursor/biochar ratios. The XRD spectra of the g-C_3N_4 prepared from melamine alone are also included as reference. g-C_3N_4 photocatalyst shows two significant peaks of graphitic carbon nitride. A diffraction peak observed at about 13° can be correlated to the in-planar structural packing, and another peak at around 27° is a typical interplanar stacking peak (Zhu et al. 2018). For the biochar-based g-C_3N_4, a characteristic peak which can be associated with the interplanar stacking peak can be observed at around 27° (Zhu et al. 2018, 2020). However, the intensity of the peak was reducing with addition of biochar. The finding from XRD results suggested the successful formation of BC/g-C_3N_4.

Effect of Various Parameters During Photocatalytic Degradation

In general, the ratio of biochar to precursor in photocatalyst would affect the properties of the produced photocatalyst. Thus, investigation on the effect of ratio of biomass to precursor was further studied to understand the effect of percentage of biochar during thermal polycondensation. As can be referred in Table 5.1, different ratio of melamine to biomass was used to produce the photocatalyst. Figure 5.4a–c show the percentage of degradation (%) of pyrene for different types of biomass waste. It was noted that BC/g-C_3N_4 derived from SB and PKS prepared from highest content of melamine generally demonstrated the higher degradation efficiency of pyrene. The highest degradation efficiency (38%) can be observed from photocatalyst derived from SB prepared with the highest percentage of melamine (BC(SB)/g-C_3N_4-M75). This was probably due to the narrow band gap of BC(SB)/g-C_3N_4-M75 as compared to other BC/g-C_3N_4 as can be referred in Table 5.2. The result is in accordance with a previous study that suggested that higher surface area could impart better degradation efficiency (Song et al. 2019). It can also be concluded that an increase in biochar content in the photocatalyst could result to the lower degradation of pyrene. This was mainly due to the higher coverage of biochar on the surface of the photocatalyst which could suppress the absorption of light and generation of electron thus reducing the effectiveness of the photocatalyst to degrade pyrene (Cui et al. 2020; Zhu et al. 2018). For the BC/g-C_3N_4 derived from RH, it can be concluded that the introduction of RH is not desirable due to low percentage of degradation of pyrene ($< 20\%$). It can also be pointed out that the selection of biomass waste for synthesis of BC/g-C_3N_4 is crucial that will influence the properties and performance of the photocatalyst.

Photocatalyst dosage plays an essential role affecting the photocatalytic degradation of pollutants. In this study, the effect of photocatalyst dosage was evaluated as shown in Fig. 5.5a–c. It can be deduced that further increase of photocatalyst dosage to more than the optimum value (around 1.0–1.5 g/L) could decrease the degradation efficiency. It can be clearly seen that an increase of photocatalyst dosage initially increased the degradation of pyrene due to desirable formation of active sites. However, further increase of photocatalyst dosage could suppress the absorption of light due to turbidity of the solution with an increase of concentration of photocatalyst in the solution. Additionally, previous work by Lu et al. (2015) also confirmed that the degradation efficiency decreased when the dosage of photocatalyst exceeded the optimum dosage. This is because the excess of g-C_3N_4 could lead to high recombination rate of electron–hole pair, thus inhibits the effectiveness of charge transfer, hence affecting the overall degradation efficiency.

The initial concentration of pollutants significantly affects the degradation efficiency of photocatalyst. Figure 5.6a–c illustrate the degradation efficiency of pyrene by BC/g-C_3N_4-M75 studied at various initial concentrations of pyrene. In general, it can be inferred that for all types of photocatalyst, the degradation efficiency decreased with increasing concentration of pyrene (more than 2 ppm) which was also in accordance with previous studies (Aisien et al. 2014; Zango et al. 2020). This

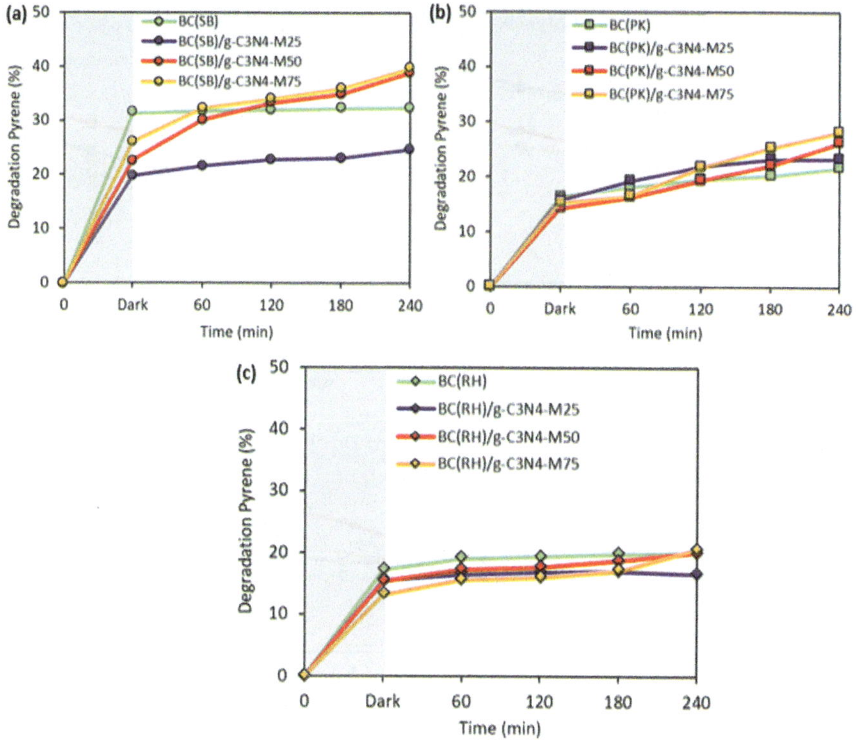

Fig. 5.4 Photocatalytic degradation of pyrene for **a** BC(SB)/g-C_3N_4, **b** BC(PKS)/g-C_3N_4 and **c** BC(RH)/g-C_3N_4

was mainly due to adsorption and deposition of pyrene onto the surface of photocatalyst which could decrease the formation of active sites, hence reducing the formation and generation of reactive radicals by the photocatalyst.

5.4 Conclusion

In this study, BC/g-C_3N_4 photocatalysts have been successfully prepared from SB, PKS and RH via the thermal polycondensation method with melamine as precursor. FTIR and XRD exhibited the main peaks of BC and g-C_3N_4 which confirms the successful formation of BC/g-C_3N_4. Photocatalyst prepared from SB demonstrated better surface area as compared to other g-C_3N_4 derived from PKS and RH, which could explain the better degradation efficiency of g-C_3N_4 derived from SB. Band gap

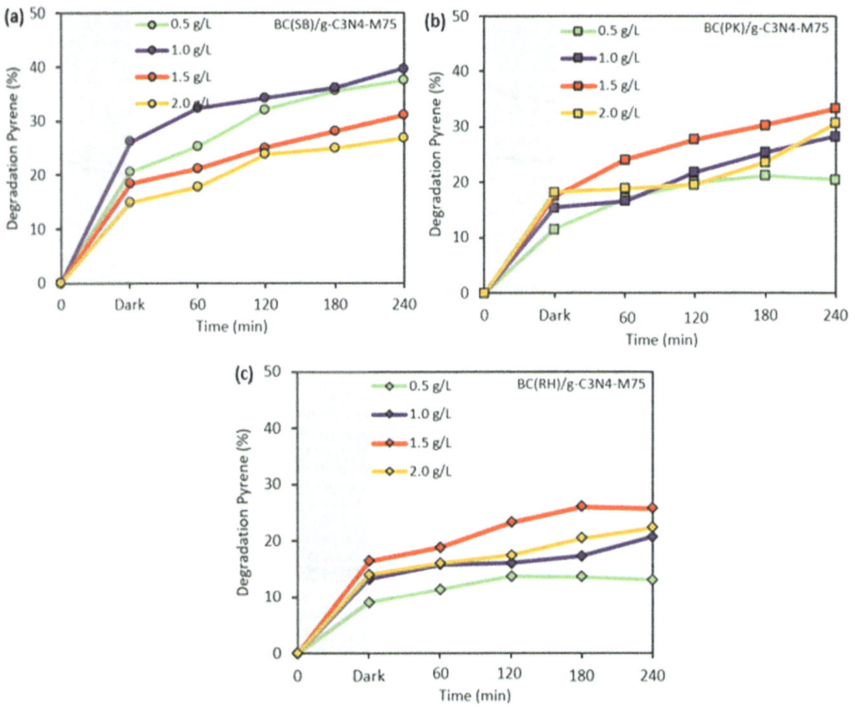

Fig. 5.5 Photocatalytic degradation of BC/g-C_3N_4-M75 derived from different biomasses (**a**: SB; **b**: PKS; **c**: RH) studied at different photocatalyst dosage

values indicated that the optical band gap is around 3.27–3.28 for all BC/g-C_3N_4. Overall results suggested that BC/g-C_3N_4 derived from SB demonstrated the highest degradation of pyrene as compared to g-C_3N_4 prepared from rice husk and palm kernel. It is suggested that the selection of biomass waste for preparation of photocatalyst is essential that will significantly affect the performance of the photocatalyst. Experimental results also suggested that the optimum temperature and duration of synthesis via thermal polycondensation were 600 °C and 3 h. Photocatalytic degradation experiments verified that the optimum parameters during photocatalytic reaction were 1.0–1.5 g/L photocatalyst dosage and 1–2 ppm of initial concentration of pyrene. This work could provide better understanding on the performance of photocatalyst prepared from biomass waste which indicated the potential of biomasses from agriculture to be utilized as low-cost photocatalyst for water and wastewater treatment.

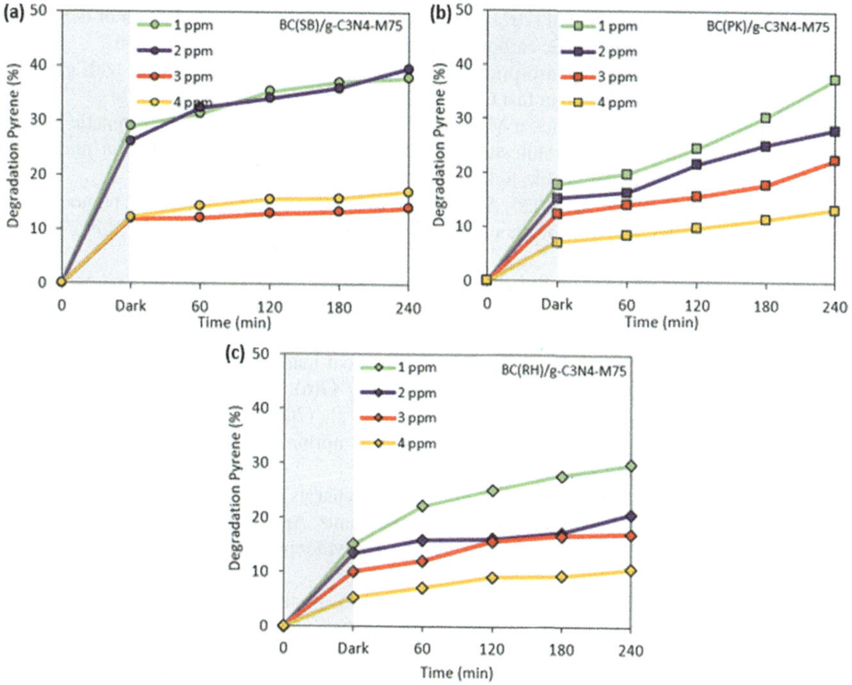

Fig. 5.6 Photocatalytic degradation of BC/g-C_3N_4-M75 derived from different biomasses (**a**: SB; **b**: PKS; **c**: RH) at different initial concentration of pyrene

Acknowledgements The authors greatly acknowledge the financial support from Ministry of Higher Education (MoHE) (Fundamental Research Grant Scheme (FRGS): FRGS/1/2019/STG07/UNIKL/02/4), FRGS/1/2020/STG05/UNIKL/02/4) and Universiti Kuala Lumpur, Kampus Cawangan Malaysian Institute of Chemical and Bioengineering Technology (UniKL MICET) for the research activities.

References

Ahmed AA (2022) Photocatalytic properties of graphitic carbon nitrides (g-C_3N_4) for sustainable green hydrogen production: recent advancement. Fuel 316:123381

Aisien F, Amenaghawon AN, Assogba M (2014) Heterogeneous photocatalytic degradation of naphthalene using periwinkle shell ash: effect of operating variables, kinetic and isotherm study. S Afr J Chem Eng 19(1):31–45

Alwin E, Koci K, Wojcieszak R et al (2020) Influence of high temperature synthesis on the structure of graphitic carbon nitride and its hydrogen generation ability. Mater 13(12):2756

Cui J, Zhang F, Li H et al (2020) Recent progress in biochar-based photocatalyst for wastewater treatment: synthesis, mechanisms, and applications. Appl Sci 10:1019

Fronczak M, Talas E, Paszti Z et al (2018) Graphitic carbon nitride doped with the s-block metals: adsorbent for the removal of methyl blue and copper(II) ions. Diam Relat Mater 34(25):7272–7283

Hasija V, Raizada P, Pardeep S et al (2021) Progress on the photocatalytic reduction of hexavalent Cr(VI) using engineered graphitic carbon nitride. Process Saf Environ Prot 202:663

Jeon P, Lee ME, Bae K (2017) Adsorption and photocatalytic activity of biochar with graphitic carbon nitride (g-C_3N_4). J Taiwan Inst Chem Eng 77:244–249

Karpuraranjith M, Chen Y, Ramadossa M et al (2021) Magnetically recyclable magnetic biochar graphitic carbon nitride nanoarchitectures for highly efficient charge separation and stable photocatalytic activity under visible-light irradiation. J Mol Liq 326:115315

Koe WS, Lee JW, Pang YL, Sim L (2020) An overview of photocatalytic degradation: photocatalyst, mechanisms, and development of photocatalytic membrane. Environ Sci Pollut Res 27:2522–2565

Lu D, Zhang G, Wan Z (2015) visible-light-driven g-C_3N_4/Ti^{3+}–TiO_2 photocatalyst co-exposed {0 0 1} and {1 0 1} facets and its enhanced photocatalytic activities for organic pollutant degradation and Cr(VI) reduction. Appl Surf Sci 358:223–230

Lu Y, Wang W, Cheng H et al (2022) Bamboo-charcoal-loaded graphitic carbon nitride for photocatalytic hydrogen evolution. Int J Hydrog Energy 47(6):3733

Nitayaphat W, Jiratumnukul N, Charuchinda S et al (2009) Mechanical properties of chitosan/bamboo charcoal composite films made with normal and surface oxidized charcoal. Carbohydr Polym 783:444–448

Pi L, Jiang R, Zhou W et al (2015) g-C_3N_4 modified biochar as an adsorptive and photocatalytic material for decontamination of aqueous organic pollutants. Appl Surf Sci 358:231–239

Qi K, Liu SY, Zada A (2020) Graphitic carbon nitride, a polymer photocatalyst. J Taiwan Inst Chem Eng 109:111–123

Shi W, Ren H, Huang X et al (2020) Low cost red mud modified graphitic carbon nitride for the removal of organic pollutants in wastewater by the synergistic effect of adsorption and photocatalysis. Sep Purif Technol 237:116477

Sliem MA, Salim AY, Mohamed GG (2018) Photocatalytic degradation of anthracene in aqueous dispersion of metal oxides nanoparticles: effect of different parameters. J Photochem Photobiol A Chem 317:327–335

Song W, Ge P, Ke Q et al (2019) Insight into the mechanisms for hexavalent chromium reduction and sulfisoxazole degradation catalyzed by graphitic carbon nitride: the yin and yang in the photo-assisted processes. Chemosphere 221:166–174

Tseng IH, You BJY, Chang PY (2022) Sugarcane bagasse supported graphitic carbon nitride for photocatalytic conversion of carbon dioxide. Catal Commun 164:106431

Wang J, Wang S (2022) A critical review on graphitic carbon nitride (g-C_3N_4)-based materials: preparation, modification and environmental application. Coord Chem Rev 453:214338

Zango ZU, Sambudi NM, Abu Bakar NHH et al (2020) Removal of pyrene from aqueous solution using Fe-based metal-organic frameworks. IOP Conf Ser Earth Environ Sci 549(1):012061

Zhu Z, Liu WFZ, Yu Y et al (2018) Fabrication of the metal-free biochar-based graphitic carbon nitride for improved 2-mercaptobenzothiazole degradation activity. J Photochem Photobiol A Chem 358:284–293

Zhu GX, Lu TL, Zhan YZ (2020) Graphitic carbon nitride (g-C_3N_4) as an efficient metal-free Fenton-like catalyst for degrading organic pollutants: the overlooked non-photocatalytic activity. Water Sci Tech 81(3):518–528

Chapter 6
Response Surface Methodology for the Optimisation of a 96-Well Microtitre Plate Coagulation Activity Assay

Farah Salma Baharudin, Mussarat Saleem, and Robert Thomas Bachmann

Abstract In order to assess the impacts of physical and chemical factors on coagulation and flocculation processes, the 1 L jar test has been employed for decades as an effective laboratory procedure. However, it is not suitable for studies involving screening of minute quantities of coagulants isolated from flora and fauna via chromatographic or solubility-based fractionation. The primary objective of this study was to develop and optimise a coagulation activity assay based on 96-well microtiter plates (MTP). The coagulation activity was determined in round bottom-shaped polystyrene MTPs due to a better signal-to-noise ratio. The response surface methodology via central composite design was used to optimise coagulation activity for shaking time (10–70 s), shaking frequency (300–1100 rpm), settling time (20–60 min) and coagulant dosage (0–140 mg/L). Alum was used as standard inorganic coagulant treating a ~ 200 μL, 100 NTU colloidal bentonite solution. MTP-based coagulation activity experiments with optimised parameters as well as conventional jar tests were also carried out with crude *Moringa oleifera* seed extracts. The coagulation activity was optimum at 40 s shaking time, 700 rpm shaking frequency, 40 min settling time and 70 mg/L alum. A further experiment on turbid river water was conducted using different dosages of crude MO seed extract with the newly developed 96 MTP coagulation activity assay. The optimum dosage was 105 mg/L confirming that the MTP assay can be applied for natural coagulants too. The coagulation activities for alum and *M. oleifera* in the MTP-based assay displayed a similar pattern. Our results demonstrate that the 96-well MTP coagulation activity assay can be used as an alternative method to the jar test approach.

F. S. Baharudin · M. Saleem · R. T. Bachmann (✉)
Green Chemistry & Sustainable Engineering Technology Cluster, Malaysian Institute of Chemical and Bioengineering Technology, Universiti Kuala Lumpur, Taboh Naning, Lot 1988, 78000 Alor Gajah, Melaka, Malaysia
e-mail: bachmann@unikl.edu.my

F. S. Baharudin
e-mail: farahsalma@gmail.com

M. Saleem
e-mail: mussarat.micet@gmail.com

A. Ismail et al. (eds.), *Advancements in Materials Science and Technology Led by Women*, Advanced Structured Materials 165,
https://doi.org/10.1007/978-3-031-21959-7_6

Keywords Coagulation activity · Microtiter plates · Response surface methodology · Alum · *Moringa oleifera*

6.1 Introduction

Coagulation–flocculation is an essential industrial process that removes colloids by using a variety of metal and polymeric coagulants such as aluminium sulphate (alum) (Aziz et al. 2007; Manda et al. 2016), poly-aluminium chloride (PACl) (Matsui et al. 1998), ferric chloride (Aziz et al. 2007; Song et al. 2015), ferrous sulphate (Aziz et al. 2007; Snodgras 1905) and ferric chloro-sulphate (Bratby 2006). Alum and PACl are among the most commonly used coagulants in the water industry (Bratby 2006). The efficiency of turbidity reduction with metallic coagulants is known to depend on the type and dosage of the coagulant (Lind 1994; Benschoten and Edzwald 1990), solution's pH (Cao et al. 2011; Lee et al. 2012; Binnie et al. 2017), mixing intensity and duration (Lee et al. 2012; Rossini et al. 1999; Young and Smith 2000) and temperature (Benschoten and Edzwald 1990; Lee et al. 2012; Guo et al. 2008).

However, metal coagulants have many shortcomings such as high carbon footprint, potential detrimental health effects and environmental pollution (Saleem and Bachmann 2019). In search for a more sustainable replacement, plant components such as seeds, leaves and barks are increasingly screened for the presence of natural coagulants (Qureshi et al. 2011; Unnisa et al. 2010; Ramamurthy et al. 2012). The greatest attention was given to seed extracts from *M. oleifera*, frequently tested at bench scale for coagulation (Ndabigengesere et al. 1995; Pritchard et al. 2010; Dezfooli et al. 2016) and known to contain coagulant-active albumin (Baptista et al. 2017). The biomolecules are classified as cationic (Gassenschmidt et al. 1995; Shebek et al. 2015) and hemagglutinating proteins (Ferreira et al. 2011; Santos et al. 2009).

The jar test has been employed for many decades in teaching, research and commercial laboratories to identify optimum process conditions for a given test water and coagulant using the one-factor-at-a-time (OFAT) method. A typical jar test unit comprises of four to six 1 L beakers each equipped with a paddle stirrer (Ramamurthy et al. 2012; Ndabigengesere et al. 1995; Dalen et al. 2010; Antov et al. 2007; Sánchez-Martín et al. 2010; Miller et al. 2008). While the jar test has proven to be useful in the water industry its use as a screening tool for new coagulants in research laboratories poses several challenges due to the comparatively large quantity of coagulants and colloidal test solution required as well as limited number of tests that can be executed concurrently. Ghebremichael et al. (2005) developed a coagulation activity assay based on 2.4 mL semi-micro-cuvettes using a micro-pipettor for mixing and measuring the absorbance of the supernatant at 500 nm. The method was adopted by a number of researchers investigating the activity of natural coagulants (Ferreira et al. 2011; Santos et al. 2009; Arunkumar et al. 2019; Gunaratna et al. 2007; Marobhe et al. 2010). This approach reduced the quantity and cost of coagulants, reagents as well as waste produced by a factor of a hundred. Nonetheless, rapid multi-factorial investigations are still not practical.

Ninety-six well microtiter plate (MTP) assays are frequently used for rapid sample screening and analysis in the field of biochemistry, biotechnology, microbiology (Hermann et al. 2003) and food chemistry (Sin et al. 2006). Its use in chemical and environmental laboratories is gaining popularity (Uttran et al. 2018) since many analytes of interest are measured spectrophotometrically. The 96-well MTP is commercially available in different designs (flat, U- and V-shaped bottom) and materials (polystyrene and polypropylene). To avoid the shortcomings of the OFAT approach and identify optimum process conditions for coagulation, statistical tools such as response surface methodology (RSM) can be combined with the MTP-based coagulation activity assay. This allows the simultaneous analysis of many samples using multiple factors. In addition, RSM enables the user to investigate the presence of synergistic or antagonistic interactions between selected process parameters while keeping the number of required experiments relatively low (Trinh and Kang 2011).

Hence, this research aims to identify the suitability of the 96-well MTP assay for coagulation activity studies. The MTP-based coagulation assay was studied via the RSM approach with alum as model coagulant and a total of four parameters (shaking frequency, shaking time, coagulant dosage and settling time). The coagulation activity of crude *M. oleifera* seed extracts was also investigated and compared with alum.

6.2 Methodology

6.2.1 Synthetic Turbid Water Samples

Synthetic turbid water samples were prepared with bentonite clay (R&M Chemicals, Essex, UK) using the Dezfooli et al. (2016) method. All water samples were of 100 $\pm$ 2 NTU and pH 6.5. The pH was adjusted with 0.1 M HCL (Merck, Germany) and 0.1 M NaOH (Merck, Germany) solutions. All necessary measurements were done using a pre-calibrated turbidity (HACH, 2100P) as well as pH meter (Mettler Toledo, LP115).

6.2.2 Preparation of Coagulant Solutions

Alum as well as crude *M. oleifera* seed extracts were used as coagulants. Aluminium sulphate ($Al_2(SO_4)_3$: 18 H_2O) (R&M Chemicals, Essex, UK) stock solution was prepared by first dissolving 1 g of alum salt in 1 L of deionised water. Four different concentrations (35, 70, 105 and 140 mg/L) were subsequently prepared by dilution with deionised water. *M. oleifera* crude extract was prepared by first crushing the deshelled seeds (Borneo Company Sdn. Bhd, Malaysia) into a fine powder using pestle and mortar. Two grams of fine *M. oleifera* powder were transferred to 100 g deionised water and mixed for 8 min at room temperature using a magnetic stirrer (Bibby). The solution was then topped up to 500 mL with deionised water and mixed again for 8 min. Afterwards, the solution was allowed to stand for 30 min before being filtered with a muslin cloth (Muyibi and Evison 1995). This was labelled as 0.4% *M. oleifera* crude extract. *M. oleifera* extracts were prepared fresh prior to each day's experiment.

6.2.3 Development of the 96-Well MTP Coagulation Activity Assay

Ninety-six well coagulation activity assays were conducted using round (U-) bottom-shaped MTP (Greiner Bio-One) of polystyrene (PS) material after preliminary coagulation activity tests demonstrated better signal-to-noise ratios compared to flat bottom and V-shaped wells (Table 6.1).

MTP wells manufactured from polypropylene (PP) exhibited lower coagulation activity. This is due to the adhesion of aggregation of flocs onto the wall of PP MTP wells which affected the absorbance measurement.

Table 6.1 ANOVA of factors, shaking frequency (*A*) [rpm], shaking time (*B*) [s], alum dosage (*C*) [mg/L] and settling time (*D*) [min], in different MTP geometry of polystyrene material

Source	Coefficient		*p*-Value	
	F-bottom	*V*-bottom	*F*-bottom	*V*-bottom
Model	−1.453	−1.630	0.0498 (poor)	< 0.0001
A	$+ 3.606 \times 10^{-3}$	$+ 1.477 \times 10^{-3}$	0.6666	< 0.0001
B	+ 0.0345	+ 0.0550	0.1908	0.0126
C	+ 0.0127	+ 0.0259	0.0014	< 0.0001
D	+ 0.0415	−0.0181	0.5995	0.5245
AB	-1.534×10^{-6}	-4.811×10^{-5}	0.9193	0.0006
AC	$+ 8.165 \times 10^{-7}$	$+ 3.239 \times 10^{-6}$	0.8999	0.5102
AD	-1.599×10^{-6}	-5.846×10^{-6}	0.9438	0.7328
BC	-1.189×10^{-6}	-2.868×10^{-4}	0.9890	0.0004
BD	-2.861×10^{-5}	$+ 2.538 \times 10^{-5}$	0.9247	0.9113
CD	-2.630×10^{-6}	$+ 1.347 \times 10^{-5}$	0.9838	0.8903
A^2	-2.584×10^{-6}	$+ 9.920 \times 10^{-7}$	0.0084	0.1431
B^2	-3.613×10^{-4}	$+ 3.548 \times 10^{-5}$	0.0308	0.7601
C^2	-6.494×10^{-5}	-8.136×10^{-5}	0.0340	0.0015
D^2	-4.648×10^{-4}	$+ 2.761 \times 10^{-4}$	0.1930	0.2991
Lack of fit			0.0277	0.0039
Status			Not significant	Significant

The total working volume per well was 200 μL which comprised of 196 μL synthetic turbid water and 4 μL of alum solution. The alum solution was first added to wells with a manual pipettor (Eppendorf) and subjected to desiccation at 60 °C for 20 min (Eppendorf 5305 Concentrator Plus). This was followed by the transfer of synthetic turbid water into designated MTP wells with an 8-channel semi-automatic pipettor (Eppendorf Xplorer). In case of blank samples, synthetic turbid water was substituted with 196 μL 1 mM NaCl solution and 4 μL of deionised water. The negative controls comprised of 196 μL of synthetic turbid water and 4 μL of deionised water. The 4 μL deionised water was desiccated as described before. Multiple loaded MTPs were then shaken at different speeds and durations using an MTP shaker (Eppendorf Thermomixer) as detailed in Sect. 2.4. Subsequently, absorbance was measured at 595 nm using an MTP reader (Bio-Rad, iMark) at various settling times.

6.2.4 Experimental Design using RSM

In order to determine the optimum coagulation activity, impact and correlations of four influential factors on the coagulation activity, the central composite design (CCD) was adopted. The four independent factors chosen in this study were shaking frequency, shaking time, alum dosage time and settling (Table 6.1). Coagulation activity, CA_{MTP}, was measured as the response (Y) using Eq. (6.2.1)

$$CA_{MTP} = \frac{A_S - A_{Blank}}{A_{NC} - A_{Blank}} \tag{6.2.1}$$

where A_S represents the absorbance of the sample, A_{NC} is the absorbance of the negative control (synthetic turbid water only), and A_{Blank} is the absorbance of the blank (1 mM NaCl solution and deionised water), respectively.

The range and levels of independent variables are summarised in Table 6.2. Design-Expert software (version 9.0.1.0, Stat-Ease) was used to create 3D surface and 2D contour plots of the response models.

Table 6.2 Experimental range and levels used for the optimisation of the MTP-based coagulation activity assay

Independent variable	Range and level				
	−2	−1	0	1	2
A	300	500	700	900	1100
B	10	25	40	55	70
C	0	35	70	105	140
D	20	30	40	50	60

A shaking frequency [rpm]; *B* shaking time [s]; *C* alum dosage [mg/L]; *D* settling time [min]

Commonly, the interaction between the independent factors and response in coagulation–flocculation processes cannot be simulated with a linear function model (Trinh and Kang 2011). The quadratic model used for predicting the optimal conditions is provided in Eq. (6.2.2)

$$Y = \beta_0 + \sum_{i=1}^{k} \beta_i X_i + \sum_{i=1}^{k} \beta_{ii} X_i^2 + \sum_{i_{1<j}}^{k} \sum_{j}^{k} \beta_{ij} X_i X_j + \cdots + \in \tag{6.2.2}$$

where Y is the response to be modelled (coagulation activity), β_O is a constant, β_i is the linear coefficient, β_{ii} is the quadratic coefficient, and β_{ij} is the cross-product coefficient. X_i and X_j are levels of the independent factors, X^2 are square of the independent factors, X_iX_j are interaction factors while k equals to the number of tested factors ($k = 4$). ϵ represents the residual error observed in response Y.

ANOVA was used to determine the effect and regression coefficients of individual linear, quadratic and interaction terms (Table 6.3). The regression coefficients were then used for statistical calculations to generate contour maps from the regression models (Sin et al. 2006). All experiments were performed with a total of three replicates.

Additional conformation experiments were performed in two trials with triplicates to validate the predicted coagulation activity under optimal conditions.

Table 6.3 ANOVA of coagulation activity at different combinations of shaking frequency (*A*) [rpm], shaking time (*B*) [s], alum dosage (*C*) [mg/L] and settling time (*D*) [min] in U-bottom MTP geometry

Source	Coefficient	*p*-Value
Model	−2.953	< 0.0001
A	$+7.625 \times 10^{-3}$	0.0003
B	+ 0.0500	0.1781
C	+ 0.0230	0.0390
D	$+8.436 \times 10^{-3}$	0.3738
AB	-1.505×10^{-5}	0.1441
AC	$+2.676 \times 10^{-6}$	0.5352
AD	-8.519×10^{-6}	0.5697
BC	-5.483×10^{-5}	0.3417
BD	$+4.119 \times 10^{-5}$	0.8359
CD	$+1.493 \times 10^{-5}$	0.8608
A^2	-4.505×10^{-6}	< 0.0001
B^2	-4.377×10^{-4}	0.0005
C^2	-1.558×10^{-4}	< 0.0001
D^2	-3.714×10^{-5}	0.8704
Lack of fit		0.0727
Status		Not significant

6.2.5 *Coagulation Activity of* M. Oleifera *Coagulant*

The optimised 96-well MTP assay was subsequently used to study the coagulation activity of *M. oleifera* seed extract in synthetic turbid water samples. The negative control and blank were similar to controls described in Sect. 2.3. The absorbance of samples was measured at 595 nm using a MTP reader (Bio-Rad, iMark). All experiments were performed in triplicates.

6.2.6 *Statistical Analysis*

The regression and graphical analysis of the experimental data was carried out using the Design-Expert 9.0.1.0 (Stat-Ease Inc., USA) software. The quadratic model quality was expressed by the coefficient of determination R^2 and its statistical significance checked by the F-test.

6.3 Results and Discussion

6.3.1 *Statistical Analysis*

The actual and predicted effects of four independent variables (shaking frequency, shaking time, alum dosage and settling time) coagulation activity are shown in Table 6.4.

The results from Table 6.4 were used to develop a quadratic model to predict coagulation activity as a function of shaking frequency, shaking time, alum dosage and settling time according to Eq. 6.2.2. The model obtained was subsequently analysed by ANOVA to identify terms that are statistically insignificant. Equation 6.3.1 illustrates the reduced quadratic model based on coded independent factors.

$$\begin{aligned}\text{Coagulation Activity}(\text{CA}_{\text{MTP}}) = {} & -2.953 + \left(7.625 \times 10^{-3}\right)A + 0.0500B \\ & + 0.0230C - \left(4.505 \times 10^{-6}\right)A^2 \\ & - \left(4.377 \times 10^{-4}\right)B^2 - \left(1.558 \times 10^{-4}\right)C^2\end{aligned} \tag{6.3.1}$$

Among the various model terms, the linear and quadratic forms of independent factors shaking frequency, shaking time and alum dosage were significant at $p = 0.05$ level (Table 6.3). The interaction terms of the quadratic model had no significant effect on the response. From Eq. 6.3.1, it can be seen that the coagulation activity increases initially with shaking frequency, shaking time and alum dosage.

Table 6.4 CCD of four independent variables and their effect on coagulation activity (CA) in U-bottom 96-well MTP geometry

Run	A	B	C	D	Coagulation activity (CA)		
					Actual	Predicted	Standard deviation
1	700	40	70	40	2.0896	2.0299	0.04
2	700	40	140	40	1.3995	1.3737	0.02
3	900	25	105	50	1.9136	1.7842	0.09
4	900	55	105	30	1.6455	1.7073	0.04
5	700	40	0	40	1.0694	1.1623	0.07
6	900	55	35	30	1.7111	1.6217	0.06
7	900	25	105	30	1.8844	1.7842	0.07
8	500	55	35	30	1.4309	1.5247	0.07
9	500	55	35	50	1.4453	1.5247	0.06
10	700	40	70	40	1.9942	2.0299	0.03
11	700	40	70	60	2.0567	2.0299	0.02
12	900	25	35	30	1.5485	1.5835	0.02
13	700	40	70	40	1.9584	2.0299	0.05
14	900	55	105	50	1.5709	1.7073	0.10
15	700	10	70	40	1.476	1.567	0.06
16	500	55	105	50	1.6988	1.5354	0.12
17	900	55	35	50	1.746	1.6217	0.09
18	700	70	70	40	1.7327	1.7091	0.02
19	500	25	105	30	1.3466	1.4315	0.06
20	300	40	70	40	1.0789	1.0863	0.01
21	500	55	105	30	1.5073	1.5354	0.02
22	700	40	70	40	2.1161	2.0299	0.06
23	500	25	105	50	1.3591	1.4315	0.05
24	500	25	35	50	1.4504	1.3057	0.10
25	700	40	70	40	2.0712	2.0299	0.03
26	1100	40	70	40	1.476	1.536	0.04
27	900	25	35	50	1.5394	1.5835	0.03
28	700	40	70	20	1.9102	2.0299	0.08
29	700	40	70	40	1.9641	2.0299	0.05
30	500	25	35	30	1.4157	1.3057	0.08

A shaking frequency [rpm]; *B* shaking time [s]; *C* alum dosage [mg/L]; *D* settling time [min]

The lack of fit (LOF) F-test illustrates the deviation of the experimental data from the fitted surface, relative to pure error. A p-value > 0.05 is preferred because it indicates that the LOF is insignificant. The large p-value presented in Table 6.3 therefore suggests that the independent factors of the reduced model and the response are significantly correlated.

The R^2 coefficient represents the fraction of the total variation in the response that is predicted by the reduced quadratic model. A value close to 1 is desirable requiring also a reasonable agreement with the adjusted R^2. The correlation coefficient of the reduced model was found to be 0.9096 and in good agreement with R^2_{adj} (0.8390).

Adequate precision (AP) refers to the signal-to-noise ratio. AP values > 4 indicate that the model has acceptable predictive power. The AP value of the reduced quadratic model is 15.448 which confirms that this model is suitable to navigate the design space as defined by CCD.

The diagnostic plot (Fig. 6.1) shows the predicted vs. actual coagulation activity and is used to ascertain the model performance. Both the experimental and predicted values are in good agreement.

The coefficient of variance (CV) defined as the ratio of the standard error of the predicted to the mean experimental value of coagulation activity reflects the reproducibility of the model. Models with CV values < 10% are considered reproducible.

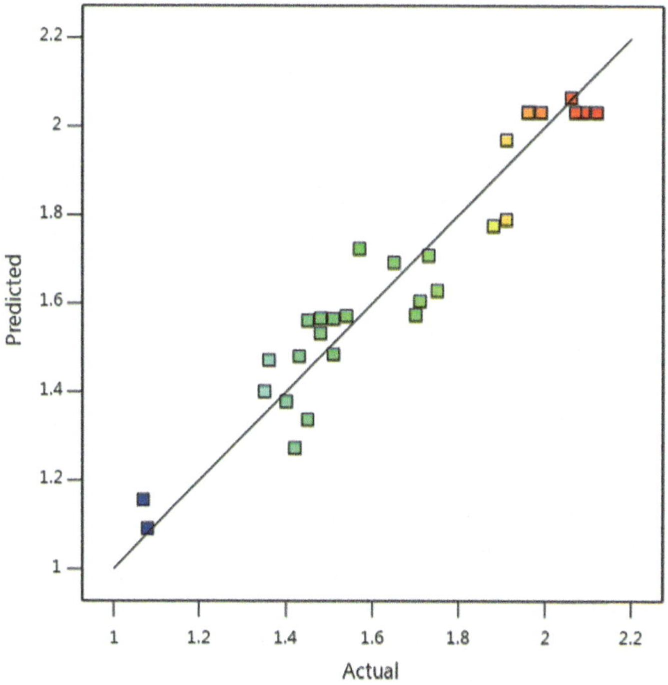

Fig. 6.1 Diagnostic plot of predicted versus actual coagulation activity of alum in U-bottom MTP

The CV value of the reduced quadratic model was found to be 6.4% demonstrating its good reproducibility.

6.3.2 Process Analysis of the MTP Coagulation Activity Assay

Settling time in this study was insignificant probably due to the mode of absorbance measurement. In conventional jar but also downscaled cuvette tests, the absorbance of the supernatent is analysed. As flocs sediment, the particle concentration in the supernatent decreases until reaching an equilibrium. In the 96-well MTP test flocs also settle but due to the small volume the sedimentation path is very short. In addition, the light in MTP assays always passes vertically through the solution from the bottom to the top making this essay insensitive to settling time. Once flocs have formed at the appropriate coagulant dosage, shaking frequency and time, the 96-well MTP coagulation activity assay can be stopped for data analysis and interpretation.

The significant factors shaking frequency, shaking time and alum dosage were analysed by contour plots and 3D surface analysis (Figs. 6.2, 6.3 and 6.4). All contour plots display almost symmetrical, circular shapes. The peaks are clearly visible in the centre of the response surfaces demonstrating that the optimal conditions for shaking frequency, shaking time and alum dosage were located within the experimental boundaries. The contour plot and 3D surface analysis of shaking frequency and shaking time in Fig. 6.2a and b suggests optimum process conditions at 700–800 rpm and 37–46 s. Shaking frequency and shaking time are important factors to ensure homogeneous particle dispersion (Micheletti et al. 2006) while providing sufficient energy for inter-particle collision to form flocs (Tissot et al. 2010). However, excessive shaking frequency and time are detrimental causing the flocs to break due to high shear forces (Yu et al. 2010). Conversely, the tilted ellipsoids in Fig. 6.2a also illustrates that a given coagulation activity can be maintained when increasing the shaking frequency while simultaneous reducing the shaking time. Flocs form faster when increasing the shaking frequency due to greater number of particle collisions. A similar trend was reported by Bratby (2006).

Figure 6.3a and b shows the effect of alum dosage and shaking frequency on coagulation activity. Optimal coagulation activity was found between 55 and 75 mg/L and 700–800 rpm. Beyond the optimum dosage, a decline in coagulation activity was noticed. The observed trend suggests that for a given shaking frequency, coagulation occurred within two alum dosages, the critical coagulant and the critical stabilisation concentration (Bratby 2006). Within these two critical concentrations, the colloids are destabilised by the adsorption and bridging mechanism. Alum dosages greater than the critical stabilisation dosage preserve individual colloids due to charge reversal until reaching a second critical coagulant concentration. At this point and beyond, alum starts to precipitate causing an enmeshment of colloidal particles in solid-phase

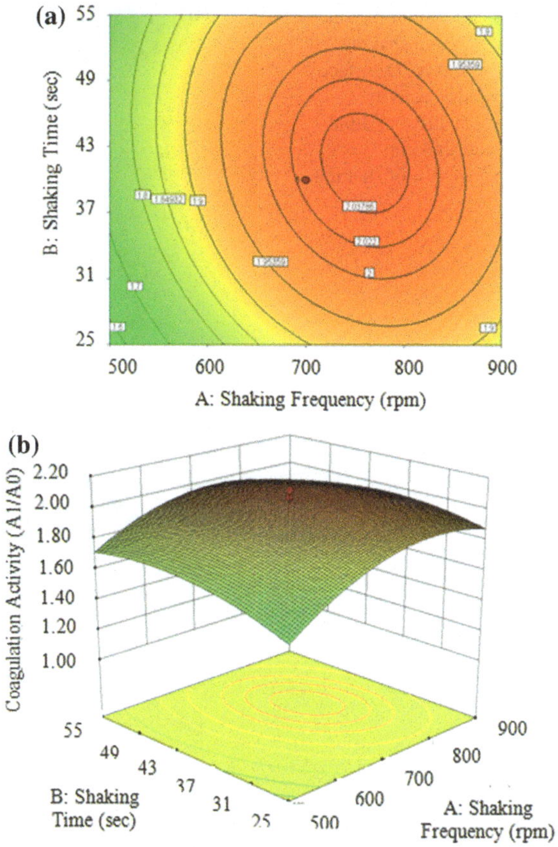

Fig. 6.2 **a** Contour plot and **b** 3D surface showing the effect of shaking time and shaking frequency on coagulation activity at 70 mg/L alum dosage

hydroxide flocs (Xiao et al. 2008). This zone is commonly referred to as the sweep-floc region. However, no renewed increase in coagulation activity was observed implying that the alum dosages used in this study did not reach the second critical coagulant concentration. It is worthwhile noting that coagulation activity was low at shaking frequencies < 550 rpm regardless of alum dosage. This stage is known as perikinetic flocculation in which microflocs form due to thermal agitation (Brownian movement) (Bratby 2006). Macrofloc formation is limited owing to the increasing energy barrier between larger flocs which cannot be overcome by Brownian movement. In contrast, similar alum dosages at shaking frequencies > 700 ppm resulted in macrofloc formation (Table 6.5) and by greater coagulation activities (Fig. 6.3). This region is known as orthokinetic flocculation in which the shaking-induced velocity gradients were greater than Brownian movement thus providing additional opportunities and energy for contact between microflocs (Bratby 2006). Our findings agree with those of Ndabigengesere and Narasiah (1996) who discovered that low-intensity mixing does not produce enough energy to sustain floc formation.

The contour plot and 3D surface analysis of alum dosage and shaking time effects on coagulation activity are shown in Fig. 6.4a and b. The optimal range of alum

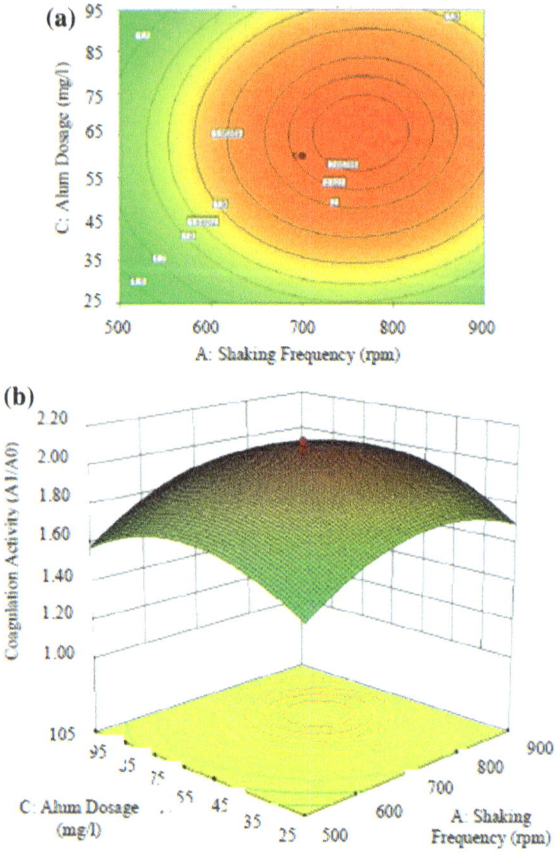

Fig. 6.3 **a** Contour plot of alum dosage and shaking frequency and **b** 3D surface of coagulation activity showing the effect of alum dosage and shaking frequency at 40 s of shaking time

dosage and shaking time appears to be between 65 and 85 mg/L as well as 37–48 s. The initially observed increase in coagulation activity with shaking time is synonymous with an increase in floc size. However, exceeding the optimal contact time results in the formation of flocs that are so large in size that they become prone to breakage. Consequently, coagulation activity declined as seen in Fig. 6.4. Similar trends were reported in the literature (Kan et al. 2002).

6.3.3 MTP Coagulation Activity Assay Optimisation and Model Verification

According to the RSM model (Sect. 3.1) and process analysis in Sect. 3.2, optimum coagulation activity of alum for round bottom MTP geometry can be achieved in the

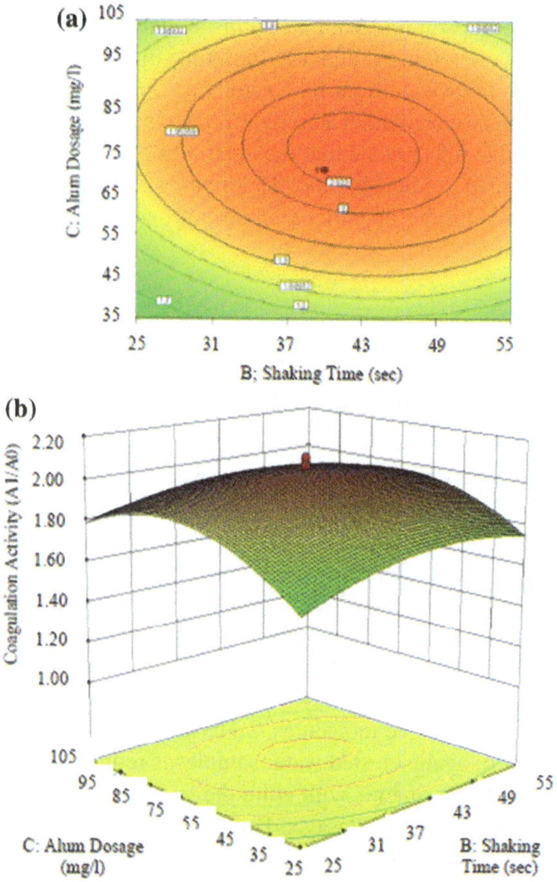

Fig. 6.4 **a** Contour plot of alum dosage and shaking time and **b** 3D surface of coagulation activity showing the effect of alum dosage and shaking time at 700 rpm of shaking frequency

Table 6.5 Verification experiments for optimum conditions of MTP-based coagulation activity assay

Replicate trial	Expected CA_{MTP}	Experimental CA_{MTP}	Image of flocs formed in 96-well MTP
Trial 1	1.808	1.80 ± 0.01	
Trial 2	1.808	1.83 ± 0.03	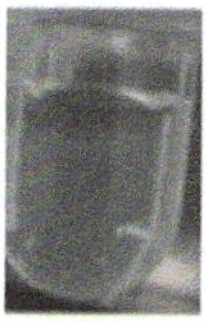

bentonite solution with a shaking frequency of 700 rpm, a shaking time of 40 s, a dosage of 40 mg/L and a settling time of 40 s.

Optimal conditions for coagulation activity were confirmed experimentally (Table 6.5). The coagulation activity was rated as desirable with a score of 0.922. The results in Table 6.5 indicate that the measured coagulation activity was reasonably close to the RSM predicted value. This finding validates the RSM approach and model for optimising the operational conditions of the coagulation activity in 96-well MTP round bottom geometry.

6.3.4 MTP Coagulation Activity Assay with Crude M. Oleifera *Seed Extracts*

The experimental coagulation activities of *M. oleifera* crude extract and alum as a model coagulant are shown in Fig. 6.5. A greater *M. oleifera* seed coagulant dosage (105–140 mg/L) was required to obtain a coagulation activity comparable to alum. *M. oleifera* seed proteins in crude and purified forms result in clay removal efficiencies ranging from 70–92% (Ferreira et al. 2011; Santos et al. 2009; Madrona et al. 2010). At an apparent dose of 15 mg/L, a coagulation activity similar to alum was attained (Ghebremichael et al. 2005). We therefore conclude that in our study the low coagulation effects at 35 mg/L is produced because of lower concentrations of coagulant-active molecules. To have similar effects to alum, a higher dosage of *M. oleifera* against tested water samples is required.

For experiments with alum the greatest coagulation activity was found between 70 and 100 mg/L dosage. This finding is also in accordance with previous results (Alley 2000). At 140 mg/L the coagulation activity of alum slightly decreased due to restablisation of colloid particles (Bratby 2006).

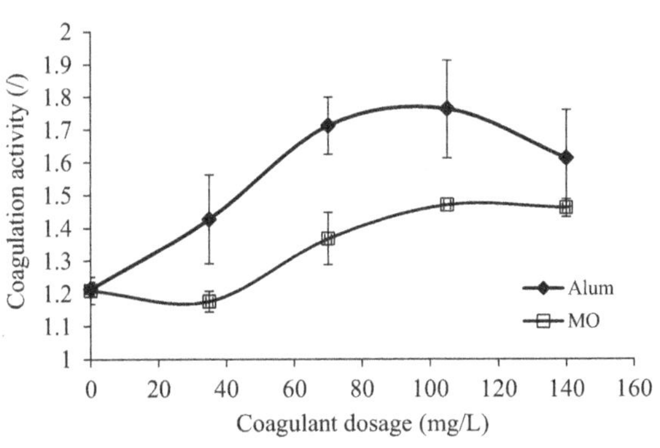

Fig. 6.5 Coagulation activity of *M. oleifera* and alum in 100 NTU synthetic turbid water based on 96-U MTPs and shaking frequency of 700 rpm, shaking time of 40 s and settling time of 40 min used as coagulation conditions

6.4 Conclusion

A 96-well MTP-based coagulation activity assay was successfully developed and optimised via RSM. U-shaped polystyrene MTP can be used to screen for coagulation activity natural and inorganic coagulants at 700 rpm shaking frequency and 40 s shaking time in 100 NTU bentonite colloidal solution of pH 6.5.

Acknowledgements Authors are thankful to the Ministry of Science, Technology and Innovation (MOSTI) for financial support of this study (Escience grant 02-02-13-SF0014).

References

Alley ER (2000) Appendix D: chemical characteristics. In: Water quality control handbook. McGraw Hill Inc

Antov MG, Šćiban MB, Adamović SR, Klašnja MT (2007) Investigation of isolation conditions and ion-exchange purification of protein coagulation components from common bean seed. Acta Periodica Technologica 38:1–19

Arunkumar P, Sadish Kumar V, Saran S, Bindun H, Devipriya SP (2019) Isolation of active coagulant protein from the seeds of *Strychnos potatorum*—a potential water treatment agent. Environ Technol 40:1624–1632

Aziz HA, Alias S, Assari F, Adlan MN (2007) The use of alum, ferric chloride and ferrous sulphate as coagulants in removing suspended solids, colour and COD from semi-aerobic landfill leachate at controlled pH. Waste Manag Res 25:556–565

Baptista ATA, Silva MO, Gomes RG, Bergamasco R, Vieira MF, Vieira AMS (2017) Protein fractionation of seeds of *Moringa oleifera* Lam and its application in superficial water treatment. Sep Purif Technol 180:114–124

Binnie C, Kimber M, Thomas H (2017) Coagulation and flocculation. In: Letterman RD (ed) Basic water treatment, 5th edn. ICE Publishing, Toronto, pp 61–83

Bratby J (2006) Coagulation and flocculation in water and wastewater treatment, second. IWA Publishing, London

Cao B, Gao B, Liu X, Wang M, Yang Z, Yue Q (2011) The impact of pH on floc structure characteristic of polyferric chloride in a low DOC and high alkalinity surface water treatment. Water Res 45:6181–6188

Dalen MB, Pam JS, Izang A, Ekele R (2010) Synergy between *Moringa oleifera* seed powder and alum in the purification of domestic. Water Sci World J 4:6–11

Dezfooli SM, Uversky VN, Saleem M, Baharudin FS, Hitam SMS, Bachmann RT (2016) A simplified method for the purification of an intrinsically disordered coagulant protein from defatted *Moringa oleifera* seeds. Process Biochem 51:1085–1091

Ferreira RS et al (2011) Coagulant and antibacterial activities of the water-soluble seed lectin from *Moringa oleifera*. Lett Appl Microbiol 53:186–192

Gassenschmidt U, Jany KD, Bernhard T, Niebergall H (1995) Isolation and characterization of a flocculating protein from *Moringa oleifera* Lam. Biochim Biophys Acta Gen Subj 1243:477–481

Ghebremichael KA, Gunaratna KR, Henriksson H, Brumer H, Dalhammar G (2005) A simple purification and activity assay of the coagulant protein from *Moringa oleifera* seed. Water Res 39:2338–2344

Gunaratna KR, Garcia B, Andersson S, Dalhammar G (2007) Screening and evaluation of natural coagulants for water treatment. Water Sci Technol Water Supply 7:19–25

Guo L et al (2008) Impacts of sterilization, microwave and ultrasonication pretreatment on hydrogen producing using waste sludge. Bioresour Technol 99:3651–3658

Hermann R, Lehmann M, Büchs J (2003) Characterization of gas–liquid mass transfer phenomena in microtiter plates. Biotechnol Bioeng 81:178–186

Kan C, Huang C, Pan JR (2002) Time requirement for rapid-mixing in coagulation. Colloids Surf A Physicochem Eng Asp 203:1–9

Lee R, Stack K, Richardson D, Lewis T, Garnier G (2012) Effect of shear, temperature and pH on the dynamics of salt induced coagulation of wood resin colloids. Colloids Surf A Physicochem Eng Asp 396:106–114

Lind C (1994) Reduce residuals with PACI coagulants. Public Work 125:43–47

Madrona GS, Serpelloni GB, Salcedo Vieira AM, Nishi L, Cardoso KC, Bergamasco R (2010) Study of the effect of saline solution on the extraction of the *Moringa oleifera* seed's active component for water treatment. Water Air Soil Pollut 211:409–415

Manda IKM, Chidya RCG, Saka JDK, Biswick TT (2016) Comparative assessment of water treatment using polymeric and inorganic coagulants. Phys Chem Earth Parts a/b/c 93:119–129

Marobhe NJ, Dalhammar G, Gunaratna KR (2010) Simple and rapid methods for purification and characterization of active coagulants from the seeds of *Vigna unguiculata* and *Parkinsonia aculeata*. Environ Technol 28:671–681

Matsui Y, Yuasa A, Furuya Y, Kamei T (1998) Dynamic analysis of coagulation with alum and PACl. J Am Water Works Assoc 90:96–106

Micheletti M et al (2006) Fluid mixing in shaken bioreactors: implications for scale-up predictions from microlitre-scale microbial and mammalian cell cultures. Chem Eng Sci 61:2939–2949

Miller SM, Fugate EJ, Craver VO, Smith JA, Zimmerman JB (2008) Toward understanding the efficacy and mechanism of *Opuntia* spp. as a natural coagulant for potential application in water treatment. Environ Sci Technol 42:4274–4279

Muyibi SA, Evison LM (1995) Optimizing physical parameters affecting coagulation of turbid water with *Morninga oleifera* seeds. Water Res 29:2689–2695

Ndabigengesere A, Narasiah KS (1996) Influence of operating parameters on turbidity removal by coagulation with *Moringa oleifera* seeds. Environ Technol 17:1103–1112

Ndabigengesere A, Narasiah KS, Talbot BG (1995) Active agents and mechanism of coagulation of turbid waters using *Moringa oleifera*. Water Res 29:703–710

Pritchard M, Craven T, Mkandawire T, Edmondson AS, O'Neill JG (2010) A comparison between *Moringa oleifera* and chemical coagulants in the purification of drinking water? An alternative sustainable solution for developing countries. Phys Chem Earth Parts a/b/c 35:798–805

Qureshi K, Bhatti I, Shaikh MS (2011) Development of biocoagulant from mango pit for the purification of turbid water. Sindh Univ Res J 43:105–110

Ramamurthy C, Maheswari MU, Selvaganabathy N, Kumar MS, Sujatha V, Thirunavakkarasu C (2012) Evaluation of eco-friendly coagulant from *Trigonella foenumgraecum* seed. Adv Biol Chem 2:58–63

Rossini M, Garrido JG, Galluzzo M (1999) Optimization of the coagulation–flocculation treatment: influence of rapid mix parameters. Water Res 33:1817–1826

Saleem M, Bachmann RT (2019) A contemporary review on plant-based coagulants for applications in water treatment. J Ind Eng Chem 72:281–297

Sánchez-Martín J, Ghebremichael K, Beltrán-Heredia J (2010) Comparison of single-step and two-step purified coagulants from *Moringa oleifera* seed for turbidity and DOC removal. Bioresour Technol 101:6259–6261

Santos AFS, Luz LA, Argolo ACC, Teixeira JA, Paiva PMG, Coelho LCBB (2009) Isolation of a seed coagulant *Moringa oleifera* lectin. Process Biochem 44:504–508

Shebek K, Schantz AB, Sines I, Lauser K, Velegol S, Kumar M (2015) The flocculating cationic polypetide from *Moringa oleifera* seeds damages bacterial cell membranes by causing membrane fusion. Langmuir 31:4496–4502

Sin HN, Yusof S, Abdul Hamid NS, Abd Rahman R (2006) Optimization of hot water extraction for sapodilla juice using response surface methodology. J Food Eng 74:352–358

Snodgras CA (1905) Purification of the St. Louis water supply by using ferrous sulphate and calcium hydroxide as a coagulant, followed by plain sedimentation. Public Heal Pap Rep 30:39–49

Song Y, Dong B, Gao N, Deng Y (2015) Comparative evaluation of aluminum sulfate and ferric sulfate-induced coagulations as pretreatment of microfiltration for treatment of surface water. Int J Environ Res Public Health 12:6700–6709

Tissot S et al (2010) Determination of a scale-up factor from mixing time studies in orbitally shaken bioreactors. Biochem Eng J 52:181–186

Trinh TK, Kang LS (2011) Response surface methodological approach to optimize the coagulation–flocculation process in drinking water treatment. Chem Eng Res Des 89:1126–1135

Unnisa SA, Deepthi P, Makkanti K (2010) Efficiency studies with *Dolichos lablab* and solar disinfection for treating turbid waters. J Environ Prot Sci 4:8–12

Uttran A, Loh SK, Kong SH, Bachmann RT (2018) Adsorption of NPK fertilizer and humic acid on palm kernel shell biochar. J Oil Palm Res 30:472–484

Van Benschoten JE, Edzwald JK (1990) Chemical aspects of coagulation using aluminum salts-II. Coagulation of fulvic acid using alum and polyaluminum chloride. Water Res 24:1527–1535

Xiao F, Ma J, Yi P, Huang JCH (2008) Effects of low temperature on coagulation of kaolinite suspensions. Water Res 42:2983–2992

Young S, Smith DW (2000) Effect of mixing on the kinetics of polymer-aided flocculation. J Water Supply Res Technol 49:1–8

Yu W, Gregory J, Campos L (2010) The effect of additional coagulant on the re-growth of alum–kaolin flocs. Sep Purif Technol 74:305–309

Chapter 7
Nutrient Recovery from Anaerobic Palm Oil Mill Effluent Using Palm Kernel Shell Biochar and Deoiled Spent Bleaching Earth and Their Effect on Oil Palm Growth

Nur Maisarah Ahmad, Nurul Shafinah Foo, Soh Kheang Loh, Khaliesah Abbas, Siew Kooi Ong, Nazatulshima Hassan, Abbas Fadhl Mubarek Al-Karkhi, Mustapha Mohammed Bello, and Robert Thomas Bachmann

Abstract Untreated palm oil mill effluent (POME) causes severe environmental pollution. Anaerobic pretreatment followed by nutrient recovery with subsequent use

---

N. M. Ahmad · N. S. Foo · K. Abbas · S. K. Ong · N. Hassan · A. F. M. Al-Karkhi · R. T. Bachmann (✉)
Green Chemistry and Sustainable Engineering Technology Cluster, Malaysian Institute of Chemical and Bio-Engineering Technology, Universiti Kuala Lumpur, 78000 Alor Gajah, Malacca, Malaysia
e-mail: bachmann@unikl.edu.my

N. M. Ahmad
e-mail: nurmaisarahmad@gmail.com

N. S. Foo
e-mail: shafinah.sharif@s.unikl.edu.my

K. Abbas
e-mail: abbas@unikl.edu.my

S. K. Ong
e-mail: skong@unikl.edu.my

N. Hassan
e-mail: nazatulshima@unikl.edu.my

A. F. M. Al-Karkhi
e-mail: abbas@unikl.edu.my

S. K. Loh
Energy and Environment Unit, Engineering and Processing Division, Malaysian Palm Oil Board, 6, Persiaran Institusi, Bandar Baru Bangi, 43000 Kajang, Selangor, Malaysia
e-mail: lohsk@mpob.gov.my

M. M. Bello
Centre for Dryland Agriculture, Bayero University, P.M.B. 3011, Kano, Nigeria
e-mail: mustiebello@gmail.com

A. Ismail et al. (eds.), *Advancements in Materials Science and Technology Led by Women*, Advanced Structured Materials 165,
https://doi.org/10.1007/978-3-031-21959-7_7

as fertiliser can avoid pollution and generate value-added products. We investigated the feasibility of recovering nutrients from anaerobically pretreated POME using palm kernel shell biochar (PKSBC) and deoiled spent bleaching earth (DSBE) as sorption agents. Nursery trials were conducted to study the effect of the nutrient-enriched PKSBC and DSBE on Haplic Acrisol soil fertility and oil palm seedling growth. About 14% of P, 43% of NH_3 and 42% of Fe content present in DSBE-pretreated POME were recovered by PKSBC, while DSBE was able to recover 47% of NH_3 and 27% of Fe from anaerobically pretreated POME. Both nutrient-enriched sorption agents enhanced soil properties, with DSBE-treated soil showing superior fertility and performance. The cation exchange capacity of Haplic Acrisol increased from 6.6 to 11.8 meq/100 g for DSBE-treated soil and to 8.7 meq/100 g for PKSBC treatment. The growth and health of oil palm seedlings, measured in terms of collar diameter and leaf colour index, was greatest for DSBE-treated soils, followed by PKSBC-treated soils and control. This study shows the reuse potential of palm oil industry by-products towards a circular palm oil economy.

Keywords Nutrient recovery · Palm oil mill effluent · Deoiled spent bleaching earth · Palm kernel shell biochar · Biofertiliser

7.1 Introduction

Palm oil is one of the major edible oils consumed worldwide and produced in countries such as Indonesia, Malaysia, Thailand, Nigeria, Colombia, Guatemala, Ghana, Papua New Guinea and Brazil (Charles et al. 2019; Tan and Lim 2019). While palm oil has contributed significantly to the economic development of the producing countries, there is a growing concern over the detrimental environmental impact of its production. The production of palm oil at the mills entails various processes, which generate by-products such as empty fruit bunches, palm kernel shells (PKS) and palm oil mill effluent (POME) (Gamaralalage et al. 2019; Kong et al. 2014). Further by-products streams are generated in palm oil refineries such as spent bleaching earth (Loh et al. 2017a).

Among these streams, POME has received wide attention due to its quantity and high organic strength. It is estimated that for every tonne of crude palm oil produced, 2.5–3.75 tonnes of POME are generated, which is characterised by high chemical oxygen demand, COD (40,000–100,000 mg/L), biological oxygen demand, BOD (25,000–65,000 mg/L), oil and grease (4000–6000 mg/L) and other organic compounds (Liew et al. 2015; Ng et al. 2016). Various technologies have been developed for the treatment of POME, ranging from biological (Ahmad Farid et al. 2019; Jalaludin et al. 2016), physico-chemical (Hossain et al. 2019; Huzir et al. 2019) to advanced oxidation processes (Charles and Cheng 2019; Bello and Abdul Raman 2017). While a lot of progress has been made in developing treatment technologies, there is a growing interest to shift from the conventional approach of treat-and-dispose to a circular economy approach of recover-and-reuse (Chojnacka et al.

Table 7.1 Physico-chemical properties of original and pyrolysed palm kernel shell (PKS)

Property	Original	PKSBC
Product yield (wt% d.b.)		
Biochar	–	29.8 ± 1.5
Bio-oil	–	6.9 ± 2.1
Syngas	–	63.2 ± 2.9
Proximate analysis		
Moisture content (wt%)	9.1 ± 3.1	3.1 ± 1.1
Volatile matter (wt% d.b.)	72.2 ± 0.9	18.9 ± 8.8
Fixed carbon (wt% d.b.)	23.9 ± 1.2	71.5 ± 6.8
Ash content (wt% d.b.)	2.0 ± 0.6	9.7 ± 4.4
VM/FC (/)	3.0 ± 0.2	0.3 ± 0.2
Ultimate analysis		
Carbon (wt% d.b.)	45.8 ± 1.4	66.2 ± 8.6
Hydrogen (wt% d.b.)	6.1 ± 0.3	2.6 ± 0.5
Nitrogen (wt% d.b.)	BDL	1.2 ± 1.5
Oxygen (wt% d.b.)	46.1 ± 1.5	20.4 ± 9.5
H/C (molar)	1.60 ± 0.08	0.48 ± 0.12
O/C (molar)	0.76 ± 0.05	0.25 ± 0.14
pH	*N/D*	9.34

PKS were pyrolysed at 500 °C for 75 min ($n = 8$; d.b.—dry basis; PKSBC)

agitator. The mixture was then left to settle for 2 h. The supernatant was decanted to collect and sundry the PKSBC.

7.2.3 Nursery Trial

A nursery trial using germinated oil palm seedlings with 4 leaves was conducted at the Malaysian Institute Chemical and Bio-Engineering Technology, Universiti Kuala Lumpur (geo coordinates 2° 27′ 14.32″ N, 102° 10′ 26.38″ E). The trial was conducted for a period of 60 days starting 9 March 2019 until 8 May 2019. Irrigation tubes were first set up, and concrete bricks were aligned in alternate manner alongside the irrigation tube and labelled. Calibration of drippers was carried out to ensure each dripper delivers a similar amount of water. The nursery trial consisted of a completely randomised design with different treatments of (1) control ($T1$), (2) soil with PKSBC addition ($T2$–$T3$), (3) soil with DSBE addition ($T4$–$T5$) and (4) soil with PKSBC-DSBE addition ($T6$–$T7$) (Table 7.2).

The Haplic Acrisol soil was mixed with nutrient-enriched PKSBC and DSBE at different concentrations and combinations (Table 7.2) to yield seven soil treatment

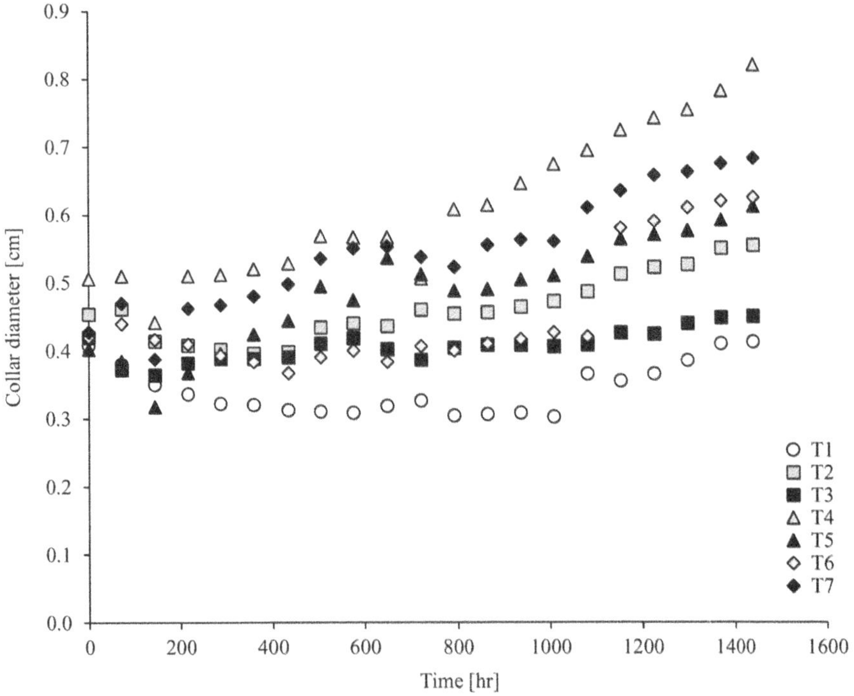

Fig. 7.1 Collar diameter measurement under different soil treatments using palm kernel shell biochar and deoiled spent bleaching earth

Table 7.2 Soil treatment composition (dry basis) containing soil and palm kernel shell biochar (PKSBC) and/or deoiled spent bleaching earth (DSBE)

Treatment	Constituent, soil:sorbent	Weight of soil, M_{soil} (kg)	Weight of BC, M_{PKSBC} (kg)	Weight of DSBE, M_{DSBE} (kg)
T1	Soil only	3.659	0	0
T2	Soil: PKSBC	3.653	0.00598	0
T3	Soil: PKSBC	3.646	0.01196	0
T4	Soil: DSBE	3.575	0	0.0802
T5	Soil: DSBE	3.490	0	0.1605
T6	Soil: PKSBC: DSBE	3.568	0.00598	0.0802
T7	Soil: PKSBC: DSBE	3.477	0.01196	0.1605

T1—soil only; *T2*—0.163 wt% PKSBC in soil; *T3*—0.327 wt% PKSBC in soil; *T4*—2.24 wt% DSBE in soil; *T5*—4.58 wt% DSBE in soil; *T6*—1.61 wt% PKSBC + 2.16 wt% DSBE in soil; *T7*—3.18 wt% PKSB + 4.27 wt% DSBE in soil

compositions designated as treatment groups *T*1–*T*7. The moisture content of each composition was predetermined to obtain the mass required on dry basis. The composition was weighed accordingly and prepared for five replicates per treatment. The PKSBC concentration in soil was 0.16 (*T*2) and 0.33 wt% (*T*3), whereas the concentration of DSBE in soil was 2.2 (*T*4) and 4.4 wt% (*T*5). The PKSBC concentration of 0.16 wt% in topsoil (30 cm) with a bulk density of 1,700 kg/m^3 is equivalent to an application rate of 8.16 MT/ha. Based on a meta-analysis carried out by Liu et al. (2013), the greatest changes in crop productivity were observed for biochar applications rate between < 10 MT/ha and 10–20 MT/ha.

Once the soil compositions were ready, the oil palm seedlings were assigned to each treatment group to form five replicates. For *T*1, for example, 2.50 kg of soil was loosely packed in a polybag and labelled. The seedling was transplanted by placing it on top of the soil and the remaining 1.16 kg of soil added to achieve the required final weight of 3.66 kg per polybag. The polybag was then labelled as *T*1*R*1 (replicate no. 1 of *T*1). The process was repeated for the rest of replicates to form *T*1*R*2, *T*1*R*3, *T*1*R*4 and *T*1*R*5. These steps were repeated for the other treatments namely *T*2 (*T*2*R*1–*T*2*R*5), *T*3 (*T*3*R*1–*T*3*R*5), *T*4 (*T*4*R*1–*T*4*R*5), *T*5 (*T*5*R*1–*T*5*R*5), *T*6 (*T*6*R*1–*T*6*R*5) and *T*7 (*T*7*R*1–*T*7*R*5).

The polybags with the transplanted oil palms were arranged at the nursery site via a completely randomised design to eliminate any external factors that might affect the growth of oil palm seedlings if arranged in groups. Watering was conducted through drippers, which were pegged onto the polybag to avoid fallout. An automated timer was set to deliver 170 mL of water in the morning (7.30 am) and 70 mL in the evening (6.30 pm) for the first month, and 90 mL of water for morning (8.30 am) and evening (7.30 pm) for the remainder of the experimental trial as suggested by Radin et al. (2018).

Collar diameter and plant health were monitored based on methods reported previously (Bachmann et al. 2018). For growth rate monitoring, the plant's collar diameter was determined using a Vernier calliper at 2 cm height from soil surface. The diameter measurements were carried out once every three days in the morning and evening before watering. The growth rate (cm/h) was determined by plotting the collar diameter measurements against time.

Plant health was monitored by a 9-scale leaf colour chart obtained from Top Fruits Sdn. Bhd (Bachmann et al. 2018). The chart scales range from 1-very healthy to 9-dead. Green leaf indicates healthy growth of the plant whereas brown shows a dying or dead plant. The score of each plant for four leaves per plant was taken, and average value was obtained. The average value of each plant representing the plant health index was taken.

7.2.4 *Analytical Techniques*

7.2.4.1 Nutrient Analysis

Nutrient recovery efficiency of DSBE and PKSBC was determined as follows. A POME sample (control) was put into a 50 mL centrifuge tube. After 2 h of mixing at 150 rpm, the final sample was taken. For the DSBE- and PKSBC-treated POME, samples were collected at initial and final stages of the 2 h mixing process. Control for POME supernatant was included as jar test for comparison with the PKSBC-treated POME. The collected POME samples were centrifuged at 3900 rpm for 30 min. Then, the supernatant was collected and preserved. The preservation medium for colour and NH_3 was sulphuric acid while nitric acid was used for P and trace elements in accordance with the standard methods (American Public Health Association 2005).

Sample analysis was carried out as described previously (Foo et al. 2020). In brief, the colour of the POME samples was determined using a colorimeter (HACH, DR900), while the NH_3 content was measured via HACH method 8155 using a colorimeter (HACH, DR900). P and trace elements analyses were conducted using a UV–Vis spectrophotometer. The concentration of P in the samples was determined following the molybdenum blue method while trace elements were analysed using the atomic absorption spectrophotometer (Perkin Elmer, AAnalyst 400).

7.2.4.2 PH and Electrical Conductivity of Soil Treatments

Soil pH and electrical conductivity were determined before and after nursery trials according to the procedure reported by Rajkovich et al. (2012). For pre-nursery soil analysis, samples were collected from soil mixing process ($T1$–$T7$) and stored in labelled bags. 1 g of sample was placed in a 250 mL Erlenmeyer flask, and 20 mL of water was added. The prepared samples were stirred using a shaker incubator at 150 rpm for 1.5 h. Afterwards, the pH measurement was carried out by inserting a pH probe metre into the sample solution. The electrical conductivity (EC) measurement was carried out via a similar method. For post-nursery soil analysis, three replicates were chosen from each treatment. Soil sample from each replicate at three different locations was obtained using a spatula and stored. The pH and EC measurements of post-nursery soils were conducted in the same manner.

7.2.4.3 Soil Analysis

Organic carbon, total organic carbon, total nitrogen, total phosphorus, potassium, calcium, magnesium, sodium, CEC and base saturation of composite samples of each treatment were determined in duplicate at a SAMM-accredited laboratory (ESPEK Research Advisory Services, Malaysia).

7.2.4.4 Statistical Analysis

Statistical analysis was carried out using MINITAB version 14.0. The number of replicates involved in the analysis for each parameter was 5. For growth rate data, a *t*-test was performed to determine if there was a significant difference in growth rate taken in the morning compared to the evening. The data taken for growth rate was also analysed using a two-way analysis of variance (ANOVA) to investigate the effect of time and treatment on the collar diameter. Tukey's test by one-way ANOVA was utilised for the post hoc test. Statistical analysis was also applied to the data of the leave health index. The effect of soil treatment and time were analysed using two-way ANOVA. Tukey's test via one-way ANOVA was also carried out for the leaf colour data. A significance level of 0.05 was used to test these effects.

7.3 Results and Discussion

7.3.1 Nutrient Recovery

DSBE and PKSBC were utilised as media for nutrient recovery from POME and mixed with soil to test their effect on oil palm seedling growth. To evaluate the nutrient recovery efficiency of DSBE and PKSBC, the nutrient concentrations in raw POME (control) and POME samples treated with DSBE and PKSBC were determined. The untreated POME served as a control to account for naturally occurring microbial degradation and physical effects (e.g. precipitation, volatilisation). Tables 7.3 and 7.4 display the results of nutrient recovery from POME using DSBE and PKSBC, respectively.

Analysis showed that anaerobic POME before nutrient recovery contained substantial macro-(N, P, K) and micronutrients (Fe, Cu, Zn and Mn) beneficial for plant growth. The concentrations of NH_3, P and Mn in POME prior to nutrient recovery agrees with values reported in literature, while Fe, Cu and Zn concentrations were below the reported range (Table 7.5). Variations in POME composition may arise due to plantation and oil palm mill-specific operation conditions as well as analytical methods deployed. Variations observed in the control during the 2 h treatment period are attributed to microbial activities, physico-chemical processes such as adsorption, desorption and precipitation as well as methodological uncertainties.

A net colour removal was achieved in both the DSBE-treated (17%) and PKSBC-treated (7%) POME samples. However, the treated POME must undergo further treatment in order to meet the final discharge limit of 200 (Zainal et al. 2017). The observed removal of colour indicates that adsorption of dissolved organic matter such as humic acid had taken place. Humic substances contain quinonoid, carboxylic and phenolic functional groups capable of attracting cationic nutrients (Wu et al. 2009). Previous studies have suggested that humic substances from POME could be involved

Table 7.3 Nutrient recovery from anaerobically pretreated palm oil mill effluent (POME) using deoiled spent bleaching earth (DSBE)

Parameter	Control POME (A)			DSBE-treated POME (B)			Net recovery
	Initial	Final	Change (%)	Initial	Final	Recovery (%)	(*B–A*) (%)
Colour (PtCo)	5,550 ± 259	5,133 ± 94	8	5,550 ± 259	4,150 ± 165	25	17
NH_3 (ppm)	202 ± 2	220 ± 0	−10	202 ± 2	128 ± 38	37	47
P (ppm)	30.7 ± 0.3	14.6	52	30.7 ± 0.3	24.3 ± 10	21	−31
Cu (ppm)	0.041 ± 0.025	0.029	29	0.041 ± 0.025	0.043 ± 0.03	−5	−34
Zn (ppm)	0.09 ± 0.00	0.20	−122	0.09 ± 0.00	0.29 ± 0.01	−222	−100
Mn (ppm)	1.11 ± 0.01	0.93	16	1.11 ± 0.01	0.47 ± 0.35	58	42
Fe (ppm)	1.92 ± 0.13	2.42	−26	1.92 ± 0.13	1.91 ± 0.64	1	27

Table 7.4 Nutrient recovery from deoiled spent bleaching earth (DSBE) pretreated palm oil mill effluent (POME) using palm kernel shell biochar (PKSBC)

Parameter	DSBE pretreated POME as control (*A*)			PKSBC-treated POME (*B*)			Net recovery
	Initial	Final	Reduction (%)	Initial	Final	Recovery (%)	*B–A* (%)
Colour (PtCo)	4,150 ± 165	3233	22	4,150 ± 165	2,950	29	7
NH_3 (ppm)	128 ± 38	145	−13	128 ± 38	90	30	43
P (ppm)	24.3 ± 10	18.1	26	24.3 ± 10	14.5	40	14
Cu (ppm)	0.043 ± 0.03	0.02	54	0.043 ± 0.03	0.02	54	0
Zn (ppm)	0.29 ± 0.01	0.34	−17	0.29 ± 0.01	0.08	72	89
Mn (ppm)	0.47 ± 0.35	0.11	77	0.47 ± 0.35	0.27	43	−34
Fe (ppm)	1.91 ± 0.64	1.08	43	1.91 ± 0.64	0.28	85	42

Table 7.5 Characteristics of raw and anaerobically digested palm oil mill effluent (POME)

Parameter	Unit	Raw POME		Digested POME		
		Loh et al. (2017b)	Lam and Lee (2011)	Zainal et al. (2017), Loh et al. (2013b)	Loh et al. (2017b)	Foo et al. (2020)
pH	–	3.3–5.7	3.4–5.2	4.7	5.8–8.9	6.6
Oil and grease	mg/L	5,614–8,812	130–18,000	4,000–6,000	1,432–2,120	NR
BOD	mg/L	18,225–23,900	10,250–43,750	25,000	236–1,266	518
COD	mg/L	45,818–54,861	15,000–100,000	51,000	3,303–6,830	2,257
Total solids	mg/L	46,304–63,192	11,500–79,000	40,000	9,891–17,089	NR
Suspended solids	mg/L	24,846 -30,920	5,000–54,000	18,000	2,539–5,713	1,567
Volatile fatty acids	mg/L	1,057–1,073	NR	NR	572–584	NR
Ammoniacal nitrogen	mg/L	77–101	4–80	35	135–165	261
Total Kjeldahl nitrogen	mg/L	670–780	180–1,400	750	288–483	480
Phosphorus, P	mg/L	109–136	NR	180	30–84	361
Potassium, K	mg/L	1,696–2,043	NR	2,270	1,569–1,879	28,356
Magnesium, Mg	mg/L	279–296	NR	615	320–334	NR
Calcium, Ca	mg/L	282–290	NR	439	212–229	NR
Iron, Fe	mg/L	65–67	NR	46.5	9–15	348
Manganese, Mn	mg/L	2.67–2.93	NR	2.0	0.90	4.56
Copper, Cu	mg/L	0.80–0.90	NR	0.89	0.22–0.24	6.15
Zinc, Zn	mg/L	1.24–2.72	NR	2.3	0.47–0.53	1.82
Chromium, Cr	mg/L	10.2	NR	NR	NR	NR
Nickel, Ni	mg/L	0.10–2.70	NR	NR	0.14–3.60	NR

NR—not reported

in urease inhibition, urea adsorption and ammonia fixation, which would be beneficial for enhanced soil fertility (Siva et al. 2000).

For NH_3, DSBE and PKSBC achieved 47 and 43% net recoveries from POME. The substantial recovery of NH_3 is important for the subsequent application in agricultural soils. The recovery of P was negative for DSBE suggesting a release of valuable P from DSBE. The presence of P in DSBE is due to its use as bleaching agent in palm oil refining removing phospholipids and other substances from the crude palm oil (Loh et al. 2013b, 2015). However, the released P could be recovered during PKSBC treatment at a net recovery of 14%. The recovery of P is highly desirable, since P is a major element required for plant growth. Rock phosphate deposits are also finite and often located in geopolitically unstable areas, thus posing a potential risk to food production security (Blackwell et al. 2019). Further research is required to understand and ultimately prevent the release of P from DSBE for maximum recovery of this mineral commodity.

For the recovery of micronutrients, different observations were made. While the DSBE could recover about 42% of the initial concentration of Mn, an increase in Cu and Zn concentrations in DSBE-pretreated POME was observed. These micronutrients possibly originated from the DSBE which could have been inherited from its previous application in palm oil refining (Loh et al. 2013b, 2015). The more reactive Mn cationic form in POME showed higher affinity to DSBE than the rest of the micronutrients. Due to higher Lewis acid strength of Zn and Cu than Mn, their ability to dissociate from DSBE would be greater in acidic POME condition. On the other hand, their presence in cationic form in DSBE-pretreated POME provided more flexible coordination geometry and lesser steric hindrance for them to bind onto the negatively charged PKSBC, hence resulted in higher nutrient recovery of Zn (89%) and Fe (42%), respectively, but none for Cu as its initial presence was considered negligible (Table 7.4). It is clear from the results presented, that the combination of DSBE and PKSBC resulted in the net recovery of most nutrients for subsequent application as fertiliser and soil modifier.

7.3.2 Effect of Nutrient-Enriched Sorbents on Soil Properties

The effect of different soil treatments on soil properties was investigated. Table 7.6 shows the amount of nutrients and CEC for the various treatments. The recommended molar ratio of carbon to nitrogen (C/N) of the soil is 9 21 for optimum plant development (Loh et al. 2013b; Li et al. 2019). A high C/N ratio is undesirable as it indicates too much carbon in the soil, which can hinder the bioavailability of N and affect plant growth. In this study, the C/N ratio of all treatment was similar (Table 7.6) and within the recommended range. The C/N ratio of the control soil ($T1$) was similar to that reported for Rengam soil (Sabrina et al. 2009).

In terms of carbon and nitrogen content, an increase was observed for all treatments compared to the control. In addition, treatments $T3$, $T5$ and $T7$ had greater carbon and nitrogen contents compared to $T2$, $T4$ and $T6$. This can be attributed to the greater concentration of PKSBC, DSBE and PKSBC + DSBE (Table 7.2).

Table 7.6 also depicts the elemental composition for each soil treatment. The concentration of P was greatest in the DSBE-treated soil, followed by PKSBC + DSBE and PKSBC. The same trend was also observed for all other monitored plant nutrients (K, Ca, Mg and Na). The findings may be attributed to the intrinsic composition of bleaching earth, the precursor of the DSBE, the nutrients present in the filter cake comprising of DSBE and suspended POME solids. Bleaching earth comprises mainly of bentonite, which is rich in elements such as Na, Ca, K and P (Loh et al. 2017a; Lim et al. 2013). PKSBC, on the other hand, did not benefit from the presence of POME solids and thus had correspondingly lower values.

Table 7.6 Soil analysis for control (*T*1) and different soil treatments (*T*2–*T*7)

Trial	wt%			C/N ratio	ppm	meq/100 g					Base saturation
	Organic carbon	Total organic carbon	Total nitrogen		Total P	K	Ca	Mg	Na	CEC	
*T*1	0.95	1.26	0.12	12	97	0.15	0.70	0.14	0.03	6.6	15.5
*T*2	0.93	1.24	0.12	12	144	0.21	1.22	0.21	0.06	7.0	23.6
*T*3	1.11	1.47	0.13	13	96	0.21	1.36	0.23	0.07	8.7	21.4
*T*4	1.02	1.36	0.13	12	302	0.23	5.96	0.63	0.05	9.6	71.8
*T*5	1.09	1.45	0.14	12	447	0.48	10.62	0.90	0.04	11.8	102
*T*6	1.05	1.39	0.13	12	227	0.28	7.13	0.70	0.06	7.5	109
*T*7	1.10	1.46	0.14	12	349	0.49	11.87	1.07	0.06	10.8	125

*T*1—soil only; *T*2—0.163 wt% PKSBC in soil; *T*3—0.327 wt% PKSBC in soil; *T*4—2.24 wt% DSBE in soil; *T*5—4.58 wt% DSBE in soil; *T*6—1.61 wt% PKSBC + 2.16 wt% DSBE in soil; *T*7—3.18 wt% PKSB + 4.27 wt% DSBE in soil

CEC of the soil measures its ability to retain important nutrients (exchangeable cations) for plant growth. Thus, soil with high CEC can hold more nutrients and water for effective plant growth. As shown in Table 7.6, the addition of DSBE resulted in a twofold increase compared to the control (*T*1). Similar effects were reported by (Loh et al. 2017a, 2013b). Although these CEC values were typical of sandy loam soils (Peats et al. 2007), the addition of DSBE and PKSBC + DSBE enhanced the CEC to a range similar to clay loam. Hence, the addition of DSBE and PKSBC could improve the soil properties, which should benefit plant growth and health.

For the base saturation, which is the percentage of the CEC occupied by the basic cations (nutrients), similar observations were made. The soil base saturation was improved with the addition of both the DSBE and PKSBC. For the control, the base saturation was 15.5%, and it increased to 21.4–23.6% for the PKSBC-treated soils. The highest base saturations were obtained for the DSBE- and (DSEB + PKSBC)-treated soils (71–125%). For *T*5, *T*6 and *T*7, the base saturations were higher than 100%. This is likely due to the unattached Ca, Mg, and/or Na that might have been technically unavoidably extracted by the soil testing process (Csuros 2018). Although there are controversies on adopting a particular optimum base saturation (Kopittke and Menzies 2007), soils with high base saturation are considered good for plant growth.

Table 7.7 shows the pH and EC for the different soil treatments. The soil used in this study was classified as Rengam soil, which has sandy loam characteristics (Sabrina et al. 2009). The pH of the control soil (*T*1) before the commencement of the nursery trial was pH 4.26, which is similar to the pH range for Rengam soil series (4.06–4.10) (Houmani et al. 2012). Addition of 0.19% and 0.37% alkaline PKSBC (Table 7.2) naturally increased the soil pH. The DSBE-treated soil (*T*4 and *T*5) experienced a greater increase in pH compared to the PKSBC-treated soils. The observed increase is in agreement with Loh et al. (2015), who reported a pH of 4.5–5.2 for spent bleaching earth. However, the pH of *T*7 (5.83) was the greatest among all the treatments. Tiemann et al. (2018) reported that 95% of oil palms in Southeast Asia are planted on soils with a pH < 5. A soil pH of 4.0–4.2 is regarded as moderately suitable while a pH > 4.2 is considered as highly suitable. In general, the initial pH values for all the soil treatments were above 4.2, indicating the suitability of the soil conditions for oil palm cultivation.

After 60 days of trial, a further increase in pH was observed for each soil treatment, ranging from 5.13 (*T*2) to 7.05 (*T*5). The increase of pH could be deduced by the presence of nutrient cations in the soil treatments, which were released from the PKSBC and DSBE over time in exchange for H^+. Microbial degradation of lignocellulosic fibres present in the biosolids was also expected to release cations bound to the organic matrix. Although the soil pH has no direct effect on the plant growth, it controls the availability of nutrients such as P and trace elements. For example, P is more available in neutral pH soil while trace elements such as Mn and Zn prefer much lower soil pH (Faloye et al. 2017). These results showed that addition of both the PKSBC and DSBE increased the soil pH, which is similar to the results reported earlier by Pandit et al. (2018).

Table 7.7 pH and electrical conductivity (EC) of soils under different treatments

Soil treatment	$pH_{initial}$ (/)	pH_{final} (/)	$EC_{initial}$ (μS/cm)	EC_{final} (μS/cm)
*T*1	4.26 ± 0.02	4.95 ± 0.14	78.4 ± 0.2	39.4 ± 5.1
*T*2	4.39 ± 0.02	5.03 ± 0.14	39.4 ± 1.4	46.5 ± 7.2
*T*3	4.54 ± 0.03	5.03 ± 0.31	57.6 ± 0.5	63 ± 33
*T*4	5.18 ± 0.01	6.71 ± 0.39	120 ± 0.3	98 ± 40
*T*5	5.67 ± 0.04	7.43 ± 0.24	171 ± 0.4	188 ± 208
*T*6	4.71 ± 0.13	7.06 ± 0.32	70.1 ± 0.5	78 ± 32
*T*7	5.83 ± 0.03	7.56 ± 0.11	189 ± 1	381 ± 254

*T*1—soil only; *T*2—0.163 wt% PKSBC in soil; *T*3—0.327 wt% PKSBC in soil; *T*4—2.24 wt% DSBE in soil; *T*5—4.58 wt% DSBE in soil; *T*6—1.61 wt% PKSBC + 2.16 wt% DSBE in soil; *T*7—3.18 wt% PKSB + 4.27 wt% DSBE in soil.

EC is a measure of dissolved cations and anions present in soil. The initial EC of the control soil (*T*1) was found to be 78 μS/cm (Table 7.7). The soil EC depends on soil type, particle size and consistency, and moisture content. The typical ranges of EC are highest in clay (100–10,000 μS/cm), intermediate in silt type (50–500 μS/cm) and lowest in sandy soil (2–40 μS/cm) (Grisso et al. 2009). The Rengam soil utilised in this study is reported to comprise 50% sand, 13% silt and 37% clay (Houmani et al. 2012). Thus, the observed EC falls under the intermediate range as expected. The addition of DSBE, which comprises primarily of bentonite clay, further increased the EC of the soil, which agrees with the study of Grisso et al. (2009). The addition of PKSBC, however, had a somewhat opposite effect, suggesting that mobile ions might have been adsorbed by the BC. The EC values for post-nursery soil ranged from 39.4 (*T*1) to 381 μS/cm (*T*7). The lower EC for *T*1 suggests that the test plants had taken up the nutrients while most other soil treatments released nutrients which could result in enhanced plant growth.

7.3.3 Effect of Nutrient-Enriched Sorbents on Growth of Oil Palm Seedlings

To investigate the effect of different soil treatments (*T*1–*T*7) on the growth of the oil palm seedlings, the collar diameters of the seedlings were measured daily over a period of 60 days. Figure 7.1 depicts the average collar diameters of each treatment against time for the observed period. The collar diameters for all treatments at the start of the trial ranged from 0.4 to 0.5 cm. The collar diameters were then monitored in the morning (am) and evening (pm). In the first week, all treatments displayed a declining trend of collar diameter, indicating that the plants were adapting to the new environment. At the end of the growth trial, the greatest collar diameter was obtained for soil treated with 2.31% DSBE (*T*4), followed by *T*7 and *T*5. The superior collar diameter in the DSBE-treated soils can be linked to the higher CEC and base saturation as discussed previously. However, the collar diameters for both the DSBE and PKSBC-treated soils were greater than the control. This indicates that both the DSBE and PKSBC have a positive effect on the plant growth, due to the enhancement of soil properties and the presence of recovered plant nutrients.

During the nursery trials, the morning-measured collar diameter was consistently greater than the evening-measured collar diameter, regardless of soil treatment (Fig. 7.2). This is expected due to the differences in evapotranspiration between the two measuring periods. During daytime, there is higher evapotranspiration due to warmer temperature and photosynthesis. Consequently, the plants were able to replenish their cells with water at night-time, resulting in the observed increase collar diameter. However, the observed differences gradually diminished. As the plants grew older, the xylem of plants became more rigid and the effect less pronounced (Zeng et al. 2018).

From the collar diameter data obtained, the growth rate of the oil palm seedlings for each soil treatment was determined. Table 7.8 displays the lag phase (*h*), growth phase (*h*) and linear growth rate (cm/h). The fastest linear growth rate among the treatments was obtained from *T*4 at 0.0003 cm/h followed by *T*1, *T*2, *T*5 and *T*6 at equal linear growth rate of 0.0002 cm/h. The linear growth rate of *T*4 was in agreement with the collar diameter and plant health index, with a correlation coefficient (R^2) of 0.9905. *T*3 experienced a slower growth compared to *T*2, indicating that a lower dosage of PKSBC may be preferred.

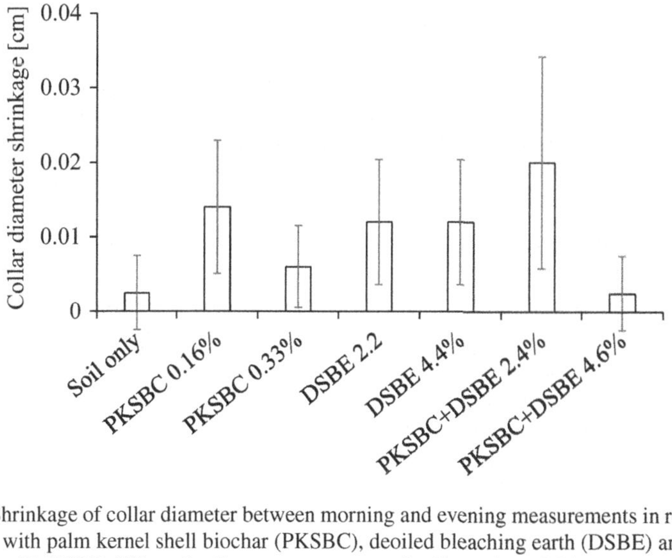

Fig. 7.2 Shrinkage of collar diameter between morning and evening measurements in response to treatments with palm kernel shell biochar (PKSBC), deoiled bleaching earth (DSBE) and mixture of PKSBC with DSBE. Effect of soil treatment was not significant ($p = 0.06$)

Table 7.8 Growth rate of oil palm seedlings

Treatment	Lag phase (*h*)	Growth phase (*h*)	Linear growth rate (cm/h)	Correlation coefficient (R^2)
*T*1	865	576	0.0002	0.8960
*T*2	793	649	0.0002	0.9788
*T*3	721	720	0.0001	0.8744
*T*4	721	720	0.0003	0.9905
*T*5	145	1,296	0.0002	0.8700
*T*6	360	1,081	0.0002	0.6211
*T*7	145	1,296	0.0001	0.9125

*T*1—soil only; *T*2—0.163 wt% PKSBC in soil; *T*3—0.327 wt% PKSBC in soil; *T*4—2.24 wt% DSBE in soil; *T*5—4.58 wt% DSBE in soil; *T*6—1.61 wt% PKSBC + 2.16 wt% DSBE in soil; *T*7—3.18 wt% PKSB + 4.27 wt% DSBE in soil

7.3.4 Comparison of Soil Treatment Performance

The collar diameter on 8 May 2019 was used to compare the effect of each treatment type on plant growth in relation to the control. The overall highest measured collar diameter was obtained in *T*4 (0.82 ± 0.11 cm), followed by *T*7 (0.68 ± 0.12 cm) and *T*2 (0.55 ± 0.12 cm). The corresponding soil pH and conductivity were 6.25 and 97.6 μS/cm (*T*4), 6.97 and 381 μS/cm (*T*7) and 5.13 and 46.5 μS/cm (*T*2). The nutrient recovery performed during soil mixing process proved that the PKSBC could recover NH_3, P and traces of Zn and Fe, which aid plant growth. Figure 7.3a compares the effect of PKSBC concentrations (*T*1 = 0, *T*2 = 0.19, and *T*3 = 0.38%) on the collar diameter. While there was no significant difference between treatments *T*1–*T*3, the highest collar diameter at *T*2 indicates detrimental effects of higher amount of PKSBC on plant growth. Similarly, the effect of lower amount of DSBE in *T*4 (2.33%) on the collar diameter was greater than *T*5 (4.62% DSBE), indicating that higher treatment of DSBE might also be detrimental to plant growth (Fig. 7.3b). Although the nutrient recovery by DSBE was lower than that of PKSBC, the fertility of soil was somewhat enhanced probably due to the presence of POME solids in the DSBE filter cake. Water holding capacity and nutrients availability are higher in soils with high CEC, thus allowing nutrient uptake by the plant roots. For the combined PKSBC and DSBE treatments (*T*6 and *T*7), a different trend was observed (Fig. 7.3c). The higher mixed concentration (*T*7 = 5% PKSBC + DSBE) resulted in a greater collar diameter than both the *T*6 (2.5% PKSBC + DSBE) and *T*1 (control). However, the effect was not statistically significant.

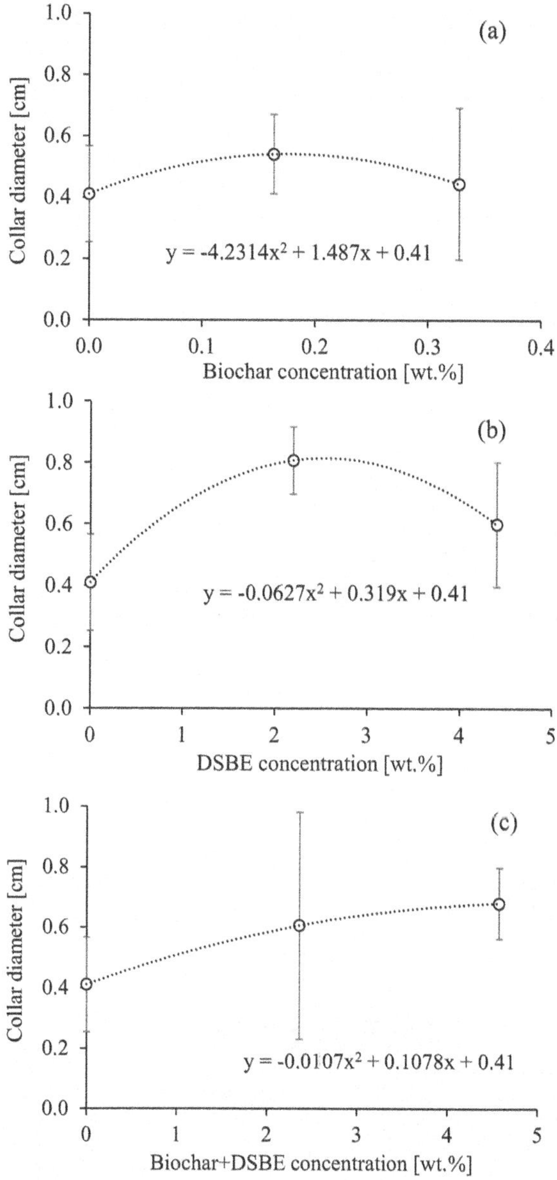

Fig. 7.3 Effect of palm kernel shell biochar (**a**), DSBE (**b**) and palm kernel shell biochar + DSBE (**c**) on the collar diameter of oil palm seedling

7.3.5 *Plant Health by Leaf Colour Data Analysis*

The effect of different soil treatments on plant health was also monitored through the leaf health index analysis. Figure 7.4 illustrates the health index for all soil treatments. In general, the health index for control and PKSBC treatments (*T*1, *T*2 and *T*3) fluctuated between scores of 4 (slightly healthy) and 5 (slightly sick) throughout the experimental period without a clear trend for *T*2. Plant health for DSBE treatment alone and in combination with PKSBC, however, had improved noticeably with time in the order of *T*2 (0.0022 1/h) > *T*6 (0.0017 1/h), *T*4 (0.0016 1/h) and *T*5 (0.0014 1/h). This finding is in good agreement with collar diameter results.

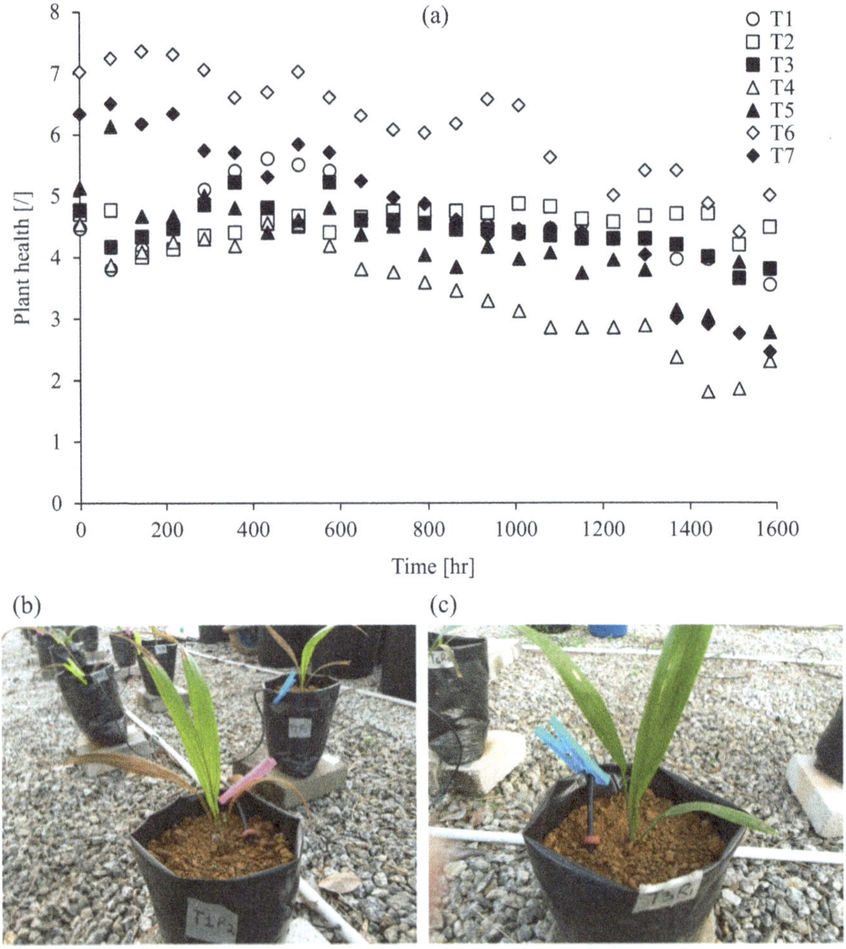

Fig. 7.4 Effect of soil treatment on leaf health index using palm kernel shell biochar and deoiled spent bleaching earth (**a**), replicate with dead leaves (**b**) and replicate with health index 4.0 (**c**)

Leaf colour data were analysed similar to collar diameter data. A two-way ANOVA was used to determine whether there was a correlation between observation time and treatments applied to the result of plant health. Although statistically there was no significant interaction between observation time and treatment types, each individual parameter had a highly significant effect on plant health index ($p < 0.001$). Overall, the findings in statistical analysis favours $T4$ as the superior treatment for this study.

7.4 Conclusion

The current interest towards circular economy necessitates the development of appropriate technologies for resources recovery. In this work, we utilised DSBE and PKSBC for nutrient recovery from POME waste stream followed by subsequent nursery trials with oil palm seedlings. DSBE and PKSBC were overall able to recover most nutrients and enhanced soil properties in terms of pH, conductivity, carbon, nitrogen and phosphor content as well as CEC. The seedlings grown in DSBE- and PKSBC-treated soils exhibited higher growth rates than those under control treatment. This study shows that reutilising DSBE and PKSBC together with high organic strength anaerobic POME could be synergistic in enhancing growth and health of oil palm seedlings due to substantial enrichment of beneficial nutrients. It serves to contribute partly towards the continued efforts in adopting a circular palm oil economy.

Acknowledgements This research did not receive any specific grant from funding agencies in the public, commercial or not-for-profit sectors.

References

Ahmad Farid MA et al (2019) A holistic treatment system for palm oil mill effluent by incorporating the anaerobic-aerobic-wetland sequential system and a convective sludge dryer. Chem Eng J 369:195–204

American Public Health Association (2005) Standard methods for the examination of water and wastewater. American Public Health Association, APHA, Washington, DC, USA

Bachmann RT, Adawiyah S, Krishnan T, Khoo B, Sian TS, Richards T (2018) Partial substitution of peat moss with biochar for sustainable cultivation of Durio zibethinus L. in nurseries. Arab J Geosci 11:426

Bello MM, Abdul Raman AA (2017) Trend and current practices of palm oil mill effluent polishing: Application of advanced oxidation processes and their future perspectives. J Environ Manage 198:170–182

Blackwell M, Darch T, Haslam R (2019) Phosphorus use efficiency and fertilizers: future opportunities for improvements. Front Agric Sci Eng 6:332

Charles A, Cheng CK (2019) Photocatalytic treatment of palm oil mill effluent by visible light-active calcium ferrite: effects of catalyst preparation technique. J Environ Manage 234:404–411

Charles A et al (2019) Facile synthesis of $CaFe_2O_4$ for visible light driven treatment of polluting palm oil mill effluent: Photokinetic and scavenging study. Sci Total Environ 661:522–530

Cheng CK, Deraman MR, Ng KH, Khan MR (2016) Preparation of titania doped argentum photocatalyst and its photoactivity towards palm oil mill effluent degradation. J Clean Prod 112:1128–1135

Chojnacka K, Gorazda K, Witek-Krowiak A, Moustakas K (2019) Recovery of fertilizer nutrients from materials - Contradictions, mistakes and future trends. Renew Sustain Energy Rev 110:485–498

Csuros M (2018) Cation Exchange Capacity. Environ Sampling Anal 4:313–315

Ding GT, Yaakob Z, Takriff MS, Salihon J, Abd Rahaman MS (2016) Biomass production and nutrients removal by a newly-isolated microalgal strain Chlamydomonas sp in palm oil mill effluent (POME). Int J Hydrogen Energy 41:4888–4895

Faloye OT, Alatise MO, Ajayi AE, Ewulo BS (2017) Synergistic effects of biochar and inorganic fertiliser on maize (zea mays) yield in an alfisol under drip irrigation. Soil Tillage Res 174:214–220

Foo S, Loh SK, Ismail K, Bachmann RT (2020) Nutrient recovery from anaerobic palm oil mill effluent with thermally regenerated spent bleaching earth using response surface methodology. J Oil Palm Res 32:245–257

Gamaralalage D, Sawai O, Nunoura T (2019) Degradation behavior of palm oil mill effluent in Fenton oxidation. J Hazard Mater 364:791–799

Gong H, Tan Z, Zhang L, Huang Q (2019) Preparation of biochar with high absorbability and its nutrient adsorption–desorption behaviour. Sci Total Environ 694:133728

Grisso R, Alley M, Holshouser D, Thomason W (2009) Precision farming tools: soil electrical conductivity. Virginia Coop Ext 442:1–6

Hossain MS, Omar F, Asis AJ, Bachmann RT, Islam Sarker MZ, Ab Kadir MO (2019) Effective treatment of palm oil mill effluent using $FeSO_4.7H_2O$ waste from titanium oxide industry: coagulation adsorption isotherm and kinetics studies. J Clean Prod 219:86–98

Houmani ZMM, Majid AA, Radiman S, Ahmad Z (2012) Effects of physico-chemical soil properties on the adsorption and transport of ^{137}Cs in Rengam and Selangor soil series. Malaysian J Anal Sci 16:94–102

Hukari S, Hermann L, Nättorp A (2016) From wastewater to fertilisers—technical overview and critical review of European legislation governing phosphorus recycling. Sci Total Environ 542:1127–1135

Huzir NM, Aziz MMA, Ismail SB, Mahmood NAN, Umor NA, Faua'ad Syed Muhammad SA, (2019) Optimization of coagulation-flocculation process for the palm oil mill effluent treatment by using rice husk ash. Ind Crops Prod 139:111482

Jalaludin N, Rahman RA, Razali F, Barghash HF, Sukri SSM (2016) Optimization of fresh palm oil mill effluent biodegradation with Aspergillus Niger and Trichoderma Virens. Arch Environ Prot 42:63–73

Kasak K et al (2018) Biochar enhances plant growth and nutrient removal in horizontal subsurface flow constructed wetlands. Sci Total Environ 639:67–74

Kasprzyk M, Gajewska M (2019) Phosphorus removal by application of natural and semi-natural materials for possible recovery according to assumptions of circular economy and closed circuit of P. Sci Total Environ 650:249–256

Kong SH, Loh SK, Bachmann RT, Rahim SA, Salimon J (2014) Biochar from oil palm biomass: a review of its potential and challenges. Renew Sustain Energy Rev 39:729–739

Kong SH, Loh SK, Bachmann RT, Zainal H, Cheong KY (2019) Palm kernel shell biochar production, characteristics and carbon sequestration potential. J Oil Palm Res 31:508–520

Kopittke PM, Menzies NW (2007) A Review of the use of the basic cation saturation ratio and the 'Ideal' soil. Soil Sci Soc Am J 71:259

Kushairi A et al (2018) Oil palm economic performance in Malaysia and R&D progress in 2017. J Oil Palm Res 30:163–195

Lam MK, Lee KT (2011) Renewable and sustainable bioenergies production from palm oil mill effluent (POME): win–win strategies toward better environmental protection. Biotechnol Adv 29:124–141

Legal research board, environmental quality Act 1974 (Act 127) and Subsidiary Legislation 1977. International law book services, Petaling Jaya, Malaysia

Li J, Duan T, Zheng C (2019) Effects of different carbon to nitrogen ratio organic fertilizers on soil microbial biomass in organic agriculture. Rev Fac Agron 36:1287–1295

Liew WL, Kassim MA, Muda K, Loh SK, Affam AC (2015) Conventional methods and emerging wastewater polishing technologies for palm oil mill effluent treatment: a review. J Environ Manage 149:222–235

Lim SC, Gomes C, Ab Kadir MZA (2013) Characterizing of bentonite with chemical, physical and electrical perspectives for improvement of electrical grounding systems. Int J Electrochem Sci 8:11429–11447

Liu X et al (2013) Biochar's effect on crop productivity and the dependence on experimental conditions—a meta-analysis of literature data. Plant Soil 373:583–594

Loh SK, Cheng SF, Choo YM, Ma AN (2006) A study of residual oils recovered from spent bleaching earth: their characteristics and applications. Am J Appl Sci 3:2063–2067

Loh SK, James S, Ngatiman M, Cheong KY, Choo YM, Lim WS (2013a) Enhancement of palm oil refinery waste—Spent bleaching earth (SBE) into bio organic fertilizer and their effects on crop biomass growth. Ind Crops Prod 49:775–781

Loh SK et al (2013b) Zero discharge treatment technology of palm oil mill effluent. J Oil Palm Res 25:273–281

Loh SK, Cheong KY, Choo YM, Salimon J (2015) Formulation and optimisation of spent bleaching earth-based bio organic fertiliser. J Oil Palm Res 27:57–66

Loh SK, Cheong KY, Salimon J (2017a) Surface-active physicochemical characteristics of spent bleaching earth on soil-plant interaction and water-nutrient uptake: A review. Appl Clay Sci 140:59–65

Loh SK et al (2017b) First Report on Malaysia's experiences and development in biogas capture and utilization from palm oil mill effluent under the economic transformation programme: current and future perspectives. Renew Sustain Energy Rev 74:1257–1274

Ng KH, Lee CH, Khan MR, Cheng CK (2016) Photocatalytic degradation of recalcitrant POME waste by using silver doped titania: Photokinetics and scavenging studies. Chem Eng J 286:282–290

Pandit NR, Mulder J, Hale SE, Martinsen V, Schmidt HP, Cornelissen G (2018) Biochar improves maize growth by alleviation of nutrient stress in a moderately acidic low-input Nepalese soil. Sci Total Environ 625:1380–1389

Peats M, Rippy JFM, Nelson PV (2007) Cation exchange capacity and base saturation variation among Alberta. North 42:349–352

Radin R, Abu Bakar R, Ishak CF, Ahmad SH, Tsong LC (2018) Biochar-compost mixture as amendment for improvement of polybag-growing media and oil palm seedlings at main nursery stage. Int J Recycl Org Waste Agric 7:11–23

Rajkovich S, Enders A, Hanley K, Hyland C, Zimmerman AR, Lehmann J (2012) Corn growth and nitrogen nutrition after additions of biochars with varying properties to a temperate soil. Biol Fertil Soils 48:271–284

Sabrina DT, Hanafi MM, Azwady Nor AA, Mahmud TMM (2009) Earthworm populations and cast properties in the soils of oil palm plantations. Malaysian J Soil Sci 13:29–42

Siva KB, Aminuddin H, Husni MHA, Manas AR (2000) Ammonia volatilization from urea as affected by humic substances derived from palm oil mill effluent (POME) and tropical peat. Trop Agric 77:13–20

Spångberg J, Tidåker P, Jönsson H (2014) Environmental impact of recycling nutrients in human excreta to agriculture compared with enhanced wastewater treatment. Sci Total Environ 493:209–219

Tan YD, Lim JS (2019) Feasibility of palm oil mill effluent elimination towards sustainable Malaysian palm oil industry. Renew Sustain Energy Rev 111:507–522
Tiemann TT et al (2018) Feeding the palm. Adv Agron 152:149–243
Wu TY, Mohammad AW, Jahim JM, Anuar N (2007) Palm oil mill effluent (POME) treatment and bioresources recovery using ultrafiltration membrane: Effect of pressure on membrane fouling. Biochem Eng J 35:309–317
Wu TY, Mohammad AW, Jahim JM, Anuar N (2009) A holistic approach to managing palm oil mill effluent (POME): Biotechnological advances in the sustainable reuse of POME. Biotechnol Adv 27:40–52
Yuan J, Ma Y, Yu F, Sun Y, Dai X, Ma J (2020) Simultaneous in situ nutrient recovery and sustainable wastewater purification based on metal anion- and cation-targeted selective adsorbents. J Hazard Mater 382:121039
Zainal NH, Jalanai NF, Mamat R, Astimar AA (2017) A review on the development of palm oil mill effluent (POME) final discharge polishing treatments. J Oil Palm Res 29:528–540
Zainal H, Loh SK, Kong SH, Bachmann RT (2018) Pilot scale biochar production from palm Kernel Shell (PKS) in a fixed bed allothermal reactor. J Oil Palm Res 30:485–494
Zeng Q, Rossi S, Yang B (2018) Effects of age and size on xylem phenology in two conifers of Northwestern China. Front Plant Sci 8:1–9

Chapter 8
Red Tilapia By-Product (*Oreochromis Sp.*) Hydrolysates: Bioactivities, Bioprocessing, and Potential Applications

Emmy Liza Anak Yaji, Nur Suraya Abdul Wahab, Norfahana Abd-Talib, Mohammad Zulkeflee Sabri, Kelly Yong Tau Len, Nadia Razali, and Khairul Faizal Pa'ee

Abstract The aquaculture and fishery industry has expanded dramatically in recent years, producing by-products that are still underutilised. This article provides a detailed review of relevant observations and studies that encompasses current research on the red tilapia by-product as a precursor for ACE-inhibitory peptides. The review includes (i) red tilapia by-product generation and its potential protein precursor for bioactive peptides, (ii) current findings on bioactive peptides and their mechanism, (iii) common and new empirical approaches to bioactive peptides

E. L. A. Yaji · N. S. A. Wahab · N. Abd-Talib · K. F. Pa'ee (✉)
Food Engineering Technology, Universiti Kuala Lumpur, Malaysian Institute of Chemical and Engineering Technology, 78000 Alor Gajah, Melaka, Malaysia
e-mail: khairulfaizal@unikl.edu.my

E. L. A. Yaji
e-mail: emmy.liza04@s.unikl.edu.my

N. S. A. Wahab
e-mail: suraya.wahab@s.unikl.edu.my

N. Abd-Talib
e-mail: norfahana@unikl.edu.my

M. Z. Sabri
Bioengineering Technology, Universiti Kuala Lumpur, Malaysian Institute of Chemical and Engineering Technology, 78000 Alor Gajah, Melaka, Malaysia
e-mail: mzulkeflee@unikl.edu.my

K. Y. T. Len
Process Engineering Technology, Universiti Kuala Lumpur, Malaysian Institute of Chemical and Engineering Technology, 78000 Alor Gajah, Melaka, Malaysia
e-mail: kytlen@unikl.edu.my

N. Razali
Environmental Engineering Technology, Universiti Kuala Lumpur, Malaysian Institute of Chemical and Engineering Technology, 78000 Alor Gajah, Melaka, Malaysia
e-mail: nadiarazali@unikl.edu.my

Centre for Women Advancement and Leadership, Universiti Kuala Lumpur, Kuala Lumpur, Malaysia

A. Ismail et al. (eds.), *Advancements in Materials Science and Technology Led by Women*, Advanced Structured Materials 165,
https://doi.org/10.1007/978-3-031-21959-7_8

production, and (iv) limitation and future outlook of bioactive peptides which includes molecular dynamics simulation and bioinformatic approach.

Keywords Red Tilapia by-product · Bioactive peptides · Bioinformatics · Functional food · Bioprocessing technology

8.1 Introductions to Red Tilapia Production and Its Processing By-Products

Over the last decade, the capture fisheries' practice has plateaued due to climate changes, environmental issues, and overfishing (Allan et al. 2005). The capture fisheries are no longer expected to meet the growing global demand in the supply of edible fishes (Subasinghe et al. 2009). The fishing industry is evolving in much the same way that man no longer hunts for meat. Consumers will increasingly rely on aquaculture or farmed fish as dependable high-nutritional-value protein sources that are responsibly sourced and available year-round in the future years. Aside from that, the aquaculture industry offers significant job and revenue opportunities, particularly in impoverished rural areas (Halwart et al. 2003).

Tilapia is the freshwater cichlid fish found in streams, wetlands, rivers, lakes, and brackish water. The group is made up of three important aquaculture genera: *Oreochromis, Sarotherodon*, and *Tilapia*. Several characteristics distinguish these three genera, but the most important may be the reproductive behaviour. Nests are built by all tilapia species, and fertilised eggs are guarded in the nest by a brooding parent. *Sarotherodon* and *Oreochromis* species are mouthbrooders; eggs are fertilised in the nest, but parents immediately pick up the eggs in their mouths and hold them during incubation and several days after hatching. Mouthbrooding is only practised by females in *Oreochromis* species, whereas mouthbrooding is practised by either the male or both the male and female in *Sarotherodon* species. The genus *Oreochromis* now encompasses all commercially farmed tilapia, with Nile tilapia accounting for more than 90% of all economically farmed tilapia. *Oreochromis aureus* (blue tilapia), *Oreochromis mossambicus* (Mozambique tilapia), and Zanzibar tilapia (*Oreochromis mossambicus*) are less typically farmed species (*O. urolepis hornorum*). Tilapia resembles sunfish or crappie in appearance, but they have an interrupted lateral line that is peculiar to the Cichlid family of fish. They have lengthy dorsal fins and a deep body and are laterally compressed. The anterior portion of the dorsal fins is heavily spined. Spines can also be found in the pelvis and anal fins. Tilapias are more resistant to high salinity, high water temperature, low dissolved oxygen, and high ammonia concentrations than most commonly farmed freshwater fish (Popma and Masser 1999). Earthen ponds, concrete tanks, and floating net cages raise the fish (wooden, galvanised, HDPE frame). Because of the encouraging aquaculture characteristics, low input costs, stable market prices, rapid maturity, and mild and sweet tastes, tilapia is the most widely cultured freshwater fish in tropical and subtropical countries (Wang and Lu 2016; Prabu et al. 2019).

The industrial activity of tilapia processing and filleting generates an enormous amount of waste which can affect the environment. Based on Medina-Medrano (Medina-Medrano et al. 2019), approximately 30% to 70% of the initial fish's weight are blood, viscera, carcass, skin, scales, and head, which will be discarded to landfill without any pre-treatment. However, emerging studies regarding the valorisation of tilapia waste such as head, bones, viscera, muscle, skin, and scales have allowed the manufacturing industry to explore alternative raw materials, gaining its popularity (Roslan et al. 2014; Coppola et al. 2021). Converting tilapia by-products into new products with good functional or biological qualities is one of the most intriguing and promising alternatives as to the by-products from tilapia processing still contain rich sources of several excellent compounds such as collagen, proteins, gelatin, lipids, amino acids, and enzymes (Zamora-Sillero et al. 2018). Several researchers reported various applications that could convert tilapia by-products into value-added products to reduce food loss and waste while also improving fish waste management strategies. Afreen & Ucak (Afreen and Ucak 2020) explored the potential of tilapia waste as animal feed ingredients by utilising several biotechnologies and biopreservation methods. Researchers extracted fish oil from various tilapia waste to make biodiesel (Oliveira et al. 2021). The moisture and free fatty acid (FFA) content of tilapia's visceral oil is low and beneficial for transesterification and conversion into high-yield biodiesel. Skin collagen from tilapia has been used as a scaffold in cartilage repair through in vitro (chondrocytes from rabbit auricula) and in vivo tests (rabbit) (Li et al. 2020). The by-products are also employed as a substrate for the ACE-inhibitory protein hydrolysate acid or alkaline hydrolysis, autolysis, and enzymatic hydrolysis (Lin et al. 2017). Fish waste is converted into a liquid mixture of hydrolysed proteins, lipids, minerals, and other nutrients easily digestible by both terrestrial and aquatic animals through the fish silage process. Other products can be made from the by-products of industrial tilapia filleting, such as surimi (minced fish) or gluten-free tapioca starch (Market 2019). As a result, a detailed analysis of its production, product features, and feasibility for larger-scale production is required, focusing on the process's cost-effectiveness and consequent reduction in pollution.

Food chemical composition is critical for human nutritional well-being and determining the future production and implementation of food materials in the food industry. Table 8.1 summarises the findings of several reports on proximate analyses of tilapia by-products. Roslan (Roslan et al. 2014) performed comparative studies on tilapia by-products and tilapia by-product hydrolysate, finding that tilapia by-products contained 14.60 $\pm$ 0.30% crude protein, 8.93 $\pm$ 0.46% ash, 66.57 $\pm$ 0.39% moisture, and 5.50 $\pm$ 0.30% fat. These findings were found to be significantly different with protein, ash, moisture, and fat content of tilapia by-product hydrolysate (62.71 $\pm$ 0.29%, 25.34 $\pm$ 0.94%, 6.48 $\pm$ 0.40%, 0.08 $\pm$ 0.05%, respectively). This high variability in protein content is due to the fact that the hydrolysate was dried, while the values for the raw material were expressed as a percentage of wet weight.

The high protein content was due to protein solubilisation during hydrolysis, the elimination of insoluble undigested non-protein molecules, and the partial removal of lipids following hydrolysis (Roslan et al. 2014; Robert et al. 2015). The proportion of solubilised protein was shown to be affected by the amount of fat in the raw

Table 8.1 Proximate composition of tilapia by-product

Sample	Moisture content	Crude protein	Fat	Ash	References
Tilapia by-product	66.57% ± 0.39	14.60% ± 0.30	5.50% ± 0.30	8.93% ± 0.46	Roslan et al. (2014)
Tilapia by-product hydrolysate powder	6.48% ± 0.40	62.71% ± 0.29	0.08% ± 0.05	25.34% ± 0.94	
Spray-dried tilapia by-product hydrolysate	3.5%	80.2%	7.4%	5.4%	Robert et al. (2015)
Powder tilapia by-product hydrolysate	2.14%	44.80%	0.32%	21.06%	Paris et al. (2016)
Dry tilapia filleting by-product	94.89% (Dry matter)	47.25%	2.27%	19.58%	Sary et al. (2017)
Freeze-dried tilapia viscera hydrolysate	16%	42.2% ± 1.1	3.6% ± 0.1	22.9% ± 0.6	Sepúlveda et al. (2021)

material, with raw materials having the highest amount of fat providing the lowest percentage of solubilised protein (Šližyte et al. 2005). During hydrolysis, muscle cell membranes begin to circle up and form insoluble vesicles. The situation led to decreased fat concentration in the tilapia by-product hydrolysate due to the removal of structured membrane lipids during centrifugation to recover soluble protein. The addition of sodium phosphate buffer during enzymatic processing may have led to the significantly high ash concentration of tilapia by-product hydrolysate. According to the data, the moisture percentage of tilapia by-product hydrolysate powder was 6.48 ± 40%, which could be due to the sample form and evaporation temperatures utilised, which led the sample to lose a lot of moisture (Roslan et al. 2014).

According to Table 8.1 (Robert et al. 2015) obtained the highest protein content in tilapia by-product hydrolysate at 80.2%, compared to other studies that found protein content ranging from 42 to 63%. Nonetheless, these results revealed that the tilapia by-product hydrolysate contained a significant amount of protein. These hydrolysates were all produced by enzymatic hydrolysis using protease enzyme; however, the parameters were different. Variations in protein content in the hydrolysate could be caused by the enzyme and parameters used during enzymatic hydrolysis. The high protein and low-fat content shows that fish hydrolysates can be used in animal or human nutrition (Robert et al. 2015). Typically, numerous separation processes are performed sequentially to purify a dietary protein. In general, the purity of a protein preparation may be enhanced by including more separation processes; however, this often results in a decrease in recovery or yield (Smith 2017).

8.2 Bioactive Peptides

Specific protein fragments known as bioactive peptides are defined as those that positively impact the body function or condition and may ultimately positively impact health (Ishak and Sarbon 2018). Hydrolysis is when proteins are broken down into smaller peptide chains, each containing specific peptide sequences with distinct bioactivities. Depending on their inherent amino acid composition and sequence, these bioactive peptides can exhibit potential bioactivities such as an angiotensin-converting enzyme (ACE) inhibitory activity, antimicrobial, antioxidative, antithrombotic, and immunomodulatory activities, among other properties (FitzGerald and Meisel 2003; Korhonen and Pihlanto 2006). The ACE-inhibitory and antioxidant activity of various protein sources was the bioactivity that had received the most attention (Murray and FitzGerald 2007; Paiva et al. 2017). Producing bioactive peptides from foods and food by-products is a feasible method of achieving this goal. Sánchez and Vázquez (Sánchez and Vázquez 2017) investigated the bioactivity of high-protein foods such as milk and dairy products, meat and various fish species, wheat, and legumes to determine their nutritional value.

The potential benefits of these compounds to human health have been investigated, and the peptide information shown in Table 8.2 can be obtained through the BIOPEP database, which is available online (Minkiewicz and Darewicz. 2019). Over the past several decades, protein-containing by-product materials have also gained popularity as a source of bioactive peptides for pharmaceutical and biotechnology applications. The most common sources of bioactive peptides have traditionally been cheese whey and meat processing by-products (Ishak and Sarbon 2018). According to Pa'ee et al. (2015), the production of hydrolysates with high potency ($IC_{50} = 206$–353 μg/mL) from acid whey with a high ionic calcium concentration in acid whey contributed to the ACE-inhibitory. The slaughter of animals produces several types of by-product materials for human consumption, including organs, limbs, bones, meat trimmings, blood, and fatty tissues, all of which are used in the meat industry. Furthermore, because of its high protein content and availability, the by-product from fish processing can be used to produce bioactive peptides due to its high protein content and availability, which accounts for approximately 50% of the total weight of fish (Ishak and Sarbon 2018).

8.2.1 Antioxidant Peptides

The antioxidant reserves are depleted due to reactive oxygen species (ROS) production by physiological processes of exogenous molecules, resulting in oxidative damage (Zamora-Sillero et al. 2018). A dynamic antioxidant protection mechanism

Table 8.2 Example of bioactive peptides with known activity available in BIOPEP (Minkiewicz and Darewicz 2019)

ID	Name of peptides	Activity	Sequence
2595	Tachyplesin III from haemocytes of horseshoe crab	Antibacterial	KWCFRVCYRGICYRRCR ~
3522	ACE inhibitor (from bovine beta-CN)	ACE inhibitor	IPP
5419	Porcine lactoferrin LFP	Antibacterial	SKCRQWSKIRRTNP
7051	Lactoferricin B, bovine (Bos taurus)	Anticancer	FKCRRWQWRMKKLGAPSITCVRRAF
7514	ACE inhibitor (from soy protein)	ACE inhibitor	NWGPLV
7570	ACE inhibitor from chicken muscle	ACE inhibitor	LAP
7678	ACE inhibitor from dried bonito	ACE inhibitor	LKPNM
7756	Anticoagulant peptide from marine echiuroid worm	Antithrombotic	GELTPESGPDLFVHFLDGNPSYSLYADAVPR
7757	Antioxidant from grass carp muscle	Antioxidative	PSKYEPFV
8110	Peptide derived from tuna backbone protein	Antioxidative	VKAGFAWTANQQLS
8118	Peptide derived from sardinella by-product proteins (*Sardinella aurita*)	Antioxidative	GALAAH

(continued)

Table 8.2 (continued)

ID	Name of peptides	Activity	Sequence
8137	Anticancer peptide from algae protein waste (*Chlorella vulgaris*)	Anticancer	VECYGPNRPQF
8165	Peptide derived from bovine k-casein (112–116)	Antithrombotic	KNQDK
8173	Peptide derived from bovine b-casein (1–28)	Immunomodulating	RELEELNVPGEIVESLSSSEESITRINK
8176	Peptide derived from ovine colostral whey	Immunomodulating	VESYVPLFP
9691	Antibacterial peptide from pig	Antibacterial	RLWKIGKVLKWI

protects organisms from such harmful effects. High levels of reactive oxygen species (ROS) can be highly harmful to cell constituents. In this case, oxidative stress occurs when the number of reactive oxygen species (ROS) generated exceeds the antioxidant defences' ability. Protein disruption, lipid peroxidation, enzyme inactivation, and DNA breakage are all linked to oxidative stress. These processes promote the development of tumours or cancer, heart disease, rheumatoid arthritis, and ageing, among other diseases or pathologies (Zamora-Sillero et al. 2018). Many research areas have addressed lipid oxidation mediated by free radicals, oxidative stress, and antioxidants (Sila and Bougatef 2016).

Several studies have demonstrated that peptides derived from fish and fish by-products are effective antioxidants. Salmon parts (Ahn et al. 2012), tilapia skin (Zhang et al. 2012), and horse mackerel (Sampath Kumar et al. 2012) were among the substrates tested. The degree of hydrolysis (DH) varies from 13 to 75% hydrolysates to exert potential antioxidative peptides. Furthermore, the molecular weight of these peptides ranged from 200 to 2000 Da (Ishak and Sarbon 2018). The majority of known antioxidant peptides are short chains of 5 to 16 polar and hydrophobic amino acids, with the amino acids Asp, Glu, Gly, Ala, Leu, His, Pro, Tyr, Phe, and Lys being the most common. Asp, Glu, Gly, Ala, Leu, His, Pro, Tyr, Phe, and Lys the precise contribution of individual amino acid residues to antioxidant activity, on the other hand, is dependent on the presence of reactive oxygen species (ROS) or free radicals, as well as the reaction medium (Udenigwe and Aluko 2012). The findings concluded that the peptides extracted from fish by-product hydrolysis are active antioxidants with low IC_{50} values. They could be used to replace synthetic antioxidants in various industries, including the pharmaceutical and food industries. It is important to conduct more experiments to produce a more conclusive argument regarding the feasibility of using peptides in vivo studies as bioactive compounds.

8.2.2 Anticancer Peptides

Resistance to synthetic anticancer drugs has recently been frequently observed; as a result, natural anticancer drugs are receiving a lot of attention. Interest in recognising and characterising natural antitumor agents has risen with the need for more efficient and less toxic pharmaceuticals (Udenigwe and Aluko 2012). Protein hydrolysates and peptides derived from seafood have been studied extensively for their potential therapeutic value in preventing and treating diseases, especially cancer. Bioactivity against breast cancer (MCF-7), colon cancer (HT-29, DLD-1), gastric cancer (AGS), glioma (U87), cervical cancer (HeLa), and prostate cancer (PC-3, DU-145, H-1299) cell lines have been reported by these studies that have examined solitary tunicate (Jumeri Kim 2011), shellfish (Kim et al. 2013), and oysters (Umayaparvathi et al. 2014). Furthermore, bioactivities were linked to amino acids noted for their anti-cancer activity, such as Pro, Gly, Lys, Arg, and Tyr. A previous study by Jumeri Kim (2011) indicated that anticancer peptides ($IC_{50} = 0.10–2.90$ mg/mL) derived from various by-product sources had low molecular weights of 300 and 1950 Da. Low-molecular-weight peptides had higher mobility and diffusivity. As a result, it enhances anticancer activity by increasing interactions with cancer cell components (Ishak and Sarbon 2018). However, the anticancer activity of fish-derived protein hydrolysates, primarily its by-product, has yet to be thoroughly investigated.

8.2.3 Antimicrobial Activity

A peptide with a molecular weight of less than 10 kDa and a chain length of less than 50 amino acids, with approximately half of them being hydrophobic, is considered an antimicrobial peptide (Najafian and Babji 2012). In the case of these peptides, their interaction with the bacterial membrane could result in the formation of pores or the blocking of membrane ion gradients, which would destroy cell constituents (Wald et al. 2016). These antimicrobial peptides may be helpful in the development of antibacterial, antiviral, antifungal, immunomodulatory, and antitumor agents (Zamora-Sillero et al. 2018). Candidates for new antibiotic discovery in the pharmaceutical industry and antimicrobial agents for use in the food industry are studied. Ennaas (Ennaas et al. 2015) used proteolytic enzymes such as Protamex, Neutrase,

Papain, and Flavourzyme to hydrolyse mackerel by-products in their research. These hydrolysates were found to inhibit both Gram-positive (*L. innocua*) and Gram-negative (*E. coli*) bacteria in the laboratory setting. Using trout pepsin, an antibacterial trout by-product protein hydrolysate was created that was effective against food contamination and fish farming pathogens (Wald et al. 2016). In the study, the hydrolysates with the highest activity against *Flavobacterium psychrophilum* and *Salmoninarum renibacterium* were those with a hydrolysis degree of 30% or higher.

Furthermore, the amino acids lysine, leucine, alanine, arginine, glycine, aspartic acid, and glutamic acid were found in the highest concentrations in this hydrolysate, as were the amino acids alanine, arginine, glycine, and aspartic acid (Zamora-Sillero et al. 2018). The Protamex enzyme was also used by Robert and colleagues (Robert et al. 2015) to hydrolyse the tilapia's head, frames, and viscera, among other things. In addition to having a well-balanced amino acid profile, these hydrolysates were found to have antibacterial activity against *Edwardsiella tarda* and *Bacillus subtilis*.

8.2.4 Angiotensin I-Converting Enzyme (ACE)-Inhibitory Peptides

A regulatory factor in the renin-angiotensin system (RAS), the angiotensin-converting enzyme (ACE; peptidyldipeptide hydrolase, EC 3.4.25.1), plays an essential role in maintaining cardiovascular homeostasis, such as blood pressure regulation. Angiotensinogen, a plasma protein synthesised in the liver, is converted into angiotensin I by the renin-angiotensin system (Fig. 8.1). Angiotensin-converting enzymes (ACE) catalyse the conversion of angiotensin I into angiotensin II, a potent vasoconstrictor that mediates the effects of RAS. In addition, ACE is a component of the kallikrein-kinin system. The enzyme deactivates bradykinin and kallidin, releasing aldosterone, increasing blood pressure (Ortiz-Chao and Jauregi 2007). As a result, inhibiting the angiotensin-converting enzyme (ACE) may lower blood pressure by preventing an angiotensin I conversion to angiotensin II. Synthetic ACE inhibitors, such as captopril, imidapril, and enalapril, are widely used to treat severe hypertension, especially in the elderly. Captopril, for example, has an IC_{50} of 0.021 $\pm$ 0.013 μM, making it a highly effective drug. Numerous side effects are associated with these medications, including dry coughing, taste disturbances, and skin rashes (Ghassem et al. 2011).

In improving the management of human diseases, bioactive peptides play critical roles in developing nutraceuticals and functional foods as naturally derived components that contribute to optimal health (Udenigwe and Aluko 2012). Alternative sources of ACE-inhibitory peptides include dairy products and fish, obtained through diet (Kim et al. 2012). It was discovered that two tripeptides that inhibit ACE, namely Ile-Pro-Pro (IPP) and Val-Pro-Pro (VPP), could be isolated from sour milk fermented with *L. helveticus* and *Saccharomyces cerevisiae* (Möller et al. 2008). Ile-Pro-Pro (IPP) and Val-Pro-Pro (VPP) were found to inhibit ACE. It has

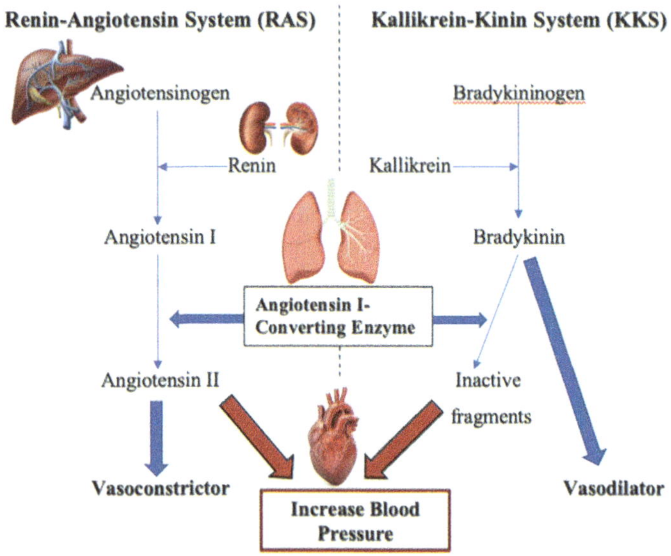

Fig. 8.1 Renin-angiotensin system (RAS) and the kallikrein-kinin system (KKS) in blood pressure regulation

been demonstrated that long-term administration of this IPP and VPP can prevent hypertension in rats and that a single oral administration can reduce blood pressure (Tuomilehto et al. 2004). The Alcalase enzyme hydrolysed rice protein, and the resulting hydrolysate was tested for ACE-inhibitory activity, yielding an IC_{50} value of 0.14 mg/mL. In another study, rice protein was hydrolysed by Alcalase, and the resulting hydrolysate was tested for ACE-inhibitory activity. Thr-Gln-Val-Tyr ($IC_{50} = 18.2\ \mu M$) is an amino acid sequence that was isolated and identified from the hydrolysate. This ACE-inhibitory peptide significantly decreased blood pressure spontaneously in hypertensive rats (SHR) when administered (Li et al. 2007). Several natural ACE inhibitor products are now available for purchase as dietary supplements, intending to maintain normal blood pressure levels in the body. As an example, the peptides bonito peptide, VasotensinR, and PeptAceTM, which are derived from the fish bonito (*Sarda orientalis*), demonstrate that the hydrolysate of bonito is sufficiently well absorbed in humans to effectively lower the systolic blood pressure of hypertensive and borderline hypertensive subjects in vivo (Fujita et al. 2001).

Natural ACE-inhibiting peptides are currently being sought in various aquatic protein sources (Table 8.3), mainly fish by-products, which are being investigated. Ahn (Ahn et al. 2012) identified salmon pectoral fin as a possible source, as well as skipjack roe (Intarasirisawat et al. 2012) and skate skin (Ngo et al. 2014). Skipjack roe, when compared to skate skin (Leu-Gly-Pro-Leu-Gly-His-Gln) ($IC_{50} = 3.09\ \mu M$) and salmon pectoral fin (Pro-Ala-Val-Pro-Leu-Thr-Gly) ($IC_{50} = 7.72\ \mu M$),

Table 8.3 Example of ACE-inhibitory peptides with known potency available in BIOPEP (Minkiewicz and Darewicz 2019)

ID	Name of peptides	Activity	IC_{50} (μM)	Sequence
3361	ACE inhibitor from sardine	ACE inhibitor	188.00	LKL
3522	ACE inhibitor (from bovine beta-CN)	ACE inhibitor	5.00	IPP
3524	ACE inhibitor (from bovine beta-CN)	ACE inhibitor	9.00	VPP
5419	ACE inhibitor from Alaskan pollack skin	ACE inhibitor	2.65	GPL
7514	ACE inhibitor (from soy protein)	ACE inhibitor	21.00	NWGPLV
7638	ACE inhibitor from tuna muscles	ACE inhibitor	330.00	LTF
7678	ACE inhibitor from dried bonito	ACE inhibitor	2.40	LKPNM
7747	ACE inhibitor from Alaska Pollack frame protein	ACE inhibitor	14.70	FGASTRGA
7751	ACE inhibitor from shark meat hydrolysate	ACE inhibitor	1.96	CF
7757	ACE inhibitor from thornback ray muscle (Raja clavata)	ACE inhibitor	170	TLKYP
7811	ACE inhibitor from tuna	ACE inhibitor	110	VWIG
7812	ACE inhibitor from blue mussel	ACE inhibitor	2.98	EVMAGNYLPG
7813	ACE inhibitor from yellowfish sole	ACE inhibitor	0.03	MIFPGAGGPEL

has shorter amino acid sequences (Met-Leu-Val-Phe-Ala-Val) and higher ACE-inhibitory activity ($IC_{50} = 3.07$ μM). Up to this point, various fish by-products have been shown to have ACE-inhibitory activity, with IC_{50} values ranging from 3.07 to 35.7 μM and ACE-inhibition rates ranging from 64%–85% reported so far. However, the IC_{50} values (3.07–35.7 μM) of these various fish by-product hydrolysate were significantly higher than those of the synthetic ACE-inhibitory inhibitor captopril, which had IC_{50} values ranging from 0.019 to 0.1 μM (Picot et al. 2010). The findings emphasised the importance that these peptides are not intended to replace synthetic drugs in the treatment of hypertension completely. Instead, they can be used as a complementary treatment in conjunction with lifestyle modifications, such as weight loss, exercise, and a controlled diet that have been reported to be some of the essential tools for effective reduction of bloating (Hermansen 2000).

8.3 The Production of ACE-Inhibitory Peptides

ACE-inhibitory peptides have been produced either through microbial fermentation, enzymatic hydrolysis, and chemical hydrolysis. Figure 8.2 outlines the general processing steps for peptide production. The production of bioactive peptides involves several steps, including pre-treatments, the hydrolysis process, and the purification and fractionation of protein hydrolysate.

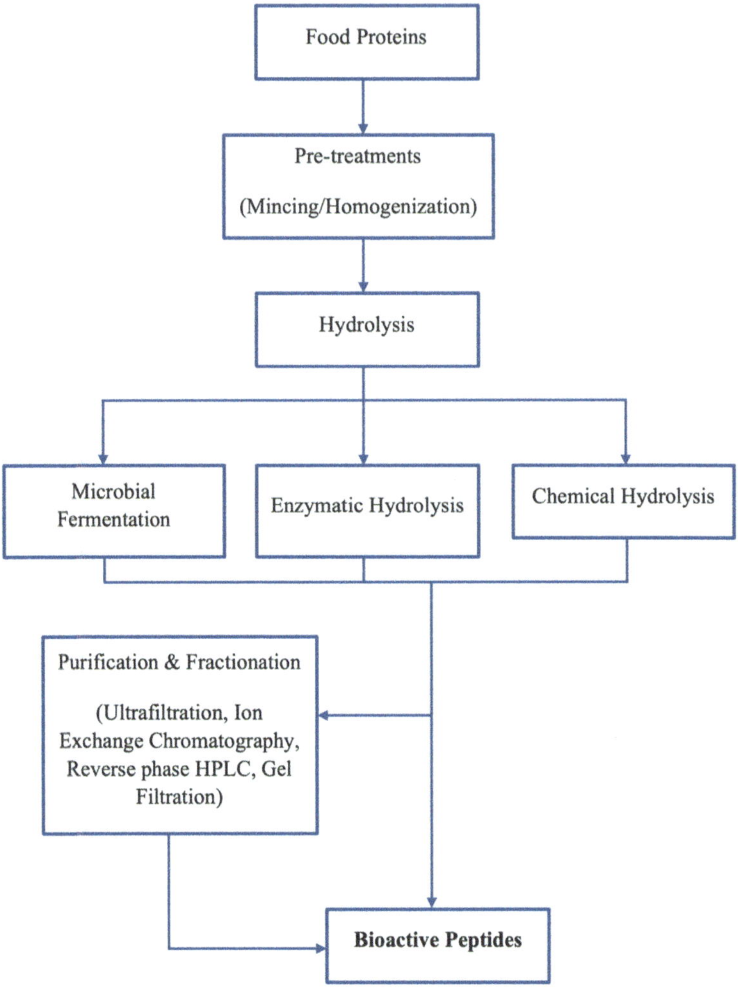

Fig. 8.2 Conversion of protein derived food item to bioactive peptides

8.3.1 Microbial Fermentation

Instead of using purified enzymes, microbial fermentation is used in several countries to produce ACE-inhibitory peptides, an important component of developing nutritious foods. Natto and tempeh are fermented soybean products that contain bioactivity peptides produced by fungal proteases. Because of the low cost and positive nutritional image associated with fermented milk drinks and yoghurt, fermentation of milk proteins using lactic acid bacteria proteolytic systems is an appealing approach for producing functional foods rich in bioactive peptides and other bioactive compounds (Hayes et al. 2007). Storage yoghurt fermented by *L. acidophilus* was examined for its fermentation properties, ACE-inhibitory activity, and content of the peptides Val-Pro-Pro (VPP) and Ile-Pro-Pro (IPP). Wang et al. (2015) investigated the *L. helveticus* isolate H9 obtained from cow, mare, and soymilk samples. All milk types and inoculation concentrations were used to make the yoghurt, and the pH and titratable acidity of the yoghurts remained stable during storage. According to the ultra-performance liquid chromatography-tandem mass spectrometry results, the ACE-inhibitory tripeptides VPP and IPP were not detected in yoghurt made from soy milk or mare milk. It has solid probiotic properties, and it may be an excellent option for fermented foods that contain probiotics. *L. helveticus* H9 is a strain of bacteria with strong probiotic properties that have been isolated from a fermented food (Wang et al. 2015).

8.3.2 Chemical Hydrolysis

Organic chemical synthesis is a crucial technique for obtaining organic molecules with specific physicochemical properties. In silico techniques and existing experimental methods in solution or solid phase allow the design and development of molecules with varying molecular complexities for use in biological research (Sánchez and Vázquez 2017). There has been a growing interest in developing potent peptide analogues and peptidomimetics with specific pharmaceutical properties using de novo design and development of novel synthetic peptides that imitate protein conformational changes. Bioactive peptides are a future innovation of biologically active regulators that can help treat various diseases and disorders and prevent oxidation and microbial deterioration of foods (Lemes et al. 2016). Some emerging technologies are currently being investigated to recover by-products from residual waste and convert them into added-value products, as well as to facilitate large-scale recovery (Agyei and Danquah 2011) and purification of peptides for future applications in the pharmaceutical (Agyei and Danquah 2011) and food industries (Korhonen and Pihlanto 2006). The progress of liquid chromatography and mass spectrometry technologies has opened up a slew of new BP detection options. However, the direct application of this method does not always allow for the identification of peptides due to signal suppression (Sánchez and Vázquez 2017).

8.3.3 Enzymatic Hydrolysis

Enzymatic hydrolysis is the most effective and reliable approach for producing ACE-inhibitory peptides, and it is widely used to make highly functional and even nutritional protein hydrolysates (Sánchez and Vázquez 2017). Processors also prefer enzymatic hydrolysis since it does not leave any residual organic solvents in the end products (Najafian and Babji 2012). The end-product hydrolysate properties are influenced by pH, temperature, time, and enzyme concentration during hydrolysis

(Intarasirisawat et al. 2012). For the various enzymes employed to hydrolyse fish protein, the effective ranges are as follows: pH (1.5–11), temperature (35–60 °C), time (10–600 min), and enzyme concentration (0.01% –5%) (Ghassem et al. 2011; Intarasirisawat et al. 2012).

Specific protease selection is necessary to achieve desired functional properties in protein hydrolysates derived from fish by-products. Proteases may be derived from a variety of sources, including plants, animals, and microorganisms. Papain (Chi et al. 2015) and bromelain (Forghani et al. 2012) are the most extensively used plant proteases, while pepsin (Jumeri Kim 2011), trypsin (Lee et al. 2011; Hatanaka et al. 2009), and α-chymotrypsin (Lee et al. 2011) are the most widely used animal proteases. Alcalase (Smith 2017; Ahn et al. 2012), protamex (Ngo et al. 2014), flavourzyme, and neutrase (Chi et al. 2015) are examples of functional microbial proteases that have been used to hydrolyse fish protein.

ACE-inhibitory hydrolysate can be made with Thermoase PC10F. The enzyme is a commercial proteolytic mixture from AMANO Enzyme Inc., a bacterial neutral metalloprotease (EC 3.4.24.27) made from *Geobacillus stearothermophilus*. The enzyme has the benefit of catalysing the hydrolysis of peptide bonds, including hydrophobic amino acids. The powerful ACE-inhibitory peptides produced from red tilapia are most likely made up of hydrophobic amino acids. However, Thermoase's data on ACE-inhibitory peptide inhibition is limited. Hatanaka et al. (2009) claimed that Thermoase PC10F could create potent ACE inhibitor peptides Val-Trp (IC_{50} = 12.9 μM) and Leu-Lys-Tyr (IC_{50} = 10.1 μM) from peeled Antarctic krill tail flesh hydrolysate. Val or Ile at the N-terminus and Pro, Phe, or Tyr at the C-terminus are typical of powerful ACE-inhibitory peptides. The most effective ACE inhibitor is dipeptide Val-Trp, a competitive inhibitor of ACE (Ortiz-Chao and Jauregi 2007). Ikeda et al. (2015) used nine enzymes to hydrolyse dwarf gulper shark (*Centrophorus atromarginatus*), including Umamizyme, Newlase F3G, Peptidase R, Papain W-40, Protease A "Amano" SD, Protease M "Amano" 3SD, Protin SD-AC10F, and Thermoase PC10F. Among these enzymes, Thermoase PC10F showed the potent ACE-inhibitory activity with an IC_{50} value of 84.2 μg mL^{-1} and Val-Trp were found as the most potent inhibitory peptide (IC_{50} = 3.3 μg/mL). Thermoase PC10F Val-Trp inhibition for ACE had a competitive mechanism. Val-Trp also has antioxidant properties, implying that bifunctional peptides can help prevent hypertension (Ikeda et al. 2015). There is currently no information about the enzyme's use in the manufacture of ACE-inhibitory peptides from tilapia by-products.

8.4 Purification of Peptides

Purification of bioactive peptides from protein hydrolysates is often accomplished using selective precipitation, membrane filtration, and chromatographic approaches (Korhonen and Marnila 2013). Table 8.4 summarises many proteins fractionation and purification methods, along with their recommended applications. A combination of several methods is often used in a trial-and-error approach to obtain peptides with the specified purity and activity during peptide purification. This method is time-consuming and mostly expensive (Agyei et al. 2016). Each bioprocess activity should strive to minimise unit operations and to achieve simple, inexpensive isolation processes.

However, without sufficient knowledge of the characteristics of bioactive peptides, it is challenging to rationally design suitable purification procedures, particularly when attempting to discover new peptides (Agyei et al. 2016). High-performance liquid chromatography (HPLC) is often employed in the laboratory and pilot plant scales to separate, identify, and purify bioactive peptides. Purification of peptides based on their molecular mass may be accomplished using the size exclusion chromatography. The RP-HPLC technique may be used to separate peptides based on their

Table 8.4 Several techniques for protein fractionation and purification

Technique	Property required	Remarks	Recommended application
Membrane ultrafiltration	Molecular size	Fractionation through a membrane with specific molecular weight cut-off (MWCO) without significant biological activity loss and produced concentrated proteins	At the beginning of a purification procedure
Ion-exchange chromatography	Charge	Separation of ionic compounds in the mobile phase by the counter ion of the opposite and protein binding capacity usually high	At the beginning of a purification procedure
Reverse-phase HPLC	Hydrophobicity	Resolution varies according to gel size and commonly used for peptide separation	Used for separation of peptides, digested purified proteins, and other applications where loss of protein's biological activity is not a concern
Gel filtration (size exclusion)	Molecular size	Usually, low resolution and provide information about protein molecular weight	At the end of a purification procedure
Preparative isoelectric focusing	Isoelectric point (pI)	Separate peptides and proteins according to their pI and proteins precipitate in the rotofor chamber	Perform subsequent analyses as RP-HPLC

hydrophobic characteristics. It was primarily used to investigate bioactive sequences' structural and functional characteristics (Pownall et al. 2010).

In the post-production of protein hydrolysate from fish waste, the desired bioactive peptides must be purified as an essential precondition to study their molecular and biological properties. Hydrolysates themselves are functional as commercial sources of bioactivity compounds, but it has been noted that biological activity is higher in purified peptides (Chi et al. 2015). Several techniques have investigated the purification of protein hydrolysates, including ultrafiltration (UF), gel filtration (GF), ion-exchange chromatography, and reverse-phase high-performance liquid chromatography (RP-HPLC) (Smith 2017). These purification steps are essential to determine the effect of peptides' specific activity, including molecular size, hydrophobicity, and surface charge on ACE-inhibitory activity.

8.4.1 Ultrafiltration (UF)

UF refines the hydrolysate to maximise its specific activity to industrialise bioactive peptides derived fish by-products for human or animal use (Picot et al. 2010). UF allows the separation of the most active peptides from a complex mixture of protein hydrolysate. As a result, the purified hydrolysate would improve its potency (low IC_{50}). Weight cut-off (MWCO) ranging from 1000 to 10,000 Da was commonly used to fractionate protein hydrolysates (Ngo et al. 2014; Picot et al. 2010). The purity increases 1.51- to 525-fold (Intarasirisawat et al. 2013). This phase has been referenced in many studies that have successfully tested the derived fractions for biological activities. Antihypertensive activity (Ghassem et al. 2011; Ngo et al. 2014), radical scavenging (Picot et al. 2010), and antioxidation (Intarasirisawat et al. 2012) are among the identified activities. Permeate of low-molecular-weight fraction displayed the highest bioactivity after sequential ultrafiltration.

Through UF, membrane separation depends on the membrane pore size, solute-membrane interactions, macromolecule structure and size, and concentration polarisation. Initially, UF membranes were employed to purify biological fluids to preserve macromolecules such as proteins. For optimum separation effectiveness, the sizes of the peptides to be separated should be at least tenfold different (Singh 2006). Additionally, since all liquid separation membranes have a predetermined pore size distribution, the molecular weight cut-off (MWCO) of the UF membrane should be at least one-half the molecular weight of the smallest solute to be removed. The phase-inversion technique is used to produce the UF membranes. Although polysulphone (PS) is the most utilised polymer, alternative polymers such as cellulose acetate (CA), polyether sulphone (PES), polyacrylonitrile (PAN), and polyvinylidene fluoride (PVDF) are increasingly being utilised. CA-based membranes, on average, have a greater flux at comparable rejections than PS membranes. While CA membranes are less prone to fouling, PS membranes are required in a wide variety of applications due to their more excellent stability (Singh 2006). PES is utilised in place of PS in the food sector when steam sterilisation is needed. Polyimide (PI) polymeric membranes are one of the most recent membranes to be used. Polyimide UF membranes have great potential due to their resistance to a wide variety of organic solvents, including acetic acid, acetone, benzene, hexane, methanol, ethyl ether, ethoxy ethanol, and chlorinated hydrocarbons (Singh and Hankins 2016).

The approach of using UF to separate protein hydrolysate is sophisticated. Numerous constraints can affect the membrane filtration, including concentration polarisation, which occurs when retained molecules accumulate near the membrane surface, protein adsorption inside the porous structure of the membrane, protein deposition on the membrane surface, and protein–protein interactions in the feed solution. Such constraints may result in membrane fouling, thus impairing membrane function (Roslan et al. 2017). The feed-in UF is directly pushed towards the filter media under applied pressure, known as dead-end filtration. The smaller molecules diffuse across the membrane to the downstream side, whereas the larger particles accumulate on the membrane surface or in the depth of the filtering medium. Considering filtration is not sustainable without eliminating accumulated particles, backwashing and filter media replacement are done frequently. Furthermore, as the filtration continues in dead-end filtration, the filter cake thickens, resulting in a lower filtration rate (at a constant transmembrane pressure) (Todaro and Vogel 2014).

Filtration needs to be discontinued when the flow or transmembrane pressure reaches a critical value to clean or replace the membrane filter. This discontinuous style of operation may be a significant drawback when dealing with process streams that include a high proportion of solids. Tangential flow filtration (TFF) may circumvent this constraint by controlling the thickness of the concentration-polarisation layer via effective fluid management. TFF is a technique of filtration in which the filtered fluid travels tangentially to the filter surface, which is composed of a permselective membrane (Musumeci et al. 2018). In TFF, the feed flows parallel to the filter medium to create shear stress and scrub the surface, while an applied pressure forces some of the fluid through the membrane to the filtrate side. Molecules that are too big to pass through the membrane's pores are held on the upstream side. Therefore, the retained components do not accumulate on the membrane's surface but are carried along by the tangential flow. TFF or "cross flow filtration" has been studied extensively in the food sector and biotechnology (Lin et al. 2015). Additionally, the pharmaceutical industry has embraced TFF during the past decade to overcome the limitations of conventional purifying techniques.

8.4.2 Chromatographic Approach

Gel filtration chromatography (GFC), also known as size exclusion chromatography, has been utilised to separate, desalt, and determine the molecular weight of peptides and proteins for over 40 years. GFC is the most straightforward and delicate of all chromatography methods, separating molecules based on size differences (Wang et al. 2017). Its separation process is based on the fact that smaller molecules enter the pores of the gel and travel a greater distance, which takes longer. In comparison, bigger molecules are excluded from the pores, resulting in much shorter retention periods. In general, size exclusion chromatography and membrane fractionation are frequently utilised as the early phase of protein hydrolysate fractionation to achieve desired chain lengths or molecular weights before further characterisation (Acquah et al. 2019). Using a gel filtration chromatographic approach, Lee et al. (2011) isolated skate skin hydrolysates using a Sephadex G-25 gel filtration column (2.5 × 70 cm). The IC_{50} of the hydrolysates with ACE-inhibitory activity decreases from 0.63 mg/ml to 0.21 mg/ml after the purification process. Unlike ion exchange and other types of chromatography, molecules do not adhere to the chromatography medium, and therefore, the buffer composition has no direct effect on the resolution. As a result, a significant benefit of GFC is that elution parameters may be adjusted to suit the kind of sample and the need for further purification, analysis, or storage without affecting the separation. GFC is well suited for biomolecules that are sensitive to changes in pH, metal ion concentrations, or cofactor concentrations, as well as harsh environmental conditions, and can be used directly following ion-exchange chromatography, as the buffer composition has hardly any effect on the final separation (Wang et al. 2017). Additionally, GFC has excellent selectivity and resolution, which are critical characteristics in a purification method.

In recent years, ion-exchange chromatography (IEX) has become increasingly popular for separating, detecting, and structural characterisation of proteins, peptides, and short nucleotides. In IEX media, charged functional groups bind molecules with a negative charge. Bonded molecules are eluted from the medium by displacement by raising the concentration of a similarly charged molecule. Many functional groups in proteins can be positively or negatively charged (Wang et al. 2017). Proteins can be separated by altering the mobile phase's pH or ionic concentration. IEX is used to capture the required protein or bulk contaminants from large amounts of material, either as an intermediate step in the purifying process or as a final step to removing impurities. Ion-exchange chromatography has been widely explored to enhance calcium-binding phosphopeptides (CPPs) from casein hydrolysates (Korhonen and Pihlanto 2006). A process-scale method for isolating high-purity CPPs was established in prior work using acid precipitation, diafiltration, and anion-exchange chromatography. Recio et al. (2000) discovered a method in which a target protein was concentrated in a chromatographic media and then hydrolysed in situ by an enzyme. The inactive peptides were washed away, while the active peptides were kept on the ion exchanger. Finally, the fraction containing the active peptides was eluted from the chromatographic medium. This method was used to identify and purify cationic antimicrobial peptides from lactoferrin and s2-casein, as well as negatively charged phosphopeptides from -casein (Korhonen and Pihlanto 2006). This technique has the advantage of not requiring the isolation of the precursor protein and allowing the recovery of the enzyme utilised in the process. This approach offers new avenues for enriching low-molecular-mass peptides, and it can readily be scaled up to yield gramme or kilogramme quantities (Recio et al. 2000).

The principle of reverse-phase high-performance liquid chromatography (RP-HPLC) is based on the interaction of a hydrophobic stationary phase with hydrophobic analytes in a polar mobile phase (Acquah et al. 2019). RP-HPLC is extensively utilised to separate peptides, digesting purified proteins, and other applications based on its hydrophobicity. The hydrophobicity of the peptides was evaluated via elution of the mobile phase's polarity changes. Increased column length may enhance the resolution of tiny peptides and proteins. Peptides are eluted according to their hydrophobicity and molecular weight during the fractionation process. Owing to their amino acid sequence, peptides with low molecular weight and low hydrophobicity have a shorter retention period on the column and therefore elute first (Acquah et al. 2019). For instance, Welderufael and Jauregi (2010) fractionated the 6-h hydrolysate permeated into eight fractions utilising RP-HPLC and analysed them for antihypertensive activity. Fractions A, B, and C eluted from RP-HPLC first and exhibited the most significant percentage of ACE inhibition, with an IC_{50} value of 71.4, 70.2, and 376 μg/ml correspondingly. This indicates that more polar peptides exhibited more significant bioactivity than less polar peptides.

To summarise, there are a growing number of high efficiency and high-resolution separation and purification technologies. As a result, researchers must choose suitable separation techniques and mediums. Although specific separation techniques remain challenging, time-consuming, and expensive, researchers are dedicated to developing more sophisticated ways to replace them.

8.5 Red Tilapia By-Product as Protein Precursor for ACE-Inhibitory Peptides

Raghavan and Kristinsson (2009) tested tilapia protein hydrolysates from its processing by-product (tilapia white muscle) for their ACE-inhibitory activity, as well as their equivalent fractionates. Rather than utilising a raw by-product, the author employed the protein extract as feed. At pH 11.0, a by-product of tilapia was solubilised, and the protein was recovered at pH 5.5. The protein was hydrolysed to 7.5% and 25% DH using Cryotin-F and Flavourzyme, respectively. The hydrolysates were ultrafiltered at three levels: > 30 kDa, 10–30 kDa, and 10 kDa. Cryotin and Flavourzyme hydrolysates containing 25% DH had the highest ACE-inhibitory action, at 71% and 73%, respectively. At a 25% degree of hydrolysis (DH), both Cryotin and Flavourzyme hydrolysates displayed some light bands between 40 and 60 kDa, but most of the peptide bands were smaller than 40 kDa. The work suggests that Protamex, Neutrase, and Alcalase can create low-molecular-weight peptides when DH levels are high. Low-molecular-weight peptides (< 40 kDa) inhibited ACE more effectively than large molecular weight peptides. All 7.5% DH Cryotin fractions and 25% DH Flavourzyme fractions showed the best percentage of ACE inhibition among fractionates.

The use of red tilapia by-products for protein hydrolysate has already been shown to have significant ACE-inhibitory activity. According to the results of BIOPEP analysis, proteins from tilapia frame and skin could potentially be precursors of

ACE-inhibitory peptides, according to Lin et al. (2017). The frequency of ACE-inhibitory peptides in tilapia skin protein isolates (SPI)-derived collagen alpha-1(I) and collagen alpha-2(I) was 0.722 and 0.756, respectively, according to BIOPEP. From tilapia frame protein isolates (FPI), the proteins (myosin heavy chain, troponin T, creatine kinase, fructose-bisphosphate aldolase A, and alpha actin) were discovered; the frequency of peptides with ACE-inhibitory action was 0.350–0.414. SPI may be a better protein source of ACE-inhibitory peptides than FPI, according to the results of the in silico research. Pepsin, papain, or bromelain were employed to hydrolyse the frame and skin in their investigation. According to the in silico findings, pepsin and papain released more ACE-inhibitory peptides from FPI than bromelain. In silico digestion of SPI with papain provided 129 and 121 ACE-inhibitory peptides, respectively, higher than the numbers produced by pepsin (54 and 51) and bromelain (97 and 85). The ACE-inhibitory action was best in the pepsin-hydrolysed frame protein isolates with the highest DH (23%) (IC_{50} = 0.57 mg/mL). In additional experiments, ultrafiltration was used to increase the potency of the hydrolysate (< 1 kDa, 1–3 kDa, 3–5 kDa, and 5–10 kDa). The peptides with the lowest molecular weight (< 1 kDa) increased the efficacy of both ACE (IC_{50} = 0.41 mg/mL) and renin activities more efficiently than those with a higher molecular weight. The < 1 kDa ultrafiltration fraction from FPHPe (FPHPe1) had the largest inhibitory effect on ACE activity, according to the results of the ACE-inhibition testing (Lin et al. 2017).

Ultrafiltration membrane is frequently utilised to fractionate fish protein hydrolysate and recover peptide fractions with high ACE-inhibitory activity. According to Roslan et al. (2017), the tilapia by-product permeates from multilayer 5/5, and 10/5 kDa membranes have stronger ACE-inhibitory activity (84.04% and 75.59%, respectively) than single membranes (5 kDa—71.83% and 10 kDa—64.32%). It could be because the permeate is high in tiny peptides. The authors investigated the use of multilayer ultrafiltration to increase the number of short peptides in the tilapia by-product protein hydrolysate separation and thereby improve selectivity. The performance of the multilayer membrane was compared to that of a single membrane system based on permeate flux and peptide transmission (5 and 10 kDa). According to selectivity analysis, most of the permeate produced contained peptides with a molecular weight of less than 1500 Da. The amount of small-sized peptides (500 Da) increased when the smaller membrane's pore size was utilised (5 kDa membrane and 5/5 multilayer membranes), demonstrating that selectivity (particularly on small-sized peptides) can be improved. As a result, using a multilayer membrane to recover strong ACE-inhibitory activity from a tilapia by-product shows promise. Fish skin, a by-product of the fishing industry, still contains a substantial amount of protein-rich substances. Proteases such as bromelain, papain, trypsin, flavourzyme, Alcalase, and neutrase were used to hydrolyse gelatin isolated from Nile tilapia skin with a yield of 20.77 $\pm$ 0.80% wet weight. The gelatin hydrolysates were tested and shown to have a high level of antihypertensive action, ranging from 89.02% to 92.55%. The inhibiting activity of the angiotensin-converting enzyme shows that all hydrolysates are effective antihypertensive agents. All tilapia skin gelatin hydrolysates investigated have strong antioxidant and antihypertensive properties. The gelatin hydrolysate from tilapia skin can be an available antihypertensive drug (Choonpicharn et al. 2015).

8.5.1 Amino Acid Compositions of Tilapia By-Products

Because of their nutritional value and impact on functional properties, the amino acid compositions of fish protein hydrolysates are important. The results of several studies on the amino acid composition of tilapia by-products are summarised in Table 8.5. According to Sepúlveda et al. (2021), the most abundant amino acids in their tilapia by-product were glutamic acid, aspartic acid, and glycine, followed in descending order alanine, leucine, and serine, while cystine was the least abundant. These results agree with other studies in tilapia (Roslan et al. 2014; Robert et al. 2015).

Table 8.5 Amino acid composition of tilapia by-product

Amino Acid	[a]Tilapia by-product (in mg/g of protein)	[a]Tilapia by-product hydrolysate powder (in mg/g of protein)	[b]Spray-dried tilapia by-product hydrolysate (% of the product)	[c]Freeze-dried tilapia viscera hydrolysate (residues/1000 residues)
Aspartic acid	9.90 ± 0.27	45.85 ± 0.50	6.26	109 ± 7
Glutamic acid	18.61 ± 0.05	79.60 ± 0.25	9.57	117 ± 7
Serine	5.74 ± 0.11	23.31 ± 0.53	3.01	74 ± 4
Glycine	15.43 ± 0.16	67.82 ± 0.96	7.00	109 ± 7
Alanine	10.38 ± 0.04	45.64 ± 0.27	5.06	95 ± 6
Cysteine	0.09 ± 0.06	0.86 ± 0.54	0.53	6 ± 0
Tyrosine	4.07 ± 0.11	15.86 ± 0.28	1.94	35 ± 2
Arginine	9.98 ± 0.13	32.22 ± 0.04	4.91	31 ± 2
Proline	9.61 ± 0.05	41.84 ± 0.48	4.82	58 ± 3
Valine	6.07 ± 0.04	24.64 ± 0.33	3.14	53 ± 3
Methionine	3.37 ± 0.08	14.84 ± 0.05	1.71	24 ± 1
Isoleucine	5.27 ± 0.13	21.47 ± 0.03	2.66	37 ± 2
Leucine	9.41 ± 0.19	39.22 ± 0.97	4.68	77 ± 5
Phenylalanine	5.51 ± 0.05	22.37 ± 0.55	2.64	43 ± 3
Histidine	2.95 ± 0.08	11.68 ± 0.69	1.53	18 ± 1
Lysine	10.53 ± 0.17	40.47 ± 0.47	5.09	58 ± 3
Threonine	5.37 ± 0.01	24.48 ± 0.73	3.03	57 ± 3

[a]Roslan et al. (2014)
[b]Robert et al. (2015)
[c]Sepúlveda et al. (2021)

The nutritional value of any component is determined by the protein's ability to fulfil the organism's needs for vital amino acids. Several factors, including raw material, enzyme source, and hydrolysis conditions, have been identified to affect the amino acid composition of fish protein hydrolysates (Roslan et al. 2014). Peptides derived from fish proteins have been extensively studied as functional foods, especially in synthesising ACE-inhibitory compounds. There is a correlation between amino acid sequences and ACE-inhibition activity. The most potent and basic peptide inhibitors have similar structures, and the C-terminal tripeptide sequence strongly influences ACE activity. Because of interactions with three subsites at the ACE active site, tripeptides with Trp, Tyr, Phe, Pro, and a hydrophobic amino acid at the C-terminus have ACE-inhibitory activity (Pihlanto-Leppälä 2000). Tilapia by-products were found to have higher concentrations of hydrophobic amino acids in these studies, leading to ACE-inhibitory activity.

8.6 Future Outlook

The most effective use of bioactive peptides for human nutrition and health poses a fascinating scientific and technical challenge while also presenting opportunities for commercial success. Bioactive peptides can be used as ingredients in functional and novel foods, dietary supplements, and pharmaceuticals to provide specific health benefits (Korhonen and Pihlanto 2006; Sánchez and Vázquez 2017). To improve health through nutrition, such personalised dietary formulations are currently being developed around the world. This strategy was primarily applied to target groups, but it will eventually be applied to individuals (Korhonen and Pihlanto 2006). However, many scientific, technical, and regulatory concerns must be addressed before these substances can be fully used for this purpose.

Much effort is being put into isolating and identifying peptides from food protein using novel high-throughput proteomic techniques. A growing number of peptides with a broader range of functionalities are expected to emerge. The majority of these are being studied in vitro, and several support their bioactivity in vivo. However, research into technologies for producing these bioactive peptides is not progressing simultaneously (Ortiz-Chao and Jauregi 2007). As a result, there is a clear need to develop novel technologies that will enable the delivery of these functions. The most common batch reactor had the advantage of being the most straightforward approach; however, the process produced a complex mixture of peptides and reduced the enzyme activity.

Thus, the process will require a longer processing time and additional enzymes, and the hydrolysate produced requires subsequent purification due to low potency. Using the integrative approach suggested by Pa'ee et al. (2015), the production and purification of ACE-inhibitory peptides occur in the same reactor. It uses the ion-exchange technique and membrane filtration to separate peptides based on their

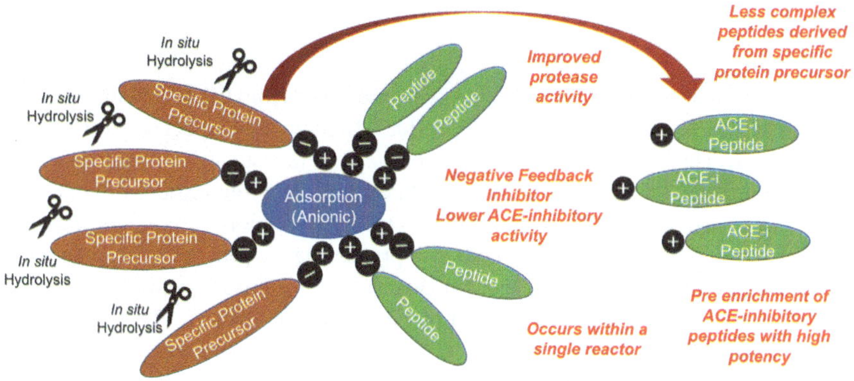

Fig. 8.3 Integrative process to produce ACE-inhibitory peptides (Abd-Talib et al. 2022)

charge and size (Fig. 8.3). The procedure revealed the possibility of being a cost-effective procedure. As a result, the next challenge would be to increase the process's scalability using available technologies.

It is critical to investigate the technical properties of active peptide fractions and to develop model foods that contain these peptides and maintain their activity for a specified period after they have been produced. Because of their lower molecular weight, peptides are more reactive than proteins, and the peptides present in the food matrix can react with other food constituents. Bioactive peptides from red tilapia by-products can produce an unpleasant odour and bitter taste (Pa'ee et al. 2021). There should be a consideration for peptide interactions with carbohydrates and lipids and the impact of processing conditions (particularly heating) on peptide activity and bioavailability, among other things (Korhonen and Pihlanto 2006). Further investigation is required due to the possibility of the formation of poisonous, allergenic, or carcinogenic substances. The development of modern analytical methods is required to investigate the safety of foods containing biologically active peptides.

Using modern computational approaches such as bioinformatics, proteomics, and molecular dynamic would expand bioactive peptides research. Such an approach would allow researchers to expand the findings on peptide sequences obtained to discern the mechanism of specific bioactivity, i.e. ACE inhibition. In the case of red tilapia by-product, the information regarding the structure implication of ACE-inhibitory peptides on ACE inhibition has not yet been unravelled due to the scarcity of information on peptide sequences. Thus, the sequence information is of importance to further elucidate the bioactivity at structural interaction behaviour.

8.7 Conclusions

The revolutionary study of protein hydrolysates derived from fish by-products would allow researchers further to discern the peptides' fundamental aspects regarding their bioactivities. The use of red tilapia by-products as feedstock has been shown to exert potential ACE-inhibitory activity. Unlike other fish protein sources, red tilapia by-product research is still in its infancy. Due to its important commodities and an effort to revolutionise the fish industrial processing by-product, researchers can expand the knowledge further. The chemical characterisation is essential to allow the researcher to predict the bioactivity and processing consideration. The hydrophobic amino acid residue, normally Pro, at the C-terminus is the most distinguishing feature of ACE-inhibitory peptides. Purification can also benefit from extending the chemical information, i.e. charged, hydrophobicity, and molecular size, as these are the main characteristics determining the bioactivity potency.

Furthermore, a cost-effective process must be of priority to allow the feasibility of the process at a commercial scale. Using freely accessible bioinformatics databases, i.e. BIOPEP, EXPaSY, UniProt, and NCBI, would enable us to search and analyse bioactive peptides in silico digestion, theoretical peptides released, sensory profile, and peptide allergenicity. Most importantly, these protein hydrolysates containing bioactive peptides can be utilised as functional food ingredients. They both eliminate environmental risks and provide job prospects, resulting in increased economic opportunities and an overall improvement in our societies' quality of living.

Acknowledgements This work was supported by the Ministry of Education (MOE) under the Fundamental Research Grant Scheme (FRGS), FRGS/1/2018/STG05/UNIKL/02/8, and Short-Term Research Grant (STR15074) awarded by Universiti Kuala Lumpur.

References

Abd-Talib N, Yaji ELA, Wahab NSA, Razali, N, Len KYT, Roslan, J, Pa'ee, KF (2022) Bioactive peptides and its alternative processes: a review. Biotechnol Bioprocess Eng. https://doi.org/10.1007/s12257-021-0160-8

Acquah C, Chan YW, Pan S, Agyei D, Udenigwe CC (2019) Structure-informed separation of bioactive peptides. J Food Biochem 43:e12765

Afreen M, Ucak I (2020) Fish processing wastes used as feed ingredient for animal feed and aquaculture feed. J. Surv. Fish. Sci. 6:55–64

Agyei D, Danquah MK (2011) Industrial-scale manufacturing of pharmaceutical-grade bioactive peptides. Biotechnol Adv 29:272–277

Agyei D, Ongkudon CM, Wei CY, Chan AS, Danquah MK (2016) Bioprocess challenges to the isolation and purification of bioactive peptides. Food Bioprod Process 98:244–256

Ahn CB, Je JY, Cho YS (2012) Antioxidant and anti-inflammatory peptide fraction from salmon by-product protein hydrolysates by peptic hydrolysis. Food Res Int 49:92–98

Allan JD, Abell R, Hogan ZEB, Revenga C, Taylor BW, Welcomme RL, Winemiller K (2005) Overfishing of inland waters. Bioscience 55:1041–1051

Chi CF, Wang B, Hu FY, Wang YM, Zhang B, Deng SG, Wu CW (2015) Purification and identification of three novel antioxidant peptides from protein hydrolysate of bluefin leatherjacket (Navodon septentrionalis) skin. Food Res Int 73:124–129

Choonpicharn S, Jaturasitha S, Rakariyatham N, Suree N, Niamsup H (2015) Antioxidant and antihypertensive activity of gelatin hydrolysate from Nile tilapia skin. J Food Sci Technol 52:3134–3139

Coppola D, Lauritano C, Palma Esposito F, Riccio G, Rizzo C, de Pascale D (2021) Fish Waste: From Problem to Valuable Resource. Mar Drugs 19:116

de Oliveira ALL, Assunção JCDC, Pascoal CVP, Bezerra MLS, Silva ACS, de Souza BV, Arruda TBMG (2021) Waste of Nile Tilapia (Oreochromis niloticus) to Biodiesel Production by Enzymatic Catalysis—Optimization Using Factorial Experimental Design. Ind Eng Chem Res 60:3554–3560

De Paris LD, Haab JCA, Sary C, Bernardi DM, Boscolo WR, Signor A (2016) Production and spray drying of protein hydrolyzate obtained from tilapia processing by-products. Acta Sci. Technol. 38:89–97

Ennaas N, Hammami R, Beaulieu L, Fliss I (2015) Production of antibacterial fraction from Atlantic mackerel (Scomber scombrus) and its processing by-products using commercial enzymes. Food Bioprod Process 96:145–153

FitzGerald RJ, Meisel H (2003) Milk Protein Hydrolysates and Bioactive Peptides. In: Fox, P.F., McSweeney, P.L.H. (eds) Advanced Dairy Chemistry—1 Proteins. Springer, Boston

Forghani B, Ebrahimpour A, Bakar J, Abdul Hamid A, Hassan Z, Saari N (2012) Enzyme hydrolysates from Stichopus horrens as a new source for angiotensin converting enzyme inhibitory peptides. J Evid Based Complement Altern Med. https://doi.org/10.1155/2012/236384

Fujita H, Yamagami T, Ohshima K (2001) Effects of an ACE-inhibitory agent, katsuobushi oligopeptide, in the spontaneously hypertensive rat and in borderline and mildly hypertensive subjects. Nutr Res 21:1149–1158

Ghassem M, Arihara K, Babji AS, Said M, Ibrahim S (2011) Purification and identification of ACE inhibitory peptides from Haruan (Channa striatus) myofibrillar protein hydrolysate using HPLC–ESI-TOF MS/MS. Food Chem 129:1770–1777

Halwart M, Funge-Smith, S, Moehl, J (2003) The role of aquaculture in rural development. Review of the State of World Aquaculture. FAO. 47–58

Hatanaka A, Miyahara H, Suzuki KI, Sato S (2009) Isolation and identification of antihypertensive peptides from antarctic krill tail meat hydrolysate. J Food Sci 74:H116–H120

Hayes M, Stanton C, Fitzgerald GF, Ross RP (2007) Putting microbes to work: dairy fermentation, cell factories and bioactive peptides. Part II: bioactive peptide functions. Biotechnology Journal: Healthcare Nutrition Technology. 2:435–449

Hermansen K (2000) Diet, blood pressure and hypertension. Br J Nutr 83:S113–S119

Ikeda A, Ichino H, Kiguchiya S, Chigwechokha P, Komatsu M, Shiozaki K (2015) Evaluation and Identification of Potent Angiotensin-I Converting Enzyme Inhibitory Peptide Derived from Dwarf Gulper Shark (C entrophorus atromarginatus). J Food Process Preserv 39:107–115

Intarasirisawat R, Benjakul S, Visessanguan W, Wu J (2012) Antioxidative and functional properties of protein hydrolysate from defatted skipjack (Katsuwonous pelamis) roe. Food Chem 135:3039–3048

Intarasirisawat R, Benjakul S, Wu J, Visessanguan W (2013) Isolation of antioxidative and ACE inhibitory peptides from protein hydrolysate of skipjack (Katsuwana pelamis) roe. J. Funct. Foods. 5:1854–1862

Ishak NH, Sarbon NM (2018) A review of protein hydrolysates and bioactive peptides deriving from wastes generated by fish processing. Food Bioprocess Tech 11:2–16

Jumeri Kim SM (2011) Antioxidant and anticancer activities of enzymatic hydrolysates of solitary tunicate (Styela clava). Food Sci. Biotechnol. 20:1075

Kim SK, Ngo DH, Vo TS (2012) Marine fish-derived bioactive peptides as potential antihypertensive agents. Adv Food Nutr Res 65:249–260

Kim EK, Kim YS, Hwang JW, Lee JS, Moon SH, Jeon BT, Park PJ (2013) Purification and characterisation of a novel anticancer peptide derived from Ruditapes philippinarum. Process Biochem 48:1086–1090

Korhonen HJ, Marnila P (2013) Milk bioactive proteins and peptides. Milk and dairy products in human nutrition: production, composition and health, 148–171

Korhonen H, Pihlanto A (2006) Bioactive peptides: Production and functionality. Int Dairy J 16:945–960

Lee JK, Jeon JK, Byun HG (2011) Effect of angiotensin I converting enzyme inhibitory peptide purified from skate skin hydrolysate. Food Chem 125:495–499

Lemes AC, Sala L, Ores JDC, Braga ARC, Egea MB, Fernandes KF (2016) A review of the latest advances in encrypted bioactive peptides from protein-rich waste. Int J Mol Sci 17:950

Li GH, Qu MR, Wan JZ, You JM (2007) Antihypertensive effect of rice protein hydrolysate with in vitro angiotensin I-converting enzyme inhibitory activity in spontaneously hypertensive rats. Asia Pac J Clin Nutr 16:275–280

Li H, Chen R, Jia Z, Wang C, Xu Y, Li C, Meng D (2020) Porous fish collagen for cartilage tissue engineering. Am. J. Transl. Res. 12:6107

Lin SY, Chiu HY, Chiang BL, Hu YC (2015) Development of EV71 virus-like particle purification processes. Vaccine 33:5966–5973

Lin HC, Alashi AM, Aluko RE, Sun Pan B, Chang, YW (2017) Antihypertensive properties of tilapia (Oreochromis spp.) frame and skin enzymatic protein hydrolysates. Food Nutr. Res. 61:1391666.

Tilapia Market. (2019). Tilapia byproducts: Commercial uses of the tilapia processing and filleting industries' waste. Tilapia Market Tilapia prices market and analysis. Retrieved from https://tilapia.market/tilapia-byproducts-commercial-uses-of-the-tilapia-processing-and-filleting-industries-waste/

Medina-Medrano JR, Quiñones-Muñoz TA, Arce-Ortíz A, Torruco-Uco JG, Hernández-Martínez R, Lizardi-Jiménez MA, Varela-Santos E (2019) Antioxidant Activity of Collagen Extracts Obtained from the Skin and Gills of Oreochromissp. J Med Food 22:722–728

Minkiewicz and Darewicz 2019

Möller NP, Scholz-Ahrens KE, Roos N, Schrezenmeir J (2008) Bioactive peptides and proteins from foods: indication for health effects. Eur J Nutr 47:171–182

Murray B, FitzGerald R (2007) Angiotensin Converting Enzyme Inhibitory Peptides Derived from Food Proteins: Biochemistry, Bioactivity and Production. Curr Pharm Des 13:773–791

Musumeci T, Leonardi A, Bonaccorso A, Pignatello R, Puglisi G (2018) Tangential flow filtration technique: an overview on nanomedicine applications. Pharm Nanotechnol. 6:48–60

Najafian L, Babji AS (2012) A review of fish-derived antioxidant and antimicrobial peptides: their production, assessment, and applications. Peptides 33:178–185

Ngo DH, Ryu B, Kim SK (2014) Active peptides from skate (Okamejei kenojei) skin gelatin diminish Angiotensin-I converting enzyme activity and intracellular free radical-mediated oxidation. Food Chem 143:246–255

Ortiz-Chao PA, Jauregi P (2007) Enzymatic production of bioactive peptides from milk and whey proteins, in: R. Rastall (ed) Novel Enzyme Technology for Food Applications, 1st edn, Woodhead Publishing, Cambridge, England

Pa'ee KF, Razali N, Sarbini SR, Ramonaran Nair SN, Yong Tau Len K, Abd-Talib N (2021) The production of collagen type I hydrolyzate derived from tilapia (*Oreochromis sp.*) skin using thermoase PC10F and its in silico analysis. Food Biotechnol 35:1–21

Pa'ee KF, Gibson T, Marakilova B, Jauregi P (2015) Production of acid whey hydrolysates applying an integrative process: effect of calcium on process performance. Process Biochem 50:302–310

Paiva L, Lima E, Neto A, Baptista J (2017) Angiotensin I-converting enzyme (ACE) inhibitory activity, antioxidant properties, phenolic content and Amino acid profiles of Fucus spiralis L. Protein Hydrolysate Fract Mar Drugs. 15:311

Picot L, Ravallec R, Fouchereau-Péron M, Vandanjon L, Jaouen P, Chaplain-Derouiniot M, Bourseau P (2010) Impact of ultrafiltration and nanofiltration of an industrial fish protein hydrolysate on its bioactive properties. J Sci Food Agric 90:1819–1826

Pihlanto-Leppälä A (2000) Bioactive peptides derived from bovine whey proteins: opioid and ace-inhibitory peptides. Trends Food Sci Technol 11:347–356
Popma T, Masser M (1999) Tilapia life history and biology Southern Regional Aquaculture Center
Pownall TL, Udenigwe CC, Aluko RE (2010) Amino acid composition and antioxidant properties of pea seed (Pisum sativum L.) enzymatic protein hydrolysate fractions. J Agric Food Chem 58:4712–4718
Prabu E, Rajagopalsamy CBT, Ahilan B, Jeevagan IJMA, Renuhadevi M (2019) Tilapia–an excellent candidate species for world aquaculture: a review. Annu Res Rev Biol 31:1–14
Raghavan S, Kristinsson HG (2009) ACE-inhibitory activity of tilapia protein hydrolysates. Food chem 117:582–588
Recio I, Floris R, Visser S (2000) Bioactive peptides from food proteins: A new isolation method. Agro Food Ind Hi Tech 11:9–11
Robert M, Zatylny-Gaudin C, Fournier V, Corre E, Le Corguillé G, Bernay B, Henry J (2015) Molecular characterization of peptide fractions of a Tilapia (Oreochromis niloticus) by-product hydrolysate and in vitro evaluation of antibacterial activity. Process Biochem 50:487–492
Roslan J, Yunos KFM, Abdullah N, Kamal SMM (2014) Characterisation of fish protein hydrolysate from tilapia (Oreochromis niloticus) by-product. Agric Agric Sci Proc 2:312–319
Roslan J, Kamal SMM, Yunos KFM, Abdullah N (2017) Assessment on multilayer ultrafiltration membrane for fractionation of tilapia by-product protein hydrolysate with angiotensin I-converting enzyme (ACE) inhibitory activity. Sep Purif Technol 173:250–257
Sampath Kumar NS, Nazeer RA, Jaiganesh R (2012) Purification and identification of antioxidant peptides from the skin protein hydrolysate of two marine fishes, horse mackerel (Magalaspis cordyla) and croaker (Otolithes ruber). Amino Acids 42:1641–1649
Sánchez A, Vázquez A (2017) Bioactive peptides: A review. Food Qual Saf 1:29–46
Sary C, Paris LDD, Bernardi DM, Lewandowski V, Signor A, Boscolo WR (2017) Tilapia by-product hydrolysate powder in diets for Nile tilapia larvae. Acta Sci 39:1–6
Sepúlveda CT, Zapata JE, Martínez-Álvarez O, Alemán A, Montero MP, Gómez-Guillén MC (2021) The preferential use of a soy-rapeseed lecithin blend for the liposomal encapsulation of a tilapia viscera hydrolysate. LWT 139:110530
Sila A, Bougatef A (2016) Antioxidant peptides from marine by-products: isolation, identification and application in food systems. A Review J Funct Foods 21:10 26
Singh R (2006) Hybrid membrane systems for water purification: technology, systems design and operations. Elsevier
Singh R, Hankins NP (2016) Introduction to membrane processes for water treatment. Emerging membrane technology for sustainable water treatment, pp 15–52
Šližyte R, Daukšas E, Falch E, Storrø I, Rustad T (2005) Yield and composition of different fractions obtained after enzymatic hydrolysis of cod (Gadus morhua) by-products. Process Biochem 40:1415–1424
Smith DM (2017) Protein separation and characterization procedures. In: Nielsen SS (eds) Food analysis. Springer, New York
Subasinghe R, Soto D, Jia J (2009) Global aquaculture and its role in sustainable development. Rev Aquac 1:2–9
Todaro CM, Vogel HC (2014) Fermentation and biochemical engineering handbook. William Andrew
Tuomilehto J, Lindström J, Hyyrynen J, Korpela R, Karhunen ML, Mikkola L, Nissinen A (2004) Effect of ingesting sour milk fermented using Lactobacillus helveticus bacteria producing tripeptides on blood pressure in subjects with mild hypertension. J Hum Hypertens 18:795–802
Udenigwe CC, Aluko RE (2012) Food protein-derived bioactive peptides: production, processing, and potential health benefits. J Food Sci 77:R11–R24
Umayaparvathi S, Meenakshi S, Vimalraj V, Arumugam M, Sivagami G, Balasubramanian T (2014) Antioxidant activity and anticancer effect of bioactive peptide from enzymatic hydrolysate of oyster (Saccostrea cucullata). Biomed Prev Nutr 4:343–353

Wald M, Schwarz K, Rehbein H, Bußmann B, Beermann C (2016) Detection of antibacterial activity of an enzymatic hydrolysate generated by processing rainbow trout by-products with trout pepsin. Food Chem 205:221–228
Wang M, Lu M (2016) Tilapia polyculture: a global review. Aquac Res 47:2363–2374
Wang J, Li C, Xue J, Yang J, Zhang Q, Zhang H, Chen Y (2015) Fermentation characteristics and angiotensin I-converting enzyme–inhibitory activity of Lactobacillus Helveticus isolate H9 in cow milk, soy milk, and mare milk. J Dairy Sci 98:3655–3664
Wang X, Yu H, Xing R, Li P (2017) Characterisation, preparation, and purification of marine bioactive peptides. BioMed Res Int. https://doi.org/10.1155/2017/9746720
Welderufael F, Jauregi P (2010) Development of an integrative process for the production of bioactive peptides from whey by proteolytic commercial mixtures. Sep Sci Technol 45:2226–2234
Zamora-Sillero J, Gharsallaoui A, Prentice C (2018) Peptides from fish by-product protein hydrolysates and its functional properties: an overview. Mar Biotechnol 20:118–130
Zhang Y, Duan X, Zhuang Y (2012) Purification and characterisation of novel antioxidant peptides from enzymatic hydrolysates of tilapia (Oreochromis niloticus) skin gelatin. Peptides 38:13–21

Chapter 9
Microhardness and Process Parameter Optimization of Friction Stir Welding on an AA5052 Thin Plate

Bakhtiar Ariff Baharudin, Mazli Mustapha, Azman Ismail, Fatin Nur Zulkipli, Fauzuddin Ayob, and Azlan Ahmad

Abstract The most important advancement in metal joining technique is FSW, which is utilized to weld high strength aluminium alloys. The joining process parameters, including tool rotational speed, travel speed, and axial force, must be tuned in order to increase the dependability and quality of the products produced by the FSW process. The process parameters for the FSW are optimized in this work using the Taguchi design of experiment (DOE) approach. The findings of the study and determination of the optimal condition using *L*9 orthogonal arrays of Taguchi were subjected to *S*/*N* analysis and analysis of variance (ANOVA) to identify the major welding factors impacting the weld quality. The relevant parameters for each welding performance criterion, such as the microhardness, were identified using ANOVA.

B. A. Baharudin (✉) · A. Ismail · F. Ayob
Universiti Kuala Lumpur Malaysian Institute of Marine Engineering Technology, 32200 Lumut, Perak, Malaysia
e-mail: bakhtiarab@unikl.edu.my

A. Ismail
e-mail: azman@unikl.edu.my

F. Ayob
e-mail: fauzuddin@unikl.edu.my

M. Mustapha · A. Ahmad
Department of Mechanical Engineering, Universiti Teknologi PETRONAS, Tronoh, Perak, Malaysia
e-mail: mazli.mustapha@utp.edu.my

A. Ahmad
e-mail: azlan.ahmad@utp.edu.my

A. Ismail
Centre for Women Advancement and Leadership, Universiti Kuala Lumpur, Jalan Sultan Ismail, Kuala Lumpur, Malaysia

F. N. Zulkipli
Faculty of Information Management, Universiti Teknologi MARA, Machang, Kelantan, Malaysia
e-mail: fatinnurzul@uitm.edu.my

A. Ismail et al. (eds.), *Advancements in Materials Science and Technology Led by Women*, Advanced Structured Materials 165,
https://doi.org/10.1007/978-3-031-21959-7_9

According to the results, the optimum settings are tool rotation speed of 350 rpm, welding speed of 550 mm/min, and axial load of 900 kg.

Keywords Taguchi's design · Friction stir welding · Microhardness

9.1 Introduction

A relatively recent method of joining metals, ferrous and non-ferrous, is friction stir welding (FSW). The method has been used to join for example ship parts and structures, aircraft structures, automobile parts, train carriages, motorcycle frames, and other parts from the energy and refrigeration industries.

In comparison with traditional fusion welds, FSW in aluminium alloys has a lower rate of cracking and evaporative loss of alloying components than typical fusion welds. This is as a result of the fact that the FSW technique used in solid-state joining results in a weld region with a finely worked or precipitated grain boundary. FSW joints are formed at temperatures below the melting point, resulting in reduced shrinkage and superior mechanical characteristics, as well as a decrease in residual stress inside the joints. In order to create a robust process design with a variety of quality features, the Taguchi dynamic experiment approach is quite useful. The primary goal of this study is to ascertain if the Taguchi dynamic technique can improve the system's robustness, high performance, and dimensional accuracy. A variety of industries have used the experimental design process known as the Taguchi method. In order to determine the ideal combination of parameters, the Taguchi approach examines the impact of varying quality standards on a sample of orthogonal arrays. These arrays create a tonne of data, even if they only require a few test runs. In this study, we show how to use the Taguchi approach to examine how FSW process factors affect the final product (Siva Rama Krishna and Pavan Kumar 2016).

This study examines the mechanical characteristics of the aluminium alloy 2250 alloy under various tool rotational speeds, transverse speeds, and axial load conditions. The tests and subsequent analyses follow the Taguchi design of experiment. The impact of various parameters on microhardness is examined using the analysis of variance (ANOVA) approach.

9.1.1 FSW Developments

Prasad and Namala experimented using the Taguchi parametric design approach for FSW of the RDE-40 aluminium alloy. The estimated optimal tensile strength value of the RDE-40 alloy friction stir welded is 303 MP (Prasad and Kumar 2018). Ghosh et al. performed experimental welding on the aluminium alloys A356 and A6061 with tool rotational rates ranging from 1000 to 1400 rpm and traversal speeds ranging from 80 to 240 mm/min. The shifting process window is what causes the fluctuation in the overall heat input and cooling volume during welding. Al-Si's eutectic network breaking and recovery recrystallization were observed in the stirring region (Ghosh et al. 2010). The FSW process parameters for the 6061 aluminium alloy were optimized by Nourani et al. using a rapid and computationally efficient method. The heat-affected zone (HAZ) region from the weld location was reduced using a thermal finite element model and the Taguchi optimization method. Additionally, during the procedure, the peak temperature was lowered (Nourani et al. 2011). Koilraj et al. experimented using the FSW technology to join two dissimilar Al-Cu alloy plates. In relation to the joint's tensile strength, the ideal process parameters were computed. The proportion of tool shoulder diameter to the pin diameter was critical in determining the joint integrity, as well as parameters such as pin shape and welding speed. The HAZ on the 5083 alloy side had the lowest hardness, where tensile failures were observed (Koilraj et al. 2012). Research shows the feasibility of using FSW to join two distinct sheets of commercially pure aluminium thickness. When welding sheets of various thicknesses, the angle of the tool is significant. The joint performance was 87.5% for the friction stir welded aluminium alloy 5052-H32 as opposed to the substance of the base. For the 5052-H32 aluminium alloy sheets that have been welded, it was also observed decreasing hardness properties in the weld region (Reddy et al. 2012). An enhanced modelling framework for the FSW of the aluminium alloys 6xxx series was developed by Simar et al. The model suite consists of a temperature in-process evolution model, a model of the evolution of microstructures, a model of strength and strain hardening, and a model of damage (Simar et al. 2012). An additional investigation examined the characteristics of two dissimilar aluminium alloys, AA5052 and AA6061, which were friction stir welded with two distinct feed rates of 28 mm per minute and 20 mm per minute and at a constant speed of 710 rpm. It is anticipated that by comparing mechanical and metallurgical characteristics, the sample's ductility was improved by welding it at a slower feed rate (RajKumar et al. 2014). As the three key impact parameters in FSW, Elanchizian et al. conducted experiments on rotational velocity, welding traversing velocity, and pressure applied on the joint. High strength to weight ratios are seen in the aluminium alloys AA6061 and AA811 (Elanchezhian et al. 2014). By combining FSW process features, Rambabu et al. created a mathematical model for the prediction of corrosion resistance of friction stir welded AA2219 aluminium alloy. The model was then optimized using the simulated annealing algorithm optimization technique (Rambabu et al. 2015). Another research on, friction stir welding aluminium alloys AA5052-H32 to AA6061-T6, each 1 mm and 1.5 mm thickness blanks was achieved. At constant tool spindle speeds of 1500 rpm, good welds were obtained. The weld produced at 63 mm/min velocity offers more ductility for both thicknesses than the weld at 98 mm/min. It was found that 1.5 mm thick tailor welded blank's elongation prior failure was 47 per cent more than that of 1 mm thickness weld (Doley and Kore 2016). Shanavas et al. during their investigation of submerged FSW of aluminium alloy 5052 achieved welded joint strength of 208.9 MPa at 700 rpm tool rotational speed and welding speed of 65 mm/min. Despite the fact that underwater welding did not produce a heat-affected region, the welded joint broke on the

Table 9.1 Taguchi's design for optimization

S. No	Factor	Level 1	Level 2	Level 3
1	Rotational speed	350	450	550
2	Welding speed	350	450	550
3	Axial load	800	900	1000

withdrawing side and the microhardness was marginally decreased close to the stir zone (Shanavas et al. 2018). A different study uses the FSW method to join plates made of the different aluminium alloys AA6063 and AA5052. The application of process parameters optimizes the tensile strength of the weld joint. The outcome demonstrates that the welding speed is the more significant parameter (Shunmugasundaram et al. 2020). Maneiah et al. performed an experiment on the FSW of a 3 mm thick aluminium 6061-T6 alloy using the H13 method steel pin and found the highest tensile strength of 191 MPa. The weld joint's 10% elongation shows that it is ductile in nature, making it suitable for use in the construction of aircraft and automobiles (Maneiah et al. 2020).

9.2 Methodology

9.2.1 Taguchi's Method

The Taguchi method is an effective resource for creating high quality systems. This approach is an efficient method for improving output and quality. The number of experiments is lowered to nine using this method. Calculating the total degree of freedom is necessary to select the best orthogonal array. The orthogonal array should have at least as many degrees of freedom as the process parameters. As a result, the *L*9 orthogonal array with 8° of freedom was selected. Using Taguchi's *L*9 orthogonal arrays, nine experimental runs were carried out. Table 9.1 displays the parameters that were chosen.

9.2.2 ANOVA Analysis

To investigate the relative importance of the complete control factor, the statistical method of ANOVA is applied. The functions of each parameter are also determined using them. Fisher's *F*-test is used as a supplementary tool for analysis. As a result, the parameters become more dominant as the F-test value increases.

9.2.3 *S/N Ratio*

Additionally, Taguchi advised utilizing the *S/N* ratio to assess the value. It uses a conceptual method that plots the influence and pinpoints the significant values.

9.2.4 *Experimental Procedure*

Sample preparation was achieved by cutting and work preparation prior to the FSW process. Using the shearing machine, the base metal sheets with dimensions of 200 mm × 100 mm × 1.5 mm were cut. The surfaces of sheared edges were ground and examined to obtain a fine finish in order to secure perpendicularity. Using mechanical clamps and backing plates, the initial joint layout secures the plates into place.

The welding direction is normal in reference to the rolling direction. Single pass welding techniques are employed to create the joints. The joints are created using a non-consumable cylindrical tool composed of high carbon steel.

To enable insertion of the tool into the protruding edges of the sheets or plates that will be crossed along the joint axis during the joining operation, the pin and shoulder are precisely designed.

The instruments' principal function is to heat the workpiece and transport material to produce joints. Friction between the tool and the workpiece causes plastic deformation. The material surrounding the pin turns softer due to the strong heat produced. The movement of materials from the front to the rear of the pin is caused by the combined effects of instrument rotational movement. The consequence is the creation of a "solid-state" joint. Fine and uniform re-crystallized grains arise as a result of the material's plastic deformation at high temperatures throughout the FSW procedure. The fine microstructure generated by the FSW technique is responsible for the workpiece's strong mechanical characteristics.

9.2.5 *Microhardness Measurement*

The Vickers hardness profile of the welded plates was measured using a HWPMT-X7B Highwood Micro-Vickers Hardness Tester under 500 gf load for 10 s at the cross-sectional thickness at the centre, parallel to the welding direction. Each sample was given 15 hardness points in accordance with ASTM: E384-17.

9.3 Results and Discussion

Table 9.2 displays the findings for microhardness at the centre of samples 1 through 9, the many tests that were run in accordance with Taguchi's *L*9 orthogonal array. The microhardness ANOVA table is shown in Table 9.3.

The signal-to-noise (*S*/*N*) ratio must be determined in order to establish which FSW process parameters have the greatest effect on the microhardness values. For *S*/*N* ratio analysis, there are three kinds of performance criteria to consider: the lower the better, the higher the better, and the nominal the better. This study considers the higher the better feature of microhardness.

The microhardness of the response characteristic is what is affected by each component, as seen by the main effects graphs. When an attribute is affected differentially by various quantities of a component, a major impact is apparent. In this study, the characteristic average for each component level is plotted using MINITAB to construct the main effects plot (Fig. 9.1). These averages correspond to those that are shown in the response table (Table 9.4). The points for each component are connected by a line. The rotational speed has the greatest impact on the microhardness, as shown by Table 9.4 and Fig. 9.1.

Table 9.2 Experimental results

Specimen No	Rotational speed (RPM)	Travel speed (mm/min)	Axial load (kg)	Hardness (Hv)
1	350	350	1000	64.30
2	350	450	800	63.50
3	350	550	900	66.60
4	450	350	800	61.50
5	450	450	900	61.20
6	450	550	1000	63.10
7	550	350	900	61.40
8	550	450	1000	58.60
9	550	550	800	58.90

Table 9.3 ANOVA of the experiment

Source	Degree of freedom	Sum of squares	Mean of squares	*F*-value	*P*-value	Contribution (%)
Rotational speed (RPM)	2	40.202	20.101	18.15	0.052	77.02
Travel speed	2	5.029	2.514	2.27	0.306	9.63
Axial load	2	4.749	2.374	2.14	0.318	9.10
Error	2	2.216	1.108			4.25
Total	8	52.196				

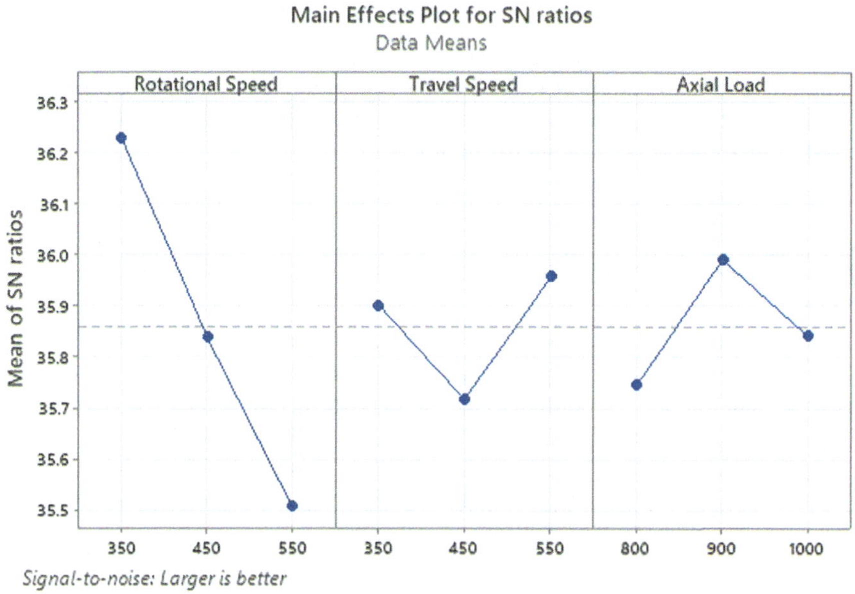

Fig. 9.1 Main effects plot for SN ratios

Table 9.4 Respond table for SN ratios

Level	Rotational speed (RPM)	Travel speed	Axial load
1	**36.23**	35.90	35.75
2	35.84	35.72	**35.99**
3	35.51	**35.96**	35.84
Delta	0.72	0.24	0.24
Rank	1	3	2

It is feasible to assess the significance of the relationship between the quality characteristic and FSW process factors using ANOVA. To do this, the entire *S*/*N* ratio variability is broken down into contributions from the error and each FSW process parameter. The total *S*/*N* ratio variability is calculated as the sum of the squared deviations from the *S*/*N* ratio's overall mean. Table 9.3 displays the ANOVA's outcomes, which were utilized to optimize the FSW procedure. To determine which of the experimental factors has a substantial influence, the Taguchi analysis's *p*-values (*P*) are utilized. The software calculates each factor's *p*-value (*P*) and compares it to the level of significance or α-level. A significant quadratic impact is present if the *p*-value for the squared effects is less than the α-level, which is often set at 0.05. It is clear from Table 9.3 that the tool rotational speed significantly affects microhardness since the *p*-value (*P*) is less than 0.05.

As shown in Table 9.3, the combined contribution of welding speed and axial load is much less than the contribution of tool rotational speed. The tool's rotating speed is a major factor of 77.02%. The response table in Table 9.4 can be used to determine the appropriate welding parameter. Table 9.4 and Fig. 9.1 show that the optimum welding conditions for achieving high microhardness are 350 rpm tool rotation speed, 550 mm/min welding speed, and 900 kg axial load. The main effects plot for microhardness is shown in Fig. 9.1.

9.4 Conclusion

In this work, the Taguchi technique was employed to create optimal FSW conditions for thin sheets of the aluminium alloy AA5052. ANOVA was used to analyse the experimental findings. According to the findings, the centre of the FSW joints had a maximum microhardness value of 66.6 Hv. Tool rotation speed of 350 rpm, welding speed of 550 mm/min, and axial load of 900 kg are the optimum welding conditions for obtaining high microhardness. With a 77.02 per cent contribution, tool rotational speed was the major factor impacting microhardness.

Acknowledgements The authors are thankful to Universiti Kuala Lumpur, Malaysian Institute of Marine Engineering Technology and Universiti Teknologi PETRONAS for providing necessary facilities and resources to complete this study for publication.

References

Bajpei T, Chelladurai H, Ansari MZ (2017) Experimental investigation and numerical analyses of residual stresses and distortions in GMA welding of thin dissimilar AA5052-AA6061 plates. J Manuf Proc 25:340–350

Doley JK, Kore SD (2016) A Study on Friction Stir Welding of Dissimilar Thin Sheets of Aluminum Alloys AA 5052-AA 6061. J Manuf Sci Eng Trans ASME 138:1–6

Elanchezhian C, Vijaya Ramnath B, Venkatesan P et al (2014) Parameter optimization of friction stir welding of AA8011-6062 using mathematical method. Procedia Eng 97:775–782

Freddi A, Salmon M (2019) Introduction to the Taguchi method. In: Springer Tracts in Mechanical Engineering. Springer, Cham.

Ghosh M, Kumar K, Kailas SV, Ray AK (2010) Optimization of friction stir welding parameters for dissimilar aluminum alloys. Mater Des 31:3033–3037

Koilraj M, Sundareswaran V, Vijayan S, Koteswara Rao SR (2012) Friction stir welding of dissimilar aluminum alloys AA2219 to AA5083 - Optimization of process parameters using Taguchi technique. Mater Des 42:1–7

Maneiah D, Mishra D, Rao KP, Raju KB (2020) Process parameters optimization of friction stir welding for optimum tensile strength in Al 6061–T6 alloy butt welded joints. Mater Today: Proc 27:904–908

Nourani M, Milani AS, Yannacopoulos S (2011) Taguchi Optimization of Process Parameters in Friction Stir Welding of 6061 Aluminum Alloy: A Review and Case Study. Eng 3:144–155

Prasad MVRD, Kumar NK (2018) Process Parameters Optimization in Friction Stir Welding by ANOVA. Mater Today: Proc 5:4824–4831

RajKumar V, VenkateshKannan M, Sadeesh P et al (2014) Studies on effect of tool design and welding parameters on the friction stir welding of dissimilar aluminium alloys AA 5052 - AA 6061. Procedia Eng 75:93–97
Rambabu G, Balaji Naik D, Venkata Rao CH et al (2015) Optimization of friction stir welding parameters for improved corrosion resistance of AA2219 aluminum alloy joints. Def Technol 11:330–337
Reddy PJ, Kailas SV, Srivatsan TS (2012) Effect of tool angle on friction stir welding of aluminum alloy 5052: Role of sheet thickness. Adv Mater Res 410:196–205
Shanavas S, Edwin Raja Dhas J, Murugan N (2018) Weldability of marine grade AA 5052 aluminum alloy by underwater friction stir welding. Int J Adv Manuf Technol 95:4535–4546
Shunmugasundaram M, Kumar AP, Sankar LP, Sivasankar S (2020) Optimization of process parameters of friction stir welded dissimilar AA6063 and AA5052 aluminum alloys by Taguchi technique. Mater Today: Proc 27:871–876
Simar A, Bréchet Y, De Meester B et al (2012) Integrated modeling of friction stir welding of 6xxx series Al alloys: Process, microstructure and properties. Prog Mater Sci 57:95–183
Siva Rama Krishna R, Pavan Kumar T (2016) Optimization of Process Parameters of Friction Stirs Welding of Aluminum Alloys (6061) Using Taguchi Method. Int J Sci Res 5:1988–1994
Ziegel ER, Ross P (1997) Taguchi Techniques for Quality Engineering. Technometrics. https://doi.org/10.2307/1270793

Chapter 10
Residual Stress Optimization of Friction Stir Welding on an AA5052 Thin Plate

Bakhtiar Ariff Baharudin, Mazli Mustapha, Azman Ismail, Fatin Nur Zulkipli, Fauzuddin Ayob, and Azlan Ahmad

Abstract Friction stir welding (FSW) is the most important advancement in metal joining technology for joining high-strength aluminium alloys. The input process parameters, including tool rotational speed, travel speed, and axial load, must be optimized in order to increase the reliability and quality of the products made by the FSW process. To identify the optimum parameters, the Taguchi design of experiment (DOE) approach was utilized. Taguchi's *L*9 orthogonal arrays were used to assess and determine the optimal condition, and the results were submitted to *S*/*N* analysis and analysis of variance (ANOVA) to identify the important welding factors impacting weld quality. The important parameters for each welding performance criterion, such

B. A. Baharudin (✉) · A. Ismail · F. Ayob
Universiti Kuala Lumpur Malaysian Institute of Marine Engineering Technology, 32200 Lumut, Perak, Malaysia
e-mail: bakhtiarab@unikl.edu.my

A. Ismail
e-mail: azman@unikl.edu.my

F. Ayob
e-mail: fauzuddin@unikl.edu.my

M. Mustapha · A. Ahmad
Department of Mechanical Engineering, Universiti Teknologi PETRONAS, Tronoh, Perak, Malaysia
e-mail: mazli.mustapha@utp.edu.my

A. Ahmad
e-mail: azlan.ahmad@utp.edu.my

A. Ismail
Centre for Women Advancement and Leadership, Universiti Kuala Lumpur, Jalan Sultan Ismail, Kuala Lumpur, Malaysia

F. N. Zulkipli
Facuty of Information Management, Universiti Teknologi MARA Campus Kelantan, Machang, Malaysia
e-mail: fatinnurzul@uitm.edu.my

A. Ismail et al. (eds.), *Advancements in Materials Science and Technology Led by Women*, Advanced Structured Materials 165,
https://doi.org/10.1007/978-3-031-21959-7_10

as residual stress, were evaluated using ANOVA. The optimal parameters obtained are 450 rpm for the tool rotation speed, 350 mm/min for the travel speed, and 900 kg for the axial load.

Keywords Taguchi's design · Friction stir welding · Residual stress

10.1 Introduction

Friction stir welding (FSW) is a technique for joining materials that are similar or dissimilar. It is an approach that is both cost-effective and environmentally friendly. It provides a number of advantages over typical fusion welding techniques. Among the most noteworthy of these advantages are low distortions and residual pressures, the absence of gases and spatters, and the absence of arc flash. It works especially well for joining alloys that are challenging or impossible to join using fusion welding since the procedure is carried out in the solid state. No filler metal or shielding gases are needed because the process is entirely self-contained. Therefore, there are no filler or slag inclusions in friction stir (FS) welded joints. For welding steels, this is crucial since they are less likely to crack because they are less likely to absorb hydrogen.

In comparison with traditional fusion welds, FSW in aluminium alloys provides several advantages, including the prevention of fracture and evaporative loss of alloying components. This is due to the solid-state joining mechanism used in the FSW process, which results in a weld zone with a finely wrought or recrystallized grain structure created by stirring and welding. FSW joints are formed at temperatures below the melting point, leading in minimal shrinkage, enhanced mechanical characteristics, and a reduction in residual stress inside the joints.

A non-replaceable rotary tool with a particular profile probe and shoulder may rotate at a specified speed in the FSW operation. Due to frictional heat, the device plunges between two materials in the shape of sheets or plates on the workpiece, generating "localized plastic deformation" in the interface region. As a result, the joint surface can be stirred down in the joining direction by the instrument. During the tool plunge operation, the "dwelling period" is the time during which the instrument spins at just one position before contact between the shoulder and the work surface occurs. During the welding process, lateral forces are generated that are parallel to the welding path (Gite et al. 2019).

The material is stirred by the pin, and the joint is created by the material's rubber deformation, while the metal is distorted by the frictional heat of the revolving instrument. The pin travels along the length of the weld while rotating, connecting the two plates. On one hand, referred to as advancing side (AS), and on the other, referred to as retreading side (RS), the tool rotation and weld orientations are comparable. As a result, an FSW joint features asymmetry, which is a distinguishing feature of the joint. Examining the mechanism's physical phenomena will reveal the important process parameters that control the joint's force (Prasad and Kumar Namala 2018).

Residual stresses are the stresses that continue to be present inside a material after they have been formed, even when external loads are not present or temperature gradients. Residual stresses can have an impact on a variety of technical aspects of industrial equipment, including fatigue life, dimension stability, corrosion resistance, and brittle fracture. There are a lot of residual stresses created by welding procedures, which are crucial industrial processes in the sector. The non-uniform thermal expansions and solidifications of welding procedures result in residual stresses.

The evaluation of mechanical equipment dependability in real-world settings, as well as the construction of structures, requires the measurement of residual stresses. The three primary categories of stress measuring techniques available are destructive, semi-destructive, and non-destructive processes. Often referred to as mechanical methods, destructive and semi-destructive procedures measure deformations brought on by residual stresses released after the specimen has been removed. For the purpose of estimating residual stresses, sectioning and the contour approach entirely destroy the specimen, whereas hole drilling, ring-core drilling, and deep-hole drilling leave microscopic holes on the surface of the material.

To assess stress-related characteristics, non-destructive techniques including ultrasonic, radiography, and magnetic particle inspection are frequently utilized. As more structural elements, including airframe structures and offshore rigs and bridges, need to be regularly inspected to prevent catastrophic damage or collapse, non-destructive stress assessment is becoming increasingly important.

10.1.1 Development of FSW

A study on the FSW of 0.8 mm thick aluminium sheets that are welded in the rolling direction using the alloys 2024-T3 and 6082-T6 was conducted. The mechanical properties of joints were exceptional. According to tensile testing, failure occurs at the welded region (Scialpi et al. 2008).

Rodriguez studied FSWs manufactured in 1 mm thick aluminium alloy plates from AA6016-*T*4 using two different tools. In terms of microstructure and mechanical qualities, the manufactured joint was examined and compared. The mechanical qualities of "hot" welds made using the fastest tool rotating speed and slowest traverse speed are enhanced (Rodrigues et al. 2009).

Another investigation was done by Zhang on the use of rotary tools without pins to join thin Al alloy plates. The FSW joint's tensile strength increases as the welding velocity for the tool with a three-spiral-flute shoulder is decreased. Around 398 MPa of tensile strength is obtained while welding at 20 mm/min and 1800 r/min, which is about 80% more than the parent material's tensile strength (Zhang et al. 2011).

Another research that employed FSW to join 1 mm thick aluminium, copper, copper-zinc, and zinc alloy plates discovered that doing so is feasible and acceptable. In actuality, the nugget's welds demonstrated strong morphological characteristics and significant grain refinement, regardless of the base material used. According to tensile and hardness testing, both welds were at least equally matched to the properties of the underlying material (Galvão et al. 2012).

At thicknesses of 1 mm and 1.5 mm, the FSW of AA5052 H32 to AA6061-*T*6 blanks was examined. Two tool traverse rates of 63 mm/min and a constant tool spindle speed of 1500 rpm were used to successfully perform the welds. The elongation before failure of a 1.5 mm thick tailor welded blank was found to be 47% larger compared to the 1 mm thick weld. Intermetallic compounds are created during welding, decreasing the joint's fracture strength (Doley and Kore 2016).

With a high rotation speed FSW, it was feasible to effectively weld thin plates of 6061-*T*6 aluminium alloy. The temperature distribution, microstructure development, and tensile properties of friction stir welded 6061-*T*6 joints were significantly impacted by high rotational and transverse speeds. The friction stir welded 6061-*T*6 joints have exceptional mechanical properties as a result of a combination of high rotational and transverse speeds. The greatest tensile strength measured was 301.8 MPa, or 85.8% of the strength of the underlying material (Liu et al. 2018a).

In a separate research, high-speed friction stir welding was used to join ultra-thin tubes made of an aluminium alloy (HS-FSW). The use of a Fe backing plate in conjunction with post-weld artificial ageing was discovered to be the most efficient way to improve the mechanical characteristics of the HS-FSW AA6061-*T*6 ultra-thin sheet joint (Liu et al. 2018b).

10.2 Methodology

10.2.1 Selection of Orthogonal Arrays

Analysing the characteristic data collected via orthogonal arrays yields the optimal process parameters (OA). To choose an acceptable orthogonal array for the experiments, the degree must be determined. As a result, Taguchi offers several linear graphs and regular orthogonal arrays. In this study, we examine three variables, namely the FSW process parameters for the three variables. Taguchi suggests a standard array at three phases. *L*9 OA (3^3) was hence chosen for this investigation, as shown in Table 10.1. Three columns and nine experimental runs are specified in this array.

Table 10.1 Taguchi's design for optimization

S. No.	Factor	Level 1	Level 2	Level 3
1	Rotational speed	350	450	550
2	Welding speed	350	450	550
3	Axial load	800	900	1000

10.2.2 ANOVA Analysis

Using the statistical technique of analysis of variance (ANOVA), the relative value of the entire control factor is explored. They are also used to figure out how much each parameter contributes. As an additional inspection technique, Fisher's F-test is employed. With a higher F-test score, the parameters become more prominent.

Taguchi also suggested utilizing the S/N ratio to analyse the values. It incorporates a conceptual method in which the effect is graphed and important values are identified.

10.2.3 Experimental Procedure

Prior to the FSW procedure, sample preparation was accomplished by cutting and work preparation. The basic metal sheets with dimensions of 200 mm × 100 mm × 1.5 mm were cut with a shearing machine. The edges of sheared faces were ground and checked for a clean finish to guarantee perpendicularity. Mechanical clamps and backing plates are used in the initial joint arrangement to secure the plates in place.

The welding direction is normal compared to the rolling direction. The joints are created using single pass welding processes. The cylindrical instrument made of high carbon steel is non-consumable. It is used to make the joints. The pin and shoulder are made to insert a non-consumable rotating tool into projecting edges of sheets or plates to be bonded and crossed along the joint axis.

The fundamental function of the instruments is to heat the workpiece. Joints are created by material movement. The workpiece deforms plastically as a result of friction between the tool and the workpiece. The material covering the pin becomes softer due to the strong heat produced. Material transfers from the front to the back of the pin as a consequence of the combined effects of instrument rotation and translation. A "solid-state" joint is thus created. The material is plastically distorted at high temperatures during the FSW process, producing small, uniform grains that have undergone re-crystallization. Strong mechanical properties of the workpiece are a result of the fine microstructure attained by the FSW process.

10.2.4 Residual Stress Measurement

The residual stress was measured using a Rigaku AutoMATE II. The centre point of the sample was measured and marked. The sample was set up as shown in Fig. 10.1. The measurement conditions are as shown in Table 10.2. The longitudinal residual stress was measured.

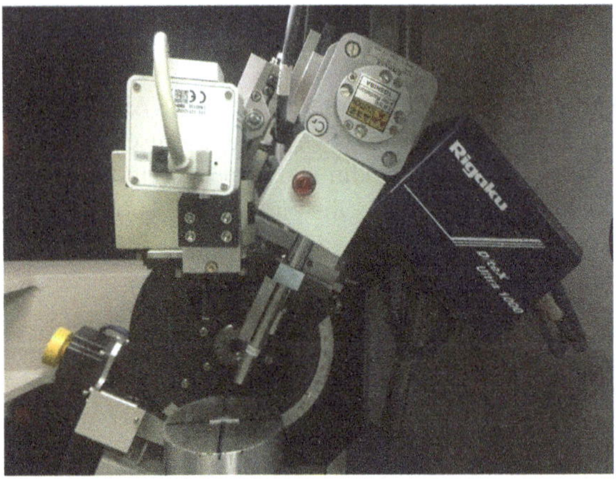

Fig. 10.1 Sample set-up in XRD chamber

Table 10.2 Residual stress measurement and analysis conditions

Radiation source	Collimator size (mm)	Measuring method	Psi angle (°)	Sample rotation	Processing range, 2theta (°)	Voltage (kV)/ Current (mA)
Cr	1.0	iso-inclination	0, 20.7, 30.0, 37.8, 45.0	No	146.14 to 166.24	40/30

10.3 Results and Discussion

Table 10.3 displays the residual stress data. The many investigations and test findings are based on Taguchi's L9 orthogonal array. The residual stress ANOVA table is shown in Table 10.4

In order to identify which FSW process factors have the largest impact on residual stress, the signal-to-noise (*S*/*N*) ratio must be calculated. There are three different types of performance criteria to take into account for *S*/*N* ratio analysis: the lower the better, the higher the better, and the nominal the better. This study takes into account residual stress's smaller-is-better characteristic.

The longitudinal residual stress, which is the response characteristic, is depicted in the main effects plots in relation to each component. When a factor's impact on a characteristic varies depending on its level, a primary effect is evident. By charting the characteristic average for each component level in this study, MINITAB is utilized to construct the major effects plot (Fig. 10.2). These averages correspond to those shown in the response (Table 10.5). The points for each component are connected

Table 10.3 Experimental results

Specimen no	Rotational speed (RPM)	Travel speed (mm/min)	Axial load (kg)	Residual stress (MPa)
1	350	350	1000	– 52.90
2	350	450	800	– 62.27
3	350	550	900	– 79.81
4	450	350	800	– 34.36
5	450	450	900	– 36.80
6	450	550	1000	4.27
7	550	350	900	43.97
8	550	450	1000	– 48.27
9	550	550	800	– 53.18

Table 10.4 ANOVA of the experiment

Source	Degree of freedom	Sum of squares	Mean of squares	*F*-value	*P*-value	Contribution (%)
Rotational speed (RPM)	2	0.25673	0.12837	0.91	0.524	34.63
Travel speed	2	0.13395	0.06698	0.47	0.679	18.07
Axial load	2	0.06776	0.03388	0.24	0.807	9.14
Error	2	0.28290	0.14145			38.16
Total	8	0.74135				

by a line. The rotational speed has the greatest impact on the residual stress, as seen in both the image and the Table 10.5.

The FSW process elements that have a substantial impact on the quality attribute are identified using an ANOVA. This is achieved by isolating the contributions from each FSW process parameter and the error, which is expressed as the sum of squared deviations from the total mean of the *S*/*N* ratio. Table 10.4 lists the ANOVA findings for the FSW process optimization. The p-values (*P*) are utilized in the Taguchi analysis to identify which experimental component has a significant effect. The computer calculates the *p*-value (*P*) of each factor and compares it to the level of significance (α-level). If the p-value for the squared effects is smaller than the α-level, then there is a significant quadratic impact (which is commonly believed to be 0.05). As shown in Table 10.4, it is evident that the combined contribution of travel speed and axial load is significantly less than that of the tool rotational speed. Tool rotational speed contributes 34.63% of the total. By using the response in Table 10.4, it is possible to determine the optimum manufacturing parameter.

According to Table 10.5 and Fig. 10.2, the optimum machining conditions for minimal residual stress are 450 rpm tool rotation speed, 350 mm/min travel speed, and 900 kg axial load. Figure 10.2 depicts the residual stress main effects plot.

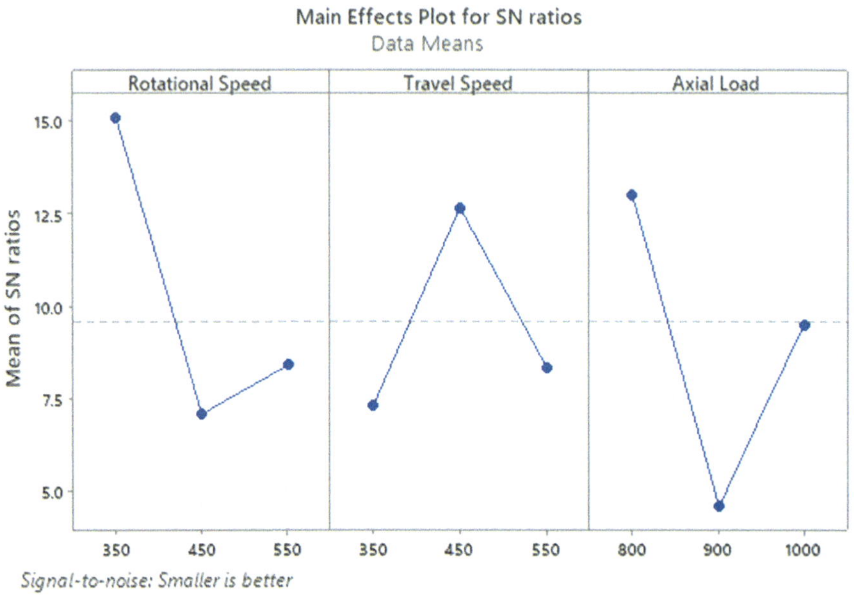

Fig. 10.2 Main effects plot for *S*/*N* ratios

Table 10.5 Respond table for *S*/*N* ratio

Level	Rotational speed (RPM)	Travel speed	Axial load
1	15.114	7.319	13.007
2	7.081	12.677	4.591
3	8.407	8.352	9.497
Delta	8.033	5.358	8.416
Rank	2	3	1

10.4 Conclusion

The Taguchi technique was employed in this study to find the best conditions for FSW of thin sheet AA5052 aluminium alloy. The results of the studies were examined using ANOVA. The FSW joints manufactured showed a minimal residual stress of −79.81 MPa, according to the data. The ideal welding conditions were determined to be 450 rpm tool rotation, 350 mm/min travel speed, and 900 kg axial load. With a contribution of 34.63%, rotational speed was the most important factor affecting residual stress.

Acknowledgements The authors are thankful to Universiti Kuala Lumpur, Malaysian Institute of Marine Engineering Technology and Universiti Teknologi PETRONAS for providing necessary facilities and resources to complete this study for publication.

References

Doley JK, Kore SD (2016) A study on friction stir welding of dissimilar thin sheets of aluminum alloys AA 5052-AA 6061. J Manuf Sci Eng Trans ASME 138:1–6

Galvão I, Leitão C, Loureiro A, Rodrigues D (2012) Friction stir welding of very thin plates. Soldag e Insp 17:2–10

Gite RA, Loharkar PK, Shimpi R (2019) Friction stir welding parameters and application: a review. Mater Today Proc 19:361–365

Liu FJ, Fu L, Chen HY (2018) Effect of high rotational speed on temperature distribution, microstructure evolution, and mechanical properties of friction stir welded 6061–T6 thin plate joints. Int J Adv Manuf Technol 96:1823–1833

Liu F, Fu L, Chen H (2018) High speed friction stir welding of ultra-thin AA6061-T6 sheets using different backing plates. J Manuf Process 33:219–227

Prasad MVRD, Kumar Namala K (2018) Process Parameters Optimization in Friction Stir Welding by ANOVA. Mater Today: Proc 5:4824–4831

Rodrigues DM, Loureiro A, Leitao C et al (2009) Influence of friction stir welding parameters on the microstructural and mechanical properties of AA 6016–T4 thin welds. Mater Des 30:1913–1921

Scialpi A, De Giorgi M, De Filippis LAC et al (2008) Mechanical analysis of ultra-thin friction stir welding joined sheets with dissimilar and similar materials. Mater Des 29:928–936

Zhang L, Ji S, Luan G et al (2011) Friction stir welding of al alloy thin plate by rotational tool without pin. J Mater Sci Technol 27(7):647–652

Chapter 11
The Effects of Fly Ash and Aluminium Trihydrate Incorporation on the Tensile and Thermal Properties of Epoxy Resin Mixtures

Wan Nursheila Wan Jusoh, Muhd Zulfadhli, Nurafiqah Naqiyah Khalid, Annur Ashran, Muhammad Hafiz, Mohamad Firzani, Rashidi Rahim, Syed Nur Azman Syed Mustaffa, and Norlaila Ramlee

Abstract This study explored the effects of adding fly ash and aluminium trihydrate (ATH) to an epoxy resin. The cementitious composite consists of epoxy resin as a binder with 5, 10, 15, and 20 (wt.%) fly ash. To enhance the fire retardant behaviour of the composite, 40 wt.% of ATH was added as a fire retardant. The tensile strength and strain were evaluated using a universal testing machine. The thermal stability of the material up to 700 °C was investigated through thermogravimetry analysis (TGA). The relative flammability of the composite was observed using the vertical

W. N. W. Jusoh (✉) · M. Zulfadhli · N. N. Khalid · A. Ashran · M. Hafiz · M. Firzani · R. Rahim
Fire Safety Research Cluster, Aerospace Section, Malaysian Institute of Aviation Technology, Universiti Kuala Lumpur, Jalan Jenderam Hulu, 43800 Dengkil, Selangor, Malaysia
e-mail: wannursheila@unikl.edu.my

M. Zulfadhli
e-mail: muhdzulfadhli@unikl.edu.my

N. N. Khalid
e-mail: naqiyahkhalid@gmail.com

A. Ashran
e-mail: annurashran96@gmail.com

M. Hafiz
e-mail: hafizzulkaple@gmail.com

M. Firzani
e-mail: mohd.firzani@yahoo.com

R. Rahim
e-mail: rashidi@unikl.edu.my

S. N. A. S. Mustaffa · N. Ramlee
School of Engineering, Manipal International University, No.1 MIU Boulevard, 71800 Putra Nilai, Nilai, Negeri Sembilan, Malaysia
e-mail: syed.azman@miu.edu.my

N. Ramlee
e-mail: norlaila.ramlee@miu.edu.my

A. Ismail et al. (eds.), *Advancements in Materials Science and Technology Led by Women*, Advanced Structured Materials 165,
https://doi.org/10.1007/978-3-031-21959-7_11

flame test. Based on the results, the tensile strength and strain of the composite declined as the amount of fly ash increased. The presence of fly ash in the mixture produced a positive but insignificant impact on the thermal stability of the composite. Nevertheless, considerable improvement was observed in the TGA and vertical flame tests of composite mixtures with ATH. The flammability of the composites was improved as the combustion rate decreased with higher fly ash content. The results suggested that fly ash and fire retardant additives such as ATH could provide sufficient fire retardant properties to mixtures.

Keywords Fly ash · Flammability · Epoxy resin · Aluminium trihydrate

11.1 Introduction

Fly ash is deposited from the combustion of coal-fired electric power plants. The coal was used as a fuel source in a boiler to generate steam for electricity. The high-pressure produce flows into a turbine, which drives a generator to generate electricity. Without proper waste management, the large quantity of fly ash in the plants can cause pollution to the air, and water, and also has been shown to endanger human health (Gajić et al. 2019; Scarlat et al. 2019; Zanoletti and Ciacci 2022).

Fly ash is viewed as a potential epoxy additive that is eco-friendly and has been a subject of interest for many researchers due to its low cost and abundance (Dharmalingam et al. 2015). Fly ash consists of aluminosilicate, calcium oxide, and ferric oxide. The size of fly ash ranges from less than 0.5 to 300 μm (Nithin Kumar et al. 2018) and can be used as fillers in composite applications, building, and polymer manufacturing (Dharmalingam et al. 2015; Ibrahim et al. 2020). Numerous polymer composite sectors are striving for environmentally friendly products while achieving the required properties. The recycling of this by-product will allow a competitive price for their final products (Sim et al. 2020).

Nagrockiene and Rutkauskas replaced cement with varying amounts of fly ash in the concrete mixture and found that 65% of fly ash produced concrete with higher density, ultrasonic pulse velocity, and low water absorption rate (Nagrockienė and Rutkauskas 2019). Gnanavel et al. observed that by adding fly ash particles in epoxy resins, the impact and hardness properties were enhanced while decreasing the tensile and flexural properties (Gnanavel et al. 2019). However, adding 5% fly ash to epoxy resin enhanced the tensile modulus slightly. Arijit Patra et al. stated that the glass transition temperature of the investigated resin was 140.4 °C, and shifted to a higher temperature of around 190 °C after fly ash was added (Patra et al. 2018). Another investigation discovered that the specific heat capacity of bagasse with fly ash particles was significantly higher than that without fly ash (Sunil and Manavendra 2017).

In general, prior work focused on the effect of fly ash on the mechanical properties of construction industry applications (Latip et al. 2020) due to the similar chemical composition and qualities of fly ash to commercial compounds used for partitioning in the construction industry (Arenas et al. 2021). As a result of the low cost and high strength of fly ash, various researchers have expressed interest in its use as a filler in polymer products (Chand 1988; Kishore et al. 2002; Kulkarni and Kishore, 2003; Garde et al. 2013; Sombatsompop et al. 2004) and the mechanical properties of polymer composite using fly ash as a filler also have been investigated (Nakamura et al. 1992a, 1992b; Mohammed Altaweel et al. 2011). Epoxy resin although it can provide good strength and stiffness to the structure, unfortunately, is a flammable material.

In this study, the influence of fly ash on the epoxy resin flammability with and without the flame retardant additive is explored. A common flame retardant additive, aluminium trihydrate (ATH), is added to explore the mixture's flame retardancy. The utilisation of ATH in the mixture was due to its low cost and effectiveness in reducing the flammability of a material (Chapple and Anandjiwala 2010). Samples were prepared by adding 0, 5, 10, 15, and 20 (wt.%) fly ash and 40 wt.% ATH to the epoxy resin. This investigation employed the thermal gravimetric analysis and vertical flame test on the composite mixture to determine the thermal properties of the mixture together with the tensile strength test to observe the changes in mechanical properties of the mixture.

11.2 Methodology

11.2.1 Preparation of Sample

The fly ash was collected from a thermal power plant company in Malaysia. The fly ash was dark brown and had a greyish tint. The epoxy resin used was an aviation-grade resin obtained from Aircraft Spruce. The ratio of resin to hardener mixture was fixed at 2:1. The samples were made by mixing the epoxy resin with 0, 5, 10, 15, and 20 wt.% fly ash. Another set of samples with 40 wt.% ATH was also prepared. The ATH was purchased from J.M. Huber Corporation. The list of sample content variations is given in Table 11.1.

The prepared sample mixtures were poured into silicone moulds. The use of silicone moulds has been widely accepted due to their flexibility and ability to withstand high temperatures without changing shape (Kang et al. 2004; Feng et al. 2020). All cured samples were trimmed to size and were stored in airtight zip lock bags.

11.2.2 Thermogravimetric Analysis

Thermogravimetric analysis (TGA) was employed to determine the thermal stability and the rate of change of the weight of the materials in the epoxy resin samples. The method is often utilised for inorganic materials, plastics, polymers, glasses,

Table 11.1 Sample composition

S. No.	Sample composition
1	0 wt.% fly ash
2	5 wt.% fly ash
3	10 wt.% fly ash
4	15 wt.% fly ash
5	20 wt.% fly ash
6	5 wt.% fly ash + ATH
7	10 wt.% fly ash + ATH
8	15 wt.% fly ash + ATH
9	20 wt.% fly ash + ATH

and composite materials (Abraham et al. 2018). The samples were analysed using a TGA4000 Perkin Elmer. The samples were exposed to a controlled nitrogen gas environment at a 10 °C/min heating rate up to 700 °C.

11.2.3 Tensile Strength Test

Tensile test was conducted using a Shimadzu AG–X 50 kN universal testing machine according to the standard test method for tensile properties of polymer matrix composite materials, American Society for Testing and Materials (ASTM) D3039. The speed of the machine was 5 mm/min for all specimens.

11.2.4 Vertical Flame Test

The vertical flame tests performed were conducted based on a standard test method for measuring the comparative burning characteristics of solid plastics in a vertical position, according to the ASTM D3801. The test determined the ability of the samples to extinguish themselves after the removal of the fire source. Schartel and Kebelmann reported that the reaction of the composite after being fed or exposed to a fire source for a short time can be observed using a vertical flame test (Schartel and Kebelmann 2019).

Each specimen was clamped at a 20° angle using a retort stand placed inside the flame test chamber. The test set-up and sample size are displayed in Fig. 11.1. The distance between the flame and the specimens was fixed at 7 cm. Before heating the specimens, the fire temperature was measured and kept constant at about 1000 °C using a thermocouple with a temperature recorder or data logger. Once the specimen caught fire, the fire source was removed to observe the flammability rate. If the

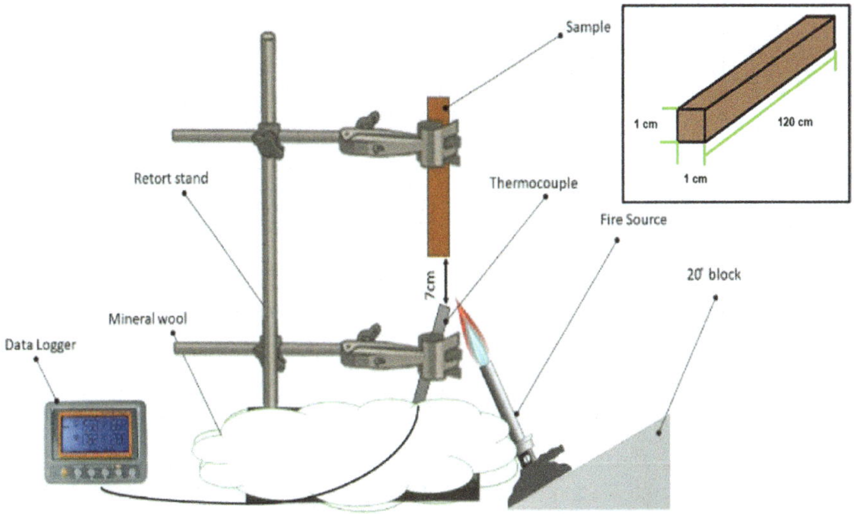

Fig. 11.1 Vertical flame test set-up

specimen did not burn, the specimen was heated for another minute, and the afterglow was observed. Any physical changes and duration of combustion were recorded.

11.3 Results and Discussion

11.3.1 Thermogravimetric Analysis

As shown in Fig. 11.2, the resin without fly ash demonstrated the highest weight decomposition, followed by the resins with 5, 10, 15, and 20 wt.% fly ash. Weight decomposition decreased with increased fly ash content. Additionally, upon heating the resins to 700 °C, the weight decomposition ranged between 69 and 85%. The samples also illustrated thermal stability and maximum decomposition within the range of 330–360 °C. The findings were consistent with a previous study that discovered that the burning rate decreased as the amount of fly ash added increased (Nguyen et al. 2019). As a result, a high amount of fly ash was preferable for decreased flammability.

The TGA curves of samples with ATH are displayed in Fig. 11.3. Based on the results, ATH promoted earlier decomposition temperatures of around 250–260 °C compared to samples without ATH. Another significant decomposition occurred around 330–360 °C, which was consistent with the results displayed in Fig. 11.1. The earlier decomposition temperature was due to the endothermic reaction of ATH after the temperature hit about 220 °C as mentioned by Chapple and Anandjiwala (2010). The samples with 20 wt.% fly ash and ATH with 30% char balance exhibited

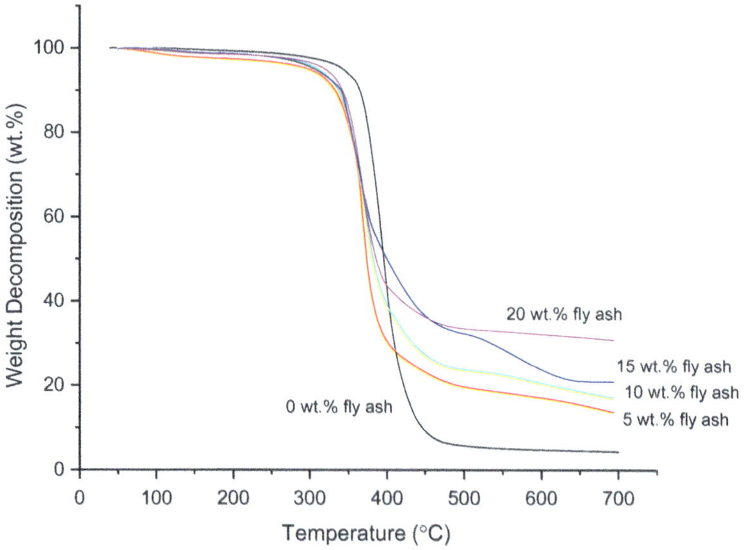

Fig. 11.2 TGA results of samples with resin and fly ash

the lowest weight decomposition as ATH and the fly ash mixture improved the thermal stability.

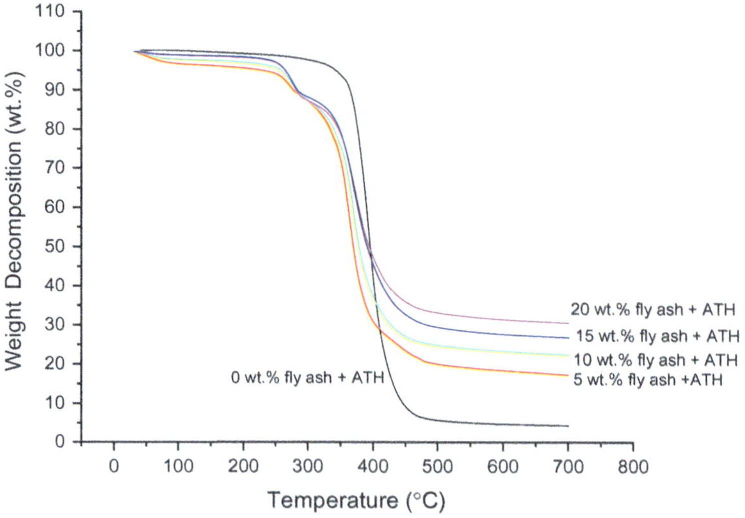

Fig. 11.3 TGA results of epoxy resins with fly ash and ATH

11.3.2 Tensile Strength Test

The effects of fly ash and ATH addition on the tensile strength of the epoxy resin are illustrated in Fig. 11.4. The addition of fly ash and ATH significantly reduced the tensile strength from 33 MPa displayed by the resin without fly ash (0 wt.%) to 9.7 MPa for 20 wt.% fly ash with ATH. The tensile strength of the composites with fly ash followed the order 10 > 5 and 15 > 20 wt.%. The tensile strengths of the composites with ATH followed the order 10 > 15 > 20 > 5 wt.%. Based on the results, fly ash and ATH reduced the ductility of the samples by changing them from polymer to composite-like mixtures. The transformation from ductile to brittle was due to a significant additive load. The samples with 10 wt.% fly ash demonstrated the highest tensile strength. However, the value was not apparent in fly ash with ATH composites.

The presented results were consistent with previous investigations, where tensile tests were conducted on fly ash embedded with epoxy (Anh et al. 2019; Nguyen and Nguyen 2020). The insufficient amount of resin to cover the fly ash and ATH particles reduced the bonding strength within the particle, decreasing the tensile strength. Therefore, the tensile strength was significantly decreased when ATH was added. The same pattern was observed for the tensile strain, illustrated in Fig. 11.5.

11.3.3 Vertical Flame Test

The vertical flame test results are illustrated in Fig. 11.6 with an insert image of the burning of a sample. The time recorded spanned the moment the samples caught fire to once the fire was completely extinguished. The samples without fly ash immediately caught fire and took two minutes to burn completely. However, an increased amount of fly ash resulted in a significantly longer burning time, ranging from six to eight minutes.

All samples burnt vigorously, dripped and produced a moderate amount of soot during the flame test with remaining char less than 1 cm. Samples with 40 wt.% ATH demonstrated the afterglow effect once the fire was removed. Figure 11.7 shows the condition of the sample after vertical flame testing. Additionally, the samples demonstrated reduced weight decomposition with increased fly ash content. According to Table 11.2, the sample with 5 wt.% fly ash and ATH exhibited the highest weight decomposition of 13%, while the sample with 20 wt.% fly ash and ATH with the least decomposition.

The sample with 5 wt.% fly ash and ATH displayed an afterglow that lasted about three seconds after one minute of exposure to fire. The afterglow of the samples with 10 and 15% wt.% fly ash and ATH lasted about two and one seconds, respectively, while the 20 wt.% fly ash and ATH sample did not demonstrate an afterglow. Based on the results, the inclusion of ATH significantly increased the fire retardancy of the composites due to the endothermic reaction of ATH upon exposure to temperatures

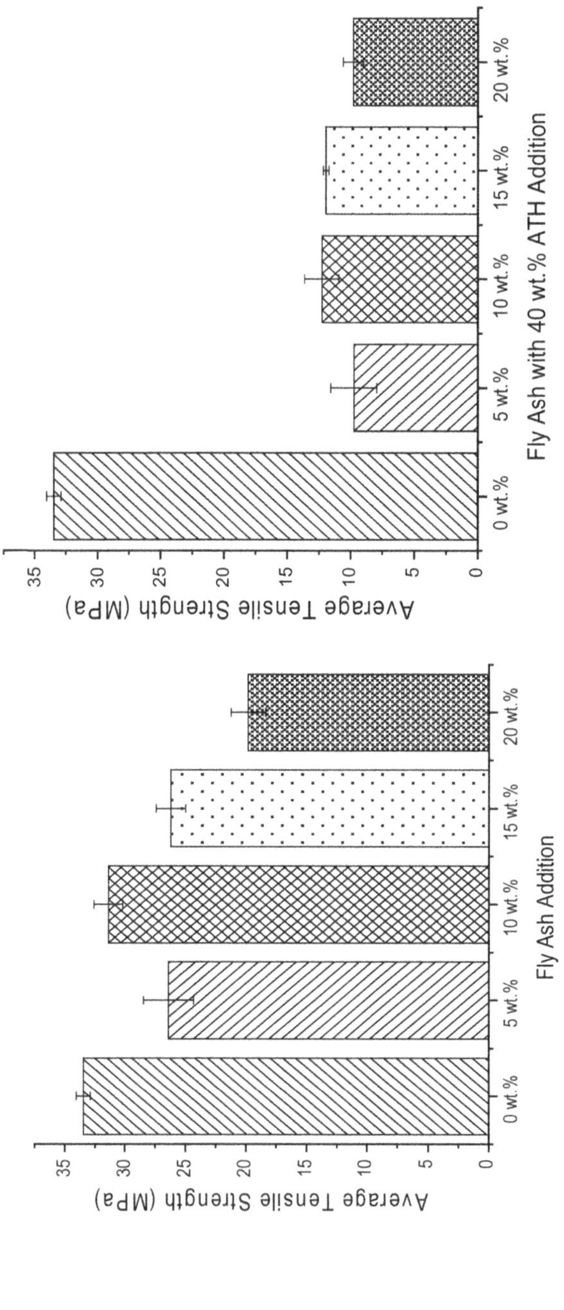

Fig. 11.4 The average tensile strength of resins with fly ash at different compositions with and without 40 wt.% ATH

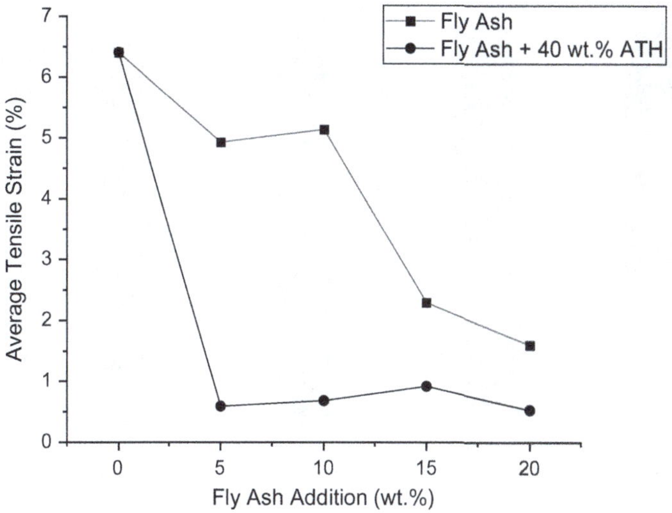

Fig. 11.5 The average tensile strain of resins with fly ash at different compositions with and without 40 wt.% ATH

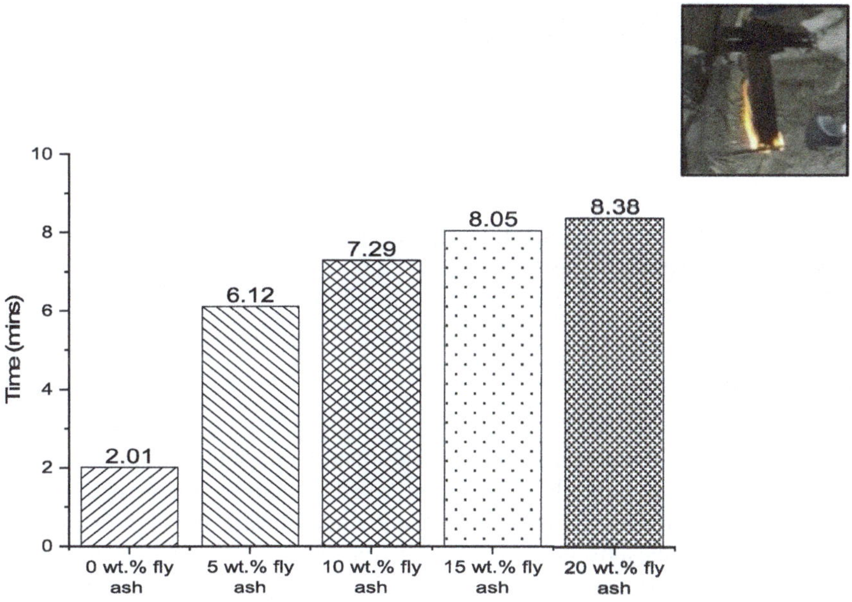

Fig. 11.6 The time taken for samples to burn completely

Fig. 11.7 Resin samples with fly ash and ATH after vertical flame test

Table 11.2 Sample weight decomposition after vertical test

Sample	Weight decomposition (%)
5 wt.% fly ash + ATH	13
10 wt.% fly ash + ATH	10
15 wt.% fly ash + ATH	4
20 wt.% fly ash + ATH	2

higher than 180 °C (Fan and Fu 2016). Hence, better fire retardancy performance on the composite mixture.

11.4 Conclusion

The remains from the decomposition reaction and flammability properties of the composites increased with fillers. However, the addition of fly ash did not improve the tensile strength or strain of the composites. Additionally, the high fly ash and ATH content significantly increased the viscosity of the mixtures, forming brittle mixtures. Only epoxy and fly ash were burnt during the flame test, considering that the total burnout time was prolonged with increased fly ash content. Samples with ATH exhibited satisfactory flammability resistance, illustrated when they did not ignite on their own when the fire source was removed. Based on the findings, the use of fly ash with ATH is recommended to promote the flammability resistance of epoxy resin. However, the low tensile strength of the composite makes it only suitable for non-load-bearing applications.

References

Abraham J, Mohammed AP, Kumar MPA, George SC, Thomas S (2018) Chapter 8-thermoanalytical techniques of nanomaterials. In: Characterization of nanomaterials: advances and key technology. Woodhead Publishing, Sawston

Anh NT, Tung NQ, Canh NX, Van Hoan N (2019) Eco-friendly flame retardant additives for epoxy resin nanocomposites. EurAsian J Biosci 13:323–332

Arenas C, Riós JD, Cifuentes H, Pecenõ B, Leiva C (2021) Experimental study of a noise reducing barrier made of fly ash. Mater Constr 71:e239

Chand N (1988) SEM observation of fractured flyash-polyester composites. J Mater Sci Lett 7:36–38

Chapple S, Anandjiwala R (2010) Flammability of natural fiber-reinforced composites and strategies for fire retardancy: a review. J Thermoplast Compos Mater 23:871–893

Dharmalingam U, Dhanasekaran M, Balasubramanian K (2015) Surface treated fly ash filled modified epoxy composites. Polimeros 25:540–546

Fan M, Fu F (2016) Advanced high strength natural fibre composites in construction. Woodhead Publishing, Sawston

Feng QK, Zhang DL, Zha J wei, Yin L juan, Dang ZM (2020) Thermal, electrical, and mechanical properties of addition-type liquid silicone rubber co-filled with Al_2O_3 particles and BN sheets. J Appl Polym Sci 137:1–9

Gajić G, Mitrović M, Pavlović P (2019) Chapter 4—Ecorestoration of fly ash deposits by native plant species at thermal power stations in Serbia. In: Phytomanagement of polluted sites. Elsevier, Amsterdam

Garde K, McGill WJ, Woolard CD (2013) Surface modification of fly ash—characterisation and evaluation as reinforcing filler in polyisoprene. Plast Rubber Compos 28:1–10

Gnanavel M, Maridurai T, Kumar K (2019) Mechanical, thermal and dielectric behaviour of C-class fly-ash coarse and fine particles reinforced epoxy resin composite. Mater Res Express 6:095507

Ibrahim NM, Ismail KN, Amat RC, Rahim NL, Rahim MA (2020) Recycling fly ash from MSWI for artificial aggregate production for concrete. IOP Conf Ser Earth Environ Sci 616:012049

Kang DW, Yeo HG, Lee KS (2004) Preparation and characteristics of liquid silicone rubber nanocomposite containing ultrafine magnesium ferrite powder. J Inorg Organomet Polym Mater 14:73–84

Kishore KSM, Sharathchandra S, Sunil D (2002) On the use of an instrumented set-up to characterize the impact behaviour of an epoxy system containing varying fly ash content. Polym Test 21:763–771

Kulkarni SM, Kishore, (2003) Effect of filler–fiber interactions on compressive strength of fly ash and short-fiber epoxy composites. J Appl Polym Sci 87:836–841

Latip NA, Sofian AH, Ali MF, Ismail SN, Idris DND (2020) Promising utilization of fly ash as an eco-friendly sustainable additive for enhancement in flame retardancy of composite production: a review. IOP Conf Ser Mater Sci Eng 736:052009

Mohammed Altaweel AMA, Ranganathaiah C, Kothandaraman B, Raj JM, Chandrashekara MN (2011) Characterization of ACS modified epoxy resin composites with fly ash and cenospheres as fillers: Mechanical and microstructural properties. Polym Compos 32:139–146

Nagrockienė D, Rutkauskas A (2019) The effect of fly ash additive on the resistance of concrete to alkali silica reaction. Constr Build Mater 201:599–609

Nakamura Y, Yamaguchi M, Kitayama A, Okubo M, Matsumoto T (1992) Effect of particle size on fracture toughness of epoxy resin filled with spherical silica. Polymer 33:3415–3426

Nakamura Y, Yamaguchi M, Okubo M, Matsumoto T (1992) Effects of particle size on mechanical and impact properties of epoxy resin filled with spherical silica. J Appl Polym Sci 45:1281–1289

Nguyen TA, Nguyen QT (2020) Study on synergies of fly ash with multiwall carbon nanotubes in manufacturing fire retardant epoxy nanocomposite. J Chem 2020:6062128

Nguyen TA, Nguyen QT, Nguyen XC, Nguyen VH (2019) Study on fire resistance ability and mechanical properties of composites based on Epikote 240 epoxy resin and thermoelectric fly ash: an ecofriendly additive. J Chem 2019:2635231

Nithin Kumar N, Siddeshchincholi HPR, Shivagiri SY, Revanasiddappa M (2018) Synthesis and characterization of fly ash /wooden fiber reinforced epoxy resin polymer composite. Mater Today Proc 5:501–507

Patra A, Das M, Anwar K, Khan B, Kamran K, Jana DR (2018) Investigation on mechanical and physical properties of fly ash reinforced epoxy resin composite. IOSR J Mech Civ Eng 15:64–68

Scarlat N, Fahl F, Dallemand JF (2019) Status and opportunities for energy recovery from municipal solid waste in europe. Waste Biomass Valorization 10:2425–2444

Schartel B, Kebelmann K (2019) Fire testing for the development of flame retardant polymeric materials. CRC Press, Florida

Sim J, Kang Y, Kim BJ, Park YH, Lee YC (2020) Preparation of fly ash/epoxy composites and its effects on mechanical properties. Polymer 12:79

Sombatsompop N, Thongsang S, Markpin T, Wimolmala E (2004) Fly ash particles and precipitated silica as fillers in rubbers. I. Untreated fillers in natural rubber and styrene-butadiene rubber compounds. J Appl Polym Sci 93:2119–2130

Sunil EM, Manavendra G (2017) Experimental investigation on thermal properties of bagasse fly ash reinforced epoxy composite. Int Res J Eng Technol 4:444–450

Zanoletti A, Ciacci L (2022) The reuse of municipal solid waste fly ash as flame retardant filler: a preliminary study. Sustain 14:2038

Chapter 12
Modeling and Simulation of Planar Micro-coils for Invasive Pressure Sensing

Norliana Yusof, Syamimi Mohd Norzeli, Siti Nurul Akmal Yusof, Noor Hidayah Mohd Yunus, and Norhayati Soin

Abstract A design modeling and simulation of a planar-based micro-coil for wireless invasive pressure sensing is presented. The rectangular-shaped micro-coil was designed using computer simulation technology (CST) software, and the finite element analysis (FEA) was conducted to analyze inductance values, quality factors and outer diameter size of the micro-coil. The width, gap, thickness, number of turns and inner diameter of the micro-coil was varied in a specific range based on the design specifications for invasive pressure sensing. The simulation results are in good agreement with the previous report and the related theory. The designed model and simulation can highly contribute to determine geometrical parameters of micro-coils for future processes, thus reducing the fabrication process and cost.

Keywords Micro-coil · Computer simulation technology (CST) · Finite element analysis (FEA) · Pressure sensing · Inductive-capacitive (LC) pressure sensor

---

N. Yusof (✉) · S. M. Norzeli · S. N. A. Yusof
Faculty of Innovative Design and Technology, Universiti Sultan Zainal Abidin, Gong Badak Campus, Gong Badak, 21300 Kuala Nerus, Terengganu, Malaysia
e-mail: norliana@unisza.edu.my

S. M. Norzeli
e-mail: syamiminorzeli@unisza.edu.my

S. N. A. Yusof
e-mail: snakmalyusof@unisza.edu.my

N. H. M. Yunus
Advanced Telecommunication Technology, Communication Technology Section, Universiti Kuala Lumpur British Malaysian Institute, Jalan Sungai Pusu, 53100 Batu 8 Gombak, Selangor, Malaysia
e-mail: noorhidayahm@unikl.edu.my

N. Soin
Department of Electrical Engineering, Faculty of Engineering, University of Malaya, 50603 Kuala Lumpur, Malaysia
e-mail: norhayatisoin@um.edu.my

A. Ismail et al. (eds.), *Advancements in Materials Science and Technology Led by Women*, Advanced Structured Materials 165,
https://doi.org/10.1007/978-3-031-21959-7_12

12.1 Introduction

The capability of wireless data transmission is one of the main advantages of inductive-capacitive (LC) pressure sensors. Sensor's measurement reading which applied the wireless concept or without physical connection allowing the LC pressure sensor to be used in measuring pressure variation detection. The main advantage of the LC pressure sensor is that no power source is required for operation. These battery-free sensors have two advantages: small size and long lifespan, making LC pressure sensors superior in conditions such as sealed environments and implantable biomedicine (Cao et al. 2013). The typical structure of the LC pressure sensor also achieves low cost. The rapid development of the Internet of things (IoT) for applications such as implantable sensors and wearable devices (Santis and Cacciotti 2020) makes LC passive wireless sensors an extensive field of research (Huang et al. 2016). The past decade has seen the rapid advancement of LC pressure sensors, especially for MEMS biomedical sensor applications (Yusof et al. 2017, 2018, 2021; Al-Nosairy et al. 2021; Farooq et al. 2020; Hu et al. 2021; Li et al. 2020a, 2020b).

In an LC pressure sensor system, two crucial components that must be developed are capacitors and inductors. The elastomer dielectric layer changes when external pressure is applied to the capacitor membrane and produces capacitance variations. The sensor circuit model is shown in Fig. 12.1, with the sensor modeled by the capacitor, *Cs* and inductor, *Ls* as well as resistance, *Rs*.

The quality factor is an essential performance parameter for measuring the effectiveness of energy transmission for LC resonance sensors. The higher the quality factor, the better the energy transmission performance. Therefore, to design micro-coils with a high-quality factor, the coils must exhibit high conductivity and low serial resistance (Zhai et al. 2010). The inductor's quality factor is the ratio of the inductive reactance to its resistance at a given frequency, which measures its efficiency. The higher the inductor's quality factor, the closer it is to the ideal inductor properties. Inductors with high-quality factors are coupled with capacitors to produce resonant circuits in radio transmitters and receivers.

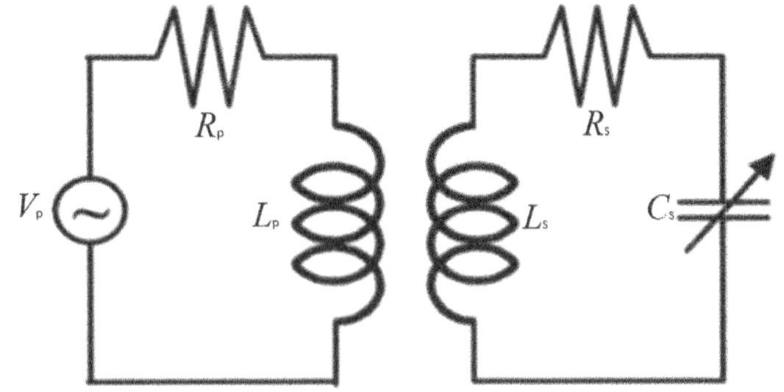

Fig. 12.1 Circuit model equivalent to telemetric reading approximation

The value of the quality factor, Q can be calculated using Eq. (12.1):

$$Q = \frac{\omega_0 L}{R} \tag{12.1}$$

where ω_0 is the angular resonance frequency ($\omega_0 = 2\pi f$), L is the inductance value and R is the coil resistance value.

For a rectangular micro-coil, the inductance value in the low-frequency range (0–20 MHz) can be calculated via Eq. (12.2) (Neagu et al. 1997):

$$L = \frac{\mu D_{out}^3}{4\pi\rho^2}(1-\alpha^2)(1-\alpha)\left[ln\frac{(1+\alpha)}{(1-\alpha)} + 0.2235\frac{(1-\alpha)}{(1+\alpha)} + 0.726\right] \tag{12.2}$$

with μ, D_{out}, α and p encompassing the relative permeability of air, the outer diameter of the coil, the ratio of the inner diameter of the coil and the outer diameter of the coil and the distance between two adjacent coil tracks. The serial resistance of the inductor can be calculated by Eq. (12.3), while the influence of the skin effect can be calculated by Eq. (12.4):

$$R = \frac{\rho l}{w\delta\left(1 - e^{-h/\delta}\right)} \tag{12.3}$$

$$\delta = \sqrt{\frac{2}{\omega\mu\sigma}} \tag{12.4}$$

where ρ is the internal resistance, l is the length of the coil, w is the width of the coil, h is the thickness of the coil, δ is the influence of the skin effect, ω is the angular frequency, μ is the magnetic permeability of the core material and σ is the conductivity of the coil. As the frequency increases, the influence of the skin effect decreases, and consequently, the parasitic serial resistance increases. An increase in such parasitic resistance reduces the ability of the sensor to make detection (Akar et al. 2001).

In this study, the design and simulation of a planar micro-coil was developed with modeling and simulation methods to determine the values of inductance, resistance and quality factor for the frequency range of 0–100 MHz.

12.2 Methodology

The rectangular micro-coil was designed using a computer simulation technology (CST) software as shown in Fig. 12.2. One of the inductors often used for MEMS pressure sensors is a rectangular micro-coil. Rectangular micro-coils were chosen because they fill the existing spatial shape most effectively compared to octagonal and circular circles (Fearday et al. 2017). Finite element analysis (FEA) was applied

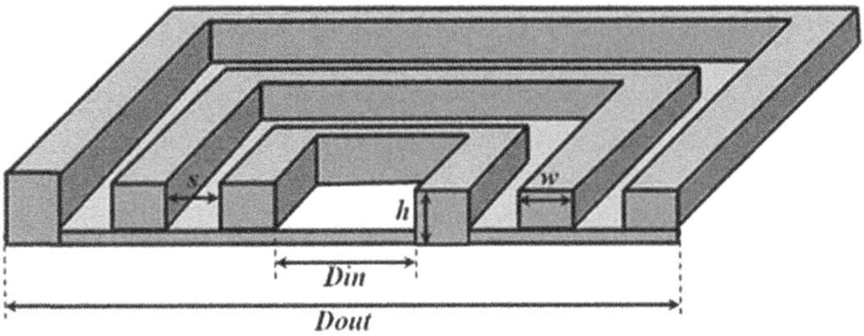

Fig. 12.2 Cross section of a rectangular micro-coil

Table 12.1 Parameters for micro-coil model simulation

Parameter	Value
Copper resistance, ρk	1.68×10^{-8} Ω.m
Young's modulus of copper, E_k	117 GPa
Reference impedance terminal, Z	50 Ω

to analyze the micro-coil's inductance value, quality factor and outer diameter size. The setting parameters for the micro-coil simulation are set as shown in Table 12.1. From the CST simulation, the values of inductance, resistance and quality factors of the rectangular micro-coils can be studied and analyzed. This study analyzes four geometric dimensional parameters: coil width, coil thickness, coil gap and the number of turns. Figure 12.2 shows a cross section of a rectangular micro-coil with width, w, of the coil, gap, s, of the coil, thickness, h, of the coil, inner diameter, D_{in}, of the coil and outer diameter, D_{out}, of the coil. The number of turns, N, also affects the performance measurement of the micro-coil.

12.3 Results and Discussion

Through the finite element analysis (FEA) method, the micro-coil parameters are set at; coil width = 50 μm, coil gap = 50 μm, coil thickness = 4 μm, inner diameter = 1000 μm and the number of turns = 15—30 with copper material. Figures 12.3, 12.4 and 12.5 shows the simulation results for the values of resistance, inductance and quality factors, respectively, for the planar micro-coils from the frequency range 0–100 MHz. Through Fig. 12.3, it can be seen that as the frequency increases, the resistance value of the coil also increases. This is due to the skin depth factor decreasing with increasing frequency which causes the resistance value of the micro-coil to decrease (Akar et al. 2001). Based on Fig. 12.3, it can be seen that the resistance value is in the range of 0–40 Ω, which proves that the micro-coils analyzed with this parameter have high electrical conductivity.

Figure 12.4 shows the simulation results of micro-coil inductance values for frequencies of 0–100 MHz. The inductance values are almost fixed with the change

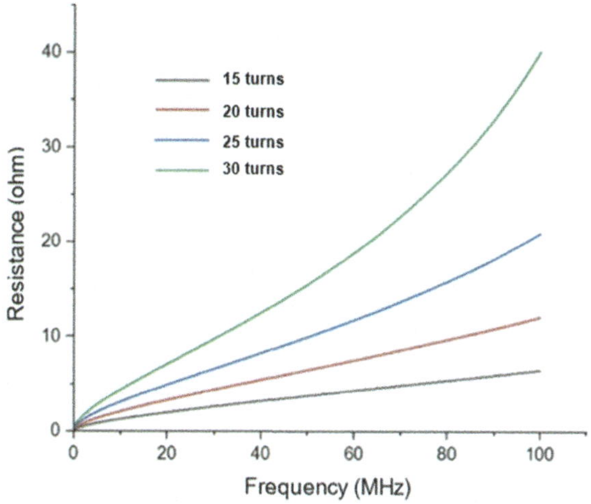

Fig. 12.3 Simulation results of micro-coil resistance values for frequencies 0–100 MHz

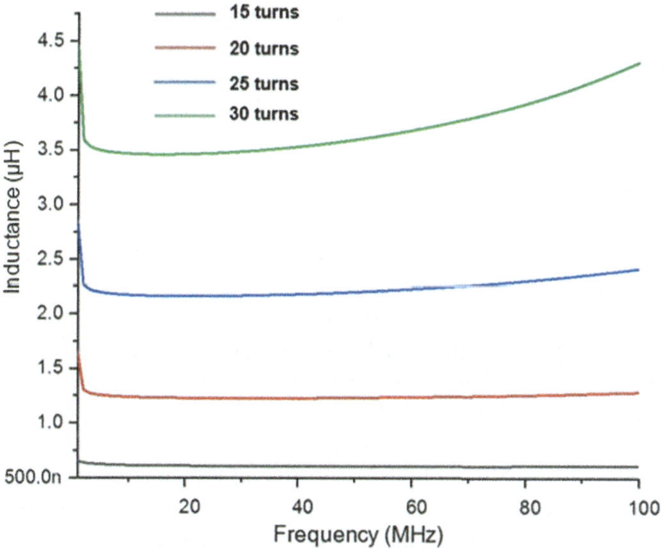

Fig. 12.4 Simulation results of micro-coil inductance values for frequencies 0–100 MHz

in frequency for the number of coil turns of 15, 20 and 25. For the number of turns of coil $N = 30$, the value of horizontal inductance for frequencies below 40 MHz and increasing for frequencies above 40 MHz. This is because the inductance value is influenced by the resistance value which refers to the inductance impedance value, $Z_L = R + X_L$, with $X_L = 2\pi fL$. Referring to Fig. 12.4, the inductance value for the simulated micro-coil is in the range of 3.5–4.25 μH for the number of turns of the coil, $N = 30$. While for the number of turns of the coil, $N = 25$, the inductance value

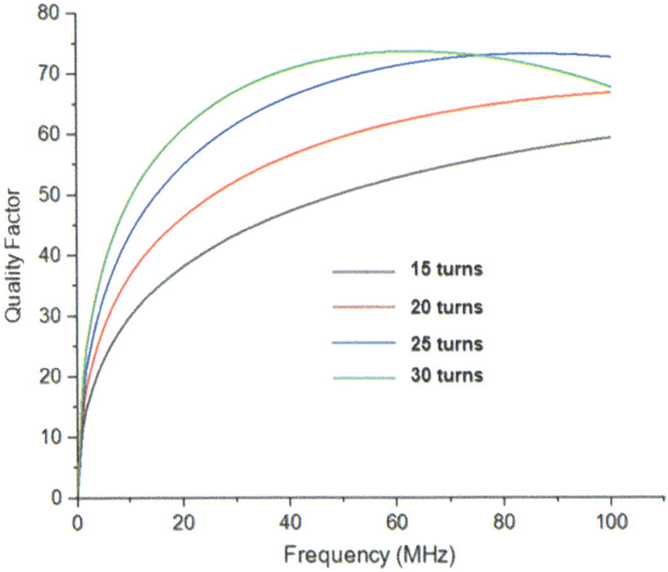

Fig. 12.5 Simulation results of the micro-coil quality factor value for the frequency 0–100 MHz

is in the range of 2.25–2.5 μH. The inductance value decreases with a decrease in the number of turns of the coil. For the number of turns of the coil, $N = 20$ and $N = 15$; the inductance values were recorded at 1.25 μH and 0.625 μH, respectively.

Figure 12.5 shows the simulation analysis results of the micro-coil quality factor against the frequency for four coil turns, $N = 15, 20, 25$ and 30. Referring to Fig. 12.5, it can be observed that the shape of the micro-coil quality factor graph is a semi-parabolic curve with factor values quality increasing with increasing frequency. The shape of the quality factor graph in the form of a parabolic curve is the effect of the resonance angle frequency, ω_o while its value is influenced by the combination of inductance and resistance values with the quality factor formula, $Q = \omega_o L/R$. Figure 12.5 also shows that the higher the value of the number of micro-coils, the higher the quality factor. The increase in the quality factor proportional to the increase in the number of coils is due to the increase in the inductance value (Akar et al. 2001). Through Fig. 12.5, it can also be observed that the maximum quality factor value is close to the value of 70 for this analysis. The high value of the quality factor adds value in improving the ability of a wireless sensor to transmit data signals better.

Referring to the sensor design specifications shown in Table 12.1, the selected frequency response is 0–100 MHz. However, the value of the self-resonance frequency (self-resonance) of the micro-coil also needs to be known to ensure that this self-resonance frequency does not interfere with the data measurement process. Figure 12.6 shows the operating frequency inductance characteristics of the simulated sensor from the frequency range of 0–1000 MHz. Figure 12.6 shows that the self-resonance frequency's value is 278.3 MHz. This confirms that the value of this

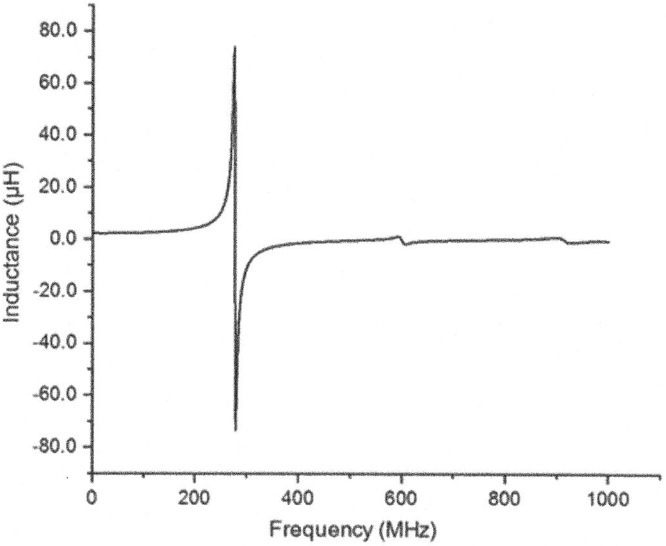

Fig. 12.6 Results of simulation of self-resonance frequency values for the frequency range 0–1000 MHz

self-resonance frequency is much higher than the sensor's operating frequency, which is at 0–100 MHz only. Therefore, this study's selection of the sensor frequency response is justified. The future experimental data measurement process is believed to be feasible without interference from the self-resonance frequency values.

12.4 Conclusion

A planar micro-coil was simulated by computer simulation technology (CST) and has been presented in this work. The geometrical parameter of simulated micro-coil was aimed to fit on the design specification for wireless invasive pressure application. The effect of geometric parameters, i.e., coil width, coil thickness, coil gap and the number of turns to the value of inductance, resistance and quality factors of micro-coil were investigated. From simulation results, the mechanical (geometrical parameters) and electrical (inductance, resistance and quality factors) behavior of micro-coil can be analyzed. Results indicate that the developed model of micro-coil from this simulation is satisfactorily compliant with the previous report and the related theory. This simulation provides benefits in terms of reducing the future fabrication cost as the geometrical design of micro-coil can be immediately determined according to the targeted performances and specifications.

References

Akar O, Akin T, Najafi K (2001) A wireless batch sealed absolute capacitive pressure sensor. Sensor Actuat A-Phys. https://doi.org/10.1016/S0924-4247(01)00753-1

Al-Nosairy KO, Bosch JJ, Pennisi V et al (2021) Use of a novel telemetric sensor to study interactions of intraocular pressure and ganglion-cell function in glaucoma. Brit J Ophthalmol 105(5):661–668

Cao H, Tata U, Landge V et al (2013) A wireless bladder volume monitoring system using a flexible capacitance-based sensor. In: IEEE topical conference on biomedical wireless technologies, networks, and sensing systems pp 34–36

Farooq M, Iqbal T, Vazquez P et al (2020) Thin-film flexible wireless pressure sensor for continuous pressure monitoring in medical applications. Sensors 20(22):1–22

Fearday C, Ward TA, Soin N et al (2017) Development of an inductor incorporated onto a carbon fiber MAV structural component. Microsyst Technol. https://doi.org/10.1007/s00542-016-2958-4

Hu S, Chen H, Jia S et al (2021) A wireless passive extra-arterial implantable blood pressure monitoring sensing system for rats. Microsyst Technol. https://doi.org/10.1007/s00542-020-05011-4

Huang Q, Dong L, Wang LF (2016) LC passive wireless sensors toward a wireless sensing platform: status, prospects, and challenges. J Microelectromech S 25(5):822–841

Li Y, Zhao M, Wei Q et al (2020a) Wireless passive intracranial pressure sensor based on vacuum packaging. Sensors. https://doi.org/10.1109/JSEN.2020.2998786

Li YT, Yang LY, Hsu WT et al (2020b) Designing and implementing an implantable wireless micromanometer system for real-time bladder pressure monitoring: a preliminary study. Sensors. https://doi.org/10.3390/s20164610

Neagu CR, Jansen HV, Smith A et al (1997) Characterization of a planar microcoil for implantable microsystems. Sensor Actuat A-Phys. https://doi.org/10.1016/S0924-4247(97)01601-4

Santis M, Cacciotti I (2020) Wireless implantable and biodegradable sensors for postsurgery monitoring: current status and future perspectives. Nanotechnology 31(25):252001

Yusof N, Bais B, Yunas J et al (2021) Fabrication of suspended PMMA-graphene membrane for high sensitivity LC-MEMS pressure sensor. Membranes (basel). https://doi.org/10.3390/membranes11120996

Yusof N, Bais B, Majlis BY et al (2017) Mechanical analysis of MEMS diaphragm for bladder pressure monitoring. In: IEEE regional symposium on micro and nanoelectronics (RSM) pp 22–25

Yusof N, Bais B, Majlis B Y et al (2018) Sensitivity and Q-factor trade-off analysis of MEMS pressure sensor for bladder implants. In: IEEE 8th international nanoelectronics conferences (INEC) pp 49–50

Zhai J, How TV, Hon B (2010) Design and modelling of a passive wireless pressure sensor. CIRP Ann. https://doi.org/10.1016/j.cirp.2010.03.103

Chapter 13
Review on Electrochemical and Biosensors and Their Application

Norilhamiah Yahya and Nur Afifah Mat Razali

Abstract Analysis of environmental, food and pharmaceutical compounds is crucial in determining the compounds content in the food products, drinking water and medicines. These compounds correspond to the health of human, thus presence of forbidden additives in the products of food, liquid samples and medicines may be harmful toward humans. Due to that, there is a high demand for analytical strategies to analyze food products and to detect harmful compounds in materials. Various studies have been done to develop low-cost, stable, high selectivity and sensitivity analytical methods. Among the analytical methods proposed, electrochemical methods show fast and selective analysis, ease of operation, portability, and they are inexpensive. On the other hand, electrochemical biosensors, which directly convert biological activities into electrical impulses, offer a different technique in analyzing the content of biological samples for the biochemical processes or biological quantification. Traditional techniques such as chronoamperometry, amperometry and potentiometry have been developed and will be further discussed in this review. Some of the challenges faced with electrochemical biosensors are instability, low signal response and selectivity affected by the presence of fouling agent and interference induced by chemicals present in the sample matrix. This shortcoming can be overcome through modification of the electrochemical electrode to improve the performance of the sensor. Detailed classification of sensors and working principle will be discussed in this review along with the review of application of electrochemical sensors in food safety, environmental analysis, glucose detection and medical diagnosis.

Keywords Sensor · Electrochemical · Chronoamperometry · Amperometry · Potentiometry

---

N. Yahya (✉) · N. A. Mat Razali
Universiti Kuala Lumpur Malaysian Institute of Chemical and Bioengineering Technology, Lot 1988, Bandar Vendor, 78000 Alor Gajah, Malacca, Malaysia
e-mail: norilhamiah@unikl.edu.my

N. A. Mat Razali
e-mail: afifah.razali22@s.unikl.edu.my

A. Ismail et al. (eds.), *Advancements in Materials Science and Technology Led by Women*, Advanced Structured Materials 165,
https://doi.org/10.1007/978-3-031-21959-7_13

13.1 Introduction

For the past few years, fabrication for analytical sensors in investigating quality of water, pharmaceutical compounds, food safety, environmental analysis has been done through spectroscopic and chromatographic methods. Although these methods are advantageous for various compounds analysis, they are not selective, reveal limitations for in toxic solvent, require time-consuming analysis and difficult operation for compounds analysis such as food and pharmaceutical samples. Due to that, low-cost analytical tools, selective and sensitive sensors have garnered attention as an alternative for spectroscopic and chromatographic methods (Karimi-Maleh et al. 2020). A sensor is a device receiving signals and responds to them (Patel et al. 2020). The state, measure or property that was sensed opposed the stimulated signals, while the output comes in the form of charge, current or voltage. Active and passive sensors are the two types of sensors where they are categorized depending on their applications, material used, cost, range or accuracy of the sensor (Patel et al. 2020). A passive sensor turns the input energy into output energy without the requirement of any additional energy sources. Thermal, electric field sensing, chemical and infrared are examples of passive sensors. On the other hand, an active sensor requires additional outside energy sources to respond toward excitation signals (Zhao and Chen 2009). Due to capability in modifying their properties in response to an exterior effect and changing them to electric signals, active sensors are called as parametric sensors. Features of a perfect sensor are response time, sensitivity, calibration, selectivity, linearity and reproducibility (Naresh and Lee 2021).

Over the past decades, research on biosensors has grown astronomically for various applications. Due to ability in designing unique reaction of chemicals by immobilizing recognition of biological components on the sensor substrate that have a precise binding affinity to the intended molecules, biosensors are highly selective. Enzymes are the most prevalent of these recognition elements for biosensors, alongside with the whole cells, nucleic acids, antibodies and receptors (Chaubey and Malhotra 2002). The surface layout of the sensor must inhibit any non-specific contact to fully utilize the specific interaction through biorecognition. Thus, various studies are carried out to find the surface modification with particular interaction capabilities for excellent sensor performance (Kasemo 2002). Evolution of biosensors had undergone three generations depending on the attachment of the biorecognition element (bioreceptor) to the transducer with the first-generation glucose oxidase (GOx) biosensor in 1962 by Clark and Lyons. Since then, GOx sensors have been widely used and undergo modifications since 1960s.

Limitation in high sensor configuration complexity and expenses is not incurred by electrochemical biosensors, as illustrated by glucose sensors (Grieshaber et al. 2008). This is due to their strong relationship to advancements in economical microelectronic circuit production and simple interference with conventional electronic readout and processing. Aside from that, they are effective analysis tools used in various applications. Fabrication of electrochemical devices garnered attention in micro- and nanotechnologies due to unique properties such as miniaturization, accessibility, low cost, ease of use, specificity, selectivity and real-time on sight monitoring capabilities (Baracu and Dinu Gugoasa 2021). Due to small volume of sample and wide active surface achieved by utilization of nanomaterials as sensing layers, electrochemical sensors exhibit significant sensitivity in detection of biomolecules (Baracu and Dinu Gugoasa 2021). For fabrication of electrochemical sensors, they can either be (1) carbon paste-based electrodes (CPE), (2) screen-printed electrodes (SPE) and (3) microfabricated electrodes (MFEs) (Baracu and Dinu Gugoasa 2021). Among this fabrication technology, MFEs show a better stability in time, easily integrated to portable devices and ease of miniaturization. On the other hand, CPEs have limitations in size-reduction and assimilation in a miniaturized device, while SPEs are not-reusable and easily damaged (Gugoasa et al. 2019; Alam et al. 2020; Moussa and Mauzeroll 2019).

Electrochemical sensors cover several classes of sensors depending on their investigated signal from the chemical response, such as (i) amperometric sensors, (ii) potentiometric sensors and (iii) conductometric sensors. In this chapter, classification of biosensors and their components will be discussed. Detailed working principle for the classes of electrochemical sensors will be discussed in the second part of this chapter, and the last of part of the chapter will review the application of electrochemical sensors in various implementations such as environmental analysis, food safety, medical diagnosis and glucose detection.

13.2 Classification of Sensors

13.2.1 Sensor

Sensors can be categorized into diverse groups, depending on physical quantity such as their energy source (either they are passive or active sensor), physical contact, comparability, digital and sensors and in terms of their signal detection toward physical, chemical, thermal and biological quantities. For instance, through biological sensors or biosensors, interactions of deoxyribonucleic acid (DNA), antibody or antigen or other biological components can be detected.

13.2.2 Biosensor

A device or probe which utilizes biological components in generating measurable component is known as a biosensor (Naresh and Lee 2021). These biosensors are available in a variety of dimensions and formations, and they can distinguish and estimate even low amounts of infections, harmful chemicals and pH values (Naresh

Table 13.1 Component of a biosensor

Components	Description
Analyte	Material of interest in which the contents are being detected such as alcohol, ammonia and glucose
Bioreceptor	Biological molecule that detects targeted substrate such as DNA or antibodies
Transducer	Important part in biosensor that converts energy from one form to another form
Electronics	Received electrical signals from transducers and transforms into the display unit
Display	Display the interpretation of the system and generate results of the biosensor. The display generates output signal in terms of numeric, graphic or tabular

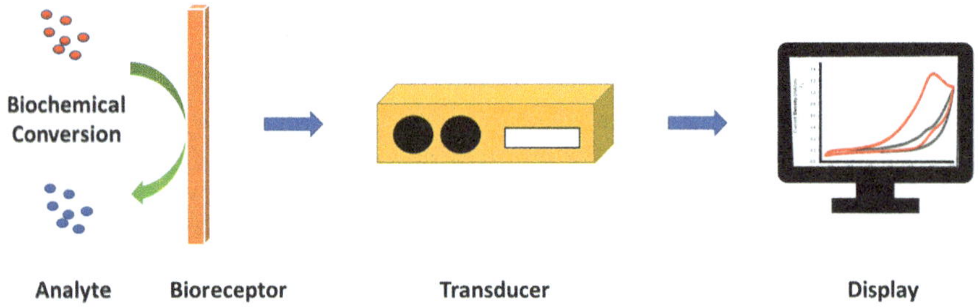

Fig. 13.1 Schematic diagram of biosensor

and Lee 2021). A biosensor consists of several components such as analyte, bioreceptor, transducer, electronics and display (Naresh and Lee 2021). Table 13.1 shows detailed explanations of the components, while a schematic diagram for a biosensor is shown in Fig. 13.1.

An ideal biosensor should possess certain requirements to achieve better performance for commercialized uses. One of the key features of biosensors is having a good selectivity to ease detection of target analyte molecule by bioreceptors in a sample that is compromised of unwanted contaminants. Aside from that, the biosensor needs to be sensitive toward the targeted analyte especially during detection of analyte in low concentrations or minimum number of steps while having ideal stability as most of biosensors require continuous monitoring (Grieshaber et al. 2008). Biosensors can be classified according to the primary component in biosensor construction; bioreceptor and transducer. Biosensors are divided into four categories according to the type of bioreceptor they use: enzymatic biosensors, immunosensors, nucleic acid-based biosensors and microbial sensors. The second division of biosensors is based on the transducer: electrochemical biosensor (further divided into potentiometric, amperometric, impedance and conductometric groups), electronic biosensor, thermal biosensor and optical biosensors (Naresh and Lee 2021; Biswas et al. 2017).

13.2.3 Biosensors Based on Bioreceptors

13.2.3.1 Enzymes-Based Biosensors

Enzymes are a conventional biocatalyst, and they are effective in boosting the biological reaction rate. This catalytic reaction and binding properties for the targeted analyte detection highly affected the function of enzyme-based sensors. The inhibition or catalysis of enzymes by the target analyte can be measured by enzymatic biosensors through several possible mechanisms (Naresh and Lee 2021; Justino et al. 2015):

(a) measurement for analyte's catalytic transformation by the enzyme determined the analyte concentration since the analyte can be metabolized by enzyme
(b) concentration of analyte is associated with reduction in formation of enzymatic production due to the enzyme can be inhibited or inhibited by the analyte
(c) monitoring of changes in enzyme properties

Since the development of enzymatic sensors in 1962 by Clark and Lyons, various biosensors are produced based on the enzyme specificity. However due to high cost and intrinsic nature of the enzyme that is sensitive, the stability, sensitivity and adaptability of this sensors are considered complicated (Naresh and Lee 2021; Justino et al. 2015). Glucose and urea biosensors are examples of the common enzyme-based biosensors. Recent studies integrate enzymes with nanomaterials which caused the high use of enzymes as recognition elements in biosensors.

13.2.3.2 Antibody-Based Biosensors

Antibodies are affinity biorecognition elements that were employed for more than two decades due to their broad application range such as environmental monitoring, medical diagnosis, food safety and clinical analysis and robust antigen–antibody interactions (Naresh and Lee 2021; Justino et al. 2015). Antibodies consist of "*Y*" formation of immunoglobulins (Ig), with two heavy and two light polypeptide chains joined by the disulfide bonds. Based on the difference in heavy chains, five kinds of antibodies have been defined: IgG, IgM, IgA, IgD and IgE antibodies (Schroeder and Cavacini 2010). Antibody implanted on biosensors that act as a ligand or perform on the antibody-antigen interaction is known as immunosensors and can be categorized as non-labeled and labeled. Detection of antigen–antibody complex is known from the calculation of physical changes due to complex's formation, and this system is defined as non-labeled immunosensors. On the other hand, labeled immunosensors are used for analysis of the antigen–antibody complex (Lima and Ahmed 2019; Lim and Ahmed 2016). Typically, estimation of proteins, biological toxins and biomarkers in numerous research domains such as food safety, clinical and environmental monitoring has been done by immunosensors (Willander et al. 2014).

13.2.3.3 Aptamer-Based Biosensors

Aptamers, single-stranded oligonucleotides (usually DNA or RNA) or peptides, can attach to their targets with high degree of specificity and affinity and can be formed into two-dimensional (2D) and three-dimensional (3D) structures (Naresh and Lee 2021; Li et al. 2019). Through these structures, it leads to high density of surface and reduced blocking of space which results in the high-binding performance of the target. Through an in vitro selection mechanisms known as systematic evolutions of Ligands by exponential enrichment (SELEX), the process is used for isolating oligonucleotide probes from wide libraries of oligonucleotides (usually RNA or single-stranded DNA) (Li et al. 2019). These biosensors are further divided into labeled and label-free aptasensors based on various transduction techniques. An example of a label-free aptasensor is surface plasmon resonance (SPR), while fluorescent dyes are an example for a label-based sensor. According to Shen et al., aptamers can be generated and screened against a wide range of targets including peptides, proteins, entire cell, small molecules and metal ions and are frequently between 25 and 90 bases long (Shen et al. 2014). Owing to the characteristics of nucleic acid of aptamers, the sensors can withstand various temperature ranges and conditions of storage, and importantly, they are functionally stable (Naresh and Lee 2021). In comparison with molecular recognition elements (MRE) such as antibodies which need a biological system to be formed, aptamers can be chemically synthesized and are stable in pH range of 2–12, and they have the potential for thermal refolding (Labib et al. 2016; Zhou et al. 2014). Another pro of aptamers is they are capable to be chemically altered in accordance with the target molecule's detection criteria.

13.2.3.4 Whole Cells-Based Biosensors

Bacteria, fungus, molds, algae, viruses and protozoa are used in designing of whole cell-based biosensors due to possession of potential biorecognition elements. Since they are self-replicating, this sensor is capable in producing recognition elements (e.g., antibodies) with the absence of extraction and purification (Gui et al. 2017; Kylilis et al. 2019). Analytical signals are recognized by analyzing the general metabolic of biological agents for whole cell-based biosensors which includes cell respiration or bacterial bioluminescence, growth inhibition, cell viability, and substrate. In comparison with enzymes, whole cell-based sensors are preferable as they reveal a high stability, low needs in purification, inexpensive production and capability detection (Carvalho 2011). With the unique characteristics possessed by these biosensors, they are being employed for environmental analysis such as for detection of heavy metals, pesticides, organic contaminants, for food analysis and drug screening (Berepiki et al. 2020).

13.2.4 Biosensors Based on Transducer

13.2.4.1 Optical Sensors

Optical glucose sensors employ fiber optics for the detection of analytes through absorption, illumination or light scattering and include various optical methods such as absorptiometry, fluorescence and surface plasmon resonance (SPR) (Bruen et al. 2017; Ju et al. 2020). An optical biosensor works by producing signals that are inversely proportional to the analyte's concentration and offers label-free and in the moment parallel detection. The transduction process in optical biosensors leads to change in absorption, reflection, transmission refraction, phase, frequency and light polarization due to physical or chemical changes produced by biorecognition elements (Naresh and Lee 2021). Enzymes, antibodies, tissues, whole cells and aptamers are examples of biorecognition components in optical biosensors. Through optical biosensors, they offer rapid reaction time, cost-saving, free electrical interference and distant sensing (Hassan et al. 2021). However, optical biosensors have their drawbacks including light interference, the narrow concentration range and necessity for powerful light sources (Kausar et al. 2015).

13.2.4.2 Thermal Sensors

Thermal biosensors employed characteristics such measuring the amount of energy consumed or emitted (exothermic or endothermic) during the process. Measurement of heat change that is directly observed for extent of process calculation for catalyst or dynamic structure of biomolecules in the dissolved state is monitored by this sensor (Danielsson 1991). Thermal biosensor can overcome the shortcomings of calorimetric-based biosensors that have long experimental procedures and are unable to specify the temperature measurement. Thermal biosensors typically utilize an immobilized enzyme reactor and flow injection analysis method, with distinctive temperature analysis across the enzyme reactor. For the conversion of a specific substrate to product, thermistors or thermopiles, common temperature sensors, are positioned across an enzyme column packed with mobilized enzymes. The heat signals produced by the catalytic reaction are inversely proportional to the substrate concentration (Sivarajasekar et al. 2019). Detection of thermometric conductivity is advantageous especially for multiple reactions since they give the sum of all enthalpies for determination of the assay sensitivity (Naresh and Lee 2021). Currently, microelectromechanical (MEMS) thermal sensors have garnered attention in monitoring metabolic applications on the basis for detection of temperature. Inexpensive integration of miniaturized devices and inexpensive batch fabrication is the benefits of these sensors. It was observed that the sensor exhibited excellent performance by improving thermal isolation, low thermal mass, high sensitivity, less consumption of power and wide linear range (Wang et al. 2008).

13.3 Electrochemical Sensors

To determine the information from the biological systems, the measurement of electrical properties is usually done electrochemically. Differs to biosensors, electrochemical detection technique is based on enzymes which may be affected from their biocatalytic activity and their binding capabilities (Naresh and Lee 2021). Typically, electrochemical sensors consist of three important components: receptor, analyte and transducer for conversion of reaction to measurable signal. During the reaction on the transducer surface, measurable current (amperometric), charge accumulation (potentiometric) or measurable in characteristic changes of conductive material (conductometric) between electrodes may be generated (Fig. 13.2). The surface of the electrode is commonly used for reaction site in which oxidation or reduction of analyte will occur. From the reaction, the current or voltage produced will be monitored to obtain data of the analyte such as the concentration of the sample. Therefore, the selection of electrode surface in terms of material, surface modification or dimensions plays a vital roles in their detection capability (Grieshaber et al. 2008). An electrochemical sensor is illustrated as shown in Fig. 13.3.

The electrodes in electrochemical sensing are known as reference electrode and usually are silver/silver chloride (Ag/AgCl), counter electrode and working electrode. The counter electrode acts as the supplier of currents toward the working electrode, while the working electrode on the other hand works as the transduction elements in the biochemical reaction. To ensure these electrodes result in excellent sensing performance, they need to be stable and conductive. The use of nanoparticles in the sensors are one of the alteration in increasing the sensor's accuracy and sensitivity while improving their thermal and electrical properties (Faridbod et al. 2011). Various studies have been conducted to explore ultra-sensitivity detection, mass production capability, and low-cost manufacturing electrochemical sensors (Abdel-Karim et al. 2020). Inexpensive nanostructured transition metals oxides (TMOs) such as copper oxide (CuO), zinc oxide (ZnO), manganese (IV) oxide (MnO_2) and iron oxide (Fe_2O_3) were made into electrochemical electrode to detect the analyte for their large surface area, selective and sensitive electrocatalytic activity and biocompatibility (Wang et al. 2018).

13.3.1 Potentiometric Biosensors

13.3.1.1 Working Principle

Combination of a biorecognition element with a transducer that senses changes in protons quantity, the accumulated analytical signal correlated with the concentration of the analyte and produces potentiometric biosensors. Typically, measurements of potentiometric are applied in determining organic and inorganic species, pesticides, ammonia and carbon dioxide (Pisoschi 2016). Information on the ion activity during electrochemical reaction is provided by potentiometry. Using the Nernst equation, the relationship of concentration and the charge potential can be observed. The equation is as follow (Skoog et al. 2016; Simões and Xavier 2017):

$$\text{EMF or } E_{\text{cell}} = E^{\circ}_{\text{cell}} - \frac{\text{RT}}{nF} \ln Q \tag{13.1}$$

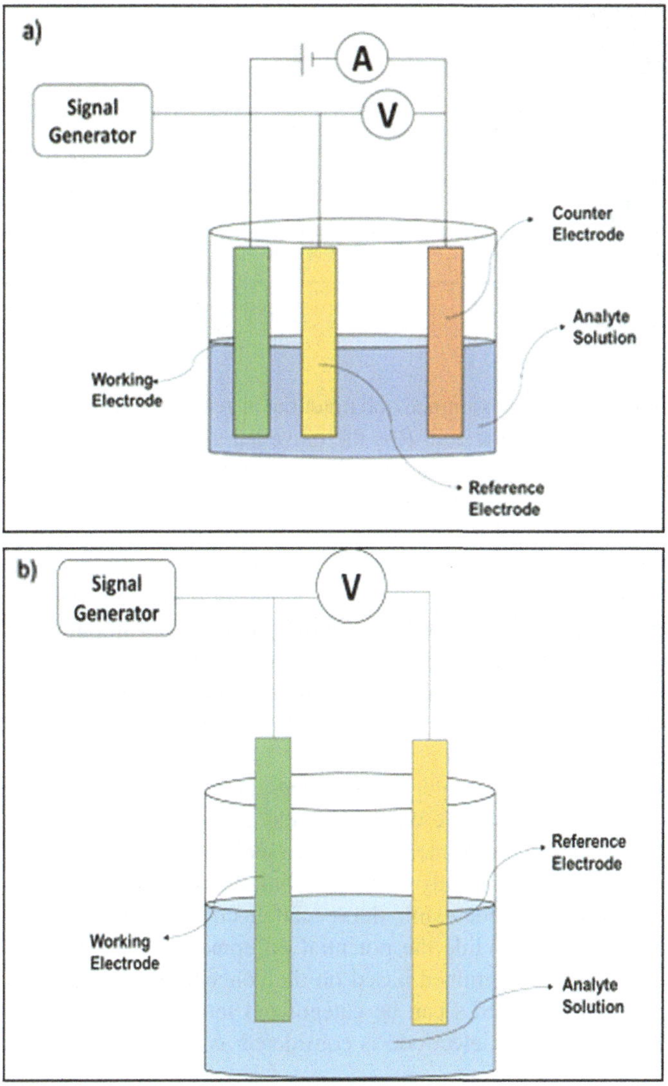

Fig. 13.2 Experimental setup for **a** potentiometric sensor, **b** amperometric sensor and **c** conductometric sensor

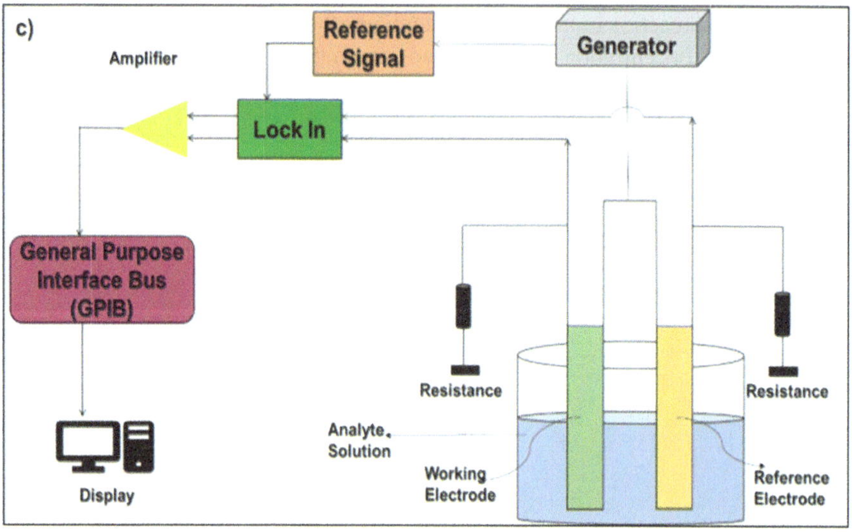

Fig. 13.2 (continued)

where EMF or E_{cell} is the examined cell potential at zero current, E°_{cell} is the constant potential contribution to the cell, R is the universal gas constant, T is the absolute temperature (Kelvin), F is the Faraday constant and Q is the ratio of ion concentration at the anode to ion concentration at the cathode.

Ion-selective electrodes for potentiometric devices are often capable in detecting the lowest limit of detection in ranges between 10^{-8} and 10^{-11} M. It is worth mentioning that potentiometric sensors are useful in observing low concentration in a limited number of samples without altering the sample. Although low detection limits are possible with potentiometric devices, detection of analytes such as nickel, mercury, nickel and arsenate ions is still limited. Potentiometry is also known as potentiometric titration which is an alternative technique in electrically determining the point in biochemical reaction.

Potentiometric consists of three basic devices, namely as ion-selective electrodes (ISE), coated wire electrodes (CWES) and field effect transistors (FETS). IES are capable in assessing the movement of certain analyte selectively (Stradiotto et al. 2003). The electrodes are usually membrane-based devices in which one electrode acts as the working electrode, while the second electrode is the reference electrode (Stradiotto et al. 2003). While the potential difference of the reference electrode is constant, it can be determined based on the concentration of the dissolved ion (Stradiotto et al. 2003). ISEs can be categorized into three groups; glass, liquid or solid, and the glass pH electrode is considered as the simplest transducer, and

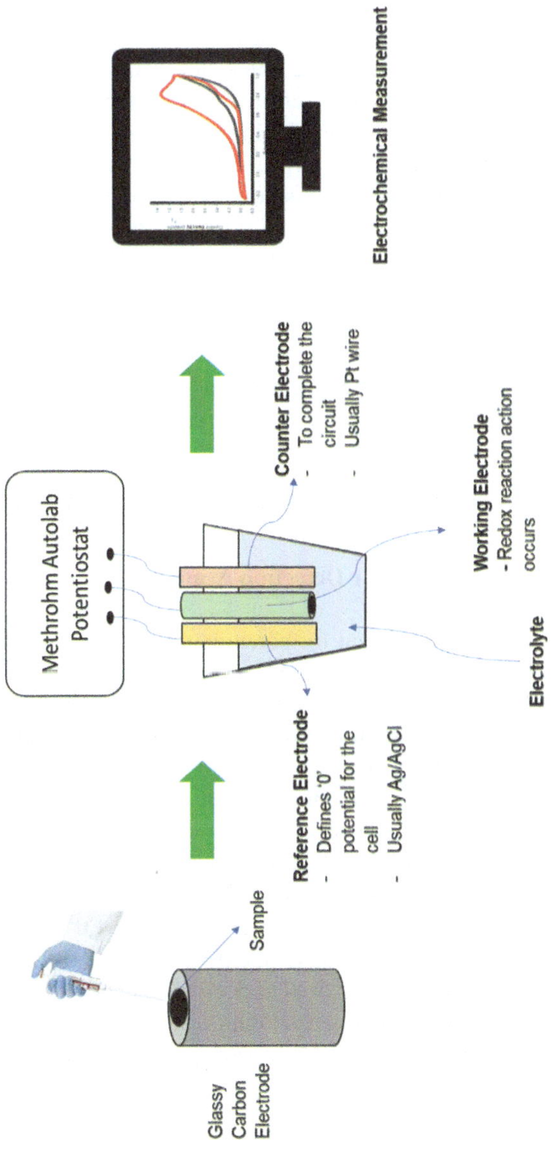

Fig. 13.3 Schematic diagram of electrochemical measurement

they can be found in assorted sizes and shapes. Through this pH electrode, the potentiometric sensors for glucose oxidase immobilization made of cellophane, nylon or nitrocellulose membranes are capable in detecting glucose in fruit juices with a linear range of 10^{-4} to 5×10^{-2} M. Despite the successfulness of the glass membrane electrode in detecting solution's pH, the limitation toward measurement in aqueous media has become a drawback for this electrode.

CWEs on the other hand fabricated the conductor with a polymer membrane in designing the electrode system that is sensitive toward the concentration of the electrolyte. Having similar response with IES in regards to detectability and range of concentration, CWEs overcome miniaturization of electrodes by eliminating the internal reference electrode, making them useful in vitro and in vivo analysis (Stradiotto et al. 2003). FETs are an extension of CWEs, and they have gained interest among researchers for their built-in potential in easing miniaturization (Menon et al. 2020). FETs semiconductor consists of three terminals; source, drain and gate in which the dielectric material gate terminal is modified with the targeted bioreceptors. Changes in surface-charge on the gate terminal through charged biomolecules (such as DNA) cause alteration of the gate voltage and thus the charge transport capabilities of the FETs channel (Menon et al. 2020; Pachauri and Ingebrandt 2016). Flaws such as semiconductor impurities become a constraint of FET-based sensors (Lee et al. 2009).

13.3.1.2 Potentiometric Sensor for Detection Application

Having fast response time is important for potentiometric sensors. To observe the response time for detection of diclofenac, a non-narcotic pain reliever, Kormosh and co-workers developed potentiometric sensors in detecting the drug, and it was observed that the sensor exhibited short response time of 2–3 s, wide concentration of 5.0×10^{-5} to 5.0×10^{-2} and range of working pH of 9.0–12.0. With the excellent performance shown by the modified sensor, pharmaceutical preparations of Diclofenac were determined (Kormosh et al. 2009). Another work by Tumur et al. (2020) a PVC-membrane potentiometric sensor was proposed for determination of Cysteamine. The developed sensors show an excellent performance with time of response of 5 s, linear concentration range of 1.0×10^{-5} to 1.0×10^{-1} mol L^{-1}, limit of detection 1.0×10^{-5} mol L^{-1} and wide pH range of 3.0–7.0. Aside from that, Gupta and colleague prepared a cobalt (II)-selective potentiometric sensor based on bridge modified calixarene. With detection limit of 0.3 ppm, wide concentration range of 5.3×10^{-6} to 1.0×10^{-1} mol L^{-1} and response time of 10 s, the developed sensor shows potential application in analyzing cobalt in real samples such as effluent and beer samples (Gupta et al. 2008).

13.3.2 Amperometric Biosensors

13.3.2.1 Working Principle

When the current is measured at constant potential, it is referred as *amperometry*, and if the current is analyzed at different controlled potential, it is referred as *voltammetry* (Grieshaber et al. 2008). Since the background charging current from different voltage is prohibited, amperometry is generally thought to be a more sensitive technique compared to voltammetry (Ryu et al. 2020). When a functioning electrode is subjected to a constant voltage, the current is measured. The target concentration relates to the cathodic or anodic current, resulting from the reduction or oxidation reaction by the applied voltage within the surface of electrode. To obtain the reduction or oxidation potential of the analyte, cyclic voltammetry is suggested (Hanrahan et al. 2004; Bard 2001). The target analyte is measured by the charge transfer at the interface. When stepped potential is given to the working electrode for a variety of applications, chronoamperometry is a highly effective sensing technique since it measures the current at the electrode as a function of time (Ryu et al. 2020). The analyte concentration is illustrated via the corresponding current change. The working electrode highly affected the performance of the amperometric sensors which resulted in various research of materials for electrode fabrication and maintenance including mercury, noble metals and carbons (Stradiotto et al. 2003). Among the suggested materials, carbons and noble metals have been suggested for fabrication of the electrode, and they are inexpensive, low background current, chemical inertness and capable to be used in various sensing and detection applications (Stradiotto et al. 2003). Nevertheless, this technique is effective in monitoring diffusion-controlled processes, and they are highly sensitive (Ryu et al. 2020).

13.3.2.2 Amperometric Sensor for Detection Application

Amperometric sensors have been developed for the usage of various fields and application especially for sensing industry by modifying the electrode for several conditions to enhance the electrode sensitivity. Yang et al. proposed a modified Fe_2O_3 nanoparticles electrochemical sensor supported by N-doped graphene for electrocatalytic oxidation and detection of L-Cys. The modified electrode exhibits effective electrocatalytic activity in oxidizing L-Cys in 0.1 M PBS at pH of 7. It was observed that the electrocatalytic current results in a direct increase with the concentration of L-Cys ranging within 0.2–400 μM, and the least detectable concentration is 0.1 μM (Yang et al. 2017). Another work of Zhad et al., proposed a silver carbon electrode (Ag/CE) and silver-silver electrode (Ag/AgE) ultramicroelectrodes (UMEs) for evaluation of the electrodes toward the reduction of nitrate. Deposit of nickel silver (AgNS) was done by pulse amperometry, while the deposition problem was overcome by ammonia acting as the complexation agent. It was observed that Ag/AgE exhibits a higher sensitivity and lower detection limit compared to Ag/CE. Through this research, it can be concluded that Ag-modified UMEs has high potential application for real-time monitoring of nitrate in the environment (Zhad and Lai 2015).

13.3.3 Conductometric Biosensors

13.3.3.1 Working Principle

Conductometric sensors depend on the electric changes in conduction of a film or a bulk material where the analyte present affected the conductivity (Stradiotto et al. 2003; Ryu et al. 2020). A small amplitude alternating current signal is applied within a wide range of frequencies to disrupt the electrode surface and measures the corresponding electrical responses, usually impedance. Impedance measurement offers information of capacitance, inductance or mass diffusion (Ryu et al. 2020). These data are then utilized in analyzing the electron charge transfer, electrolyte resistance, electroactive chemical adsorption and analyte mass transfer. This sensor is particularly useful for monitoring of target analyte adsorption on the surface of the electrode (Ryu et al. 2020). Due to the accumulation on the electrode surface, certain analytes are not capable in generating electrochemical reactions at which then impedance measurement is one of the alternative methods to overcome this drawback. Aside from that, impedance measurement can be utilized to investigate surface attributes and acquire a motion concentration correlation at various analytes (Stradiotto et al. 2003; Macdonald and Andreas 2014). In association with enzymes, changes in conductivity and ionic strength of a solution between two electrodes are the result of enzymatic reaction (Grieshaber et al. 2008). Due to that, conductometric sensors are used in enzymatic reactions study which generate variations in the concentration of charged species in a solution.

13.3.3.2 Conductometric Sensor for Detection Application

Saipiana and co-workers designed a hybrid of calixarene and the ionophore 18-crown-6, labeled 25,27-di-(5-thio-octyloxy) calix (Naresh and Lee 2021) arene-crown-6 conductometric sensor for detection of ammonium in aqueous solution. The combination of calixarenes and 18-crown-6 allows them to detect ammonium cations efficiently and highly selective. Analytical properties of the modified electrode such as their stability, selectivity toward ammonium and performance in different buffers were investigated, and it is best to mention, the modified electrode was found to be a promising instrument for detection of ammonium (Saiapina et al. 2012). Another work of Wu et al. developed a cuprous oxide (Cu_2O)/carbon nanotubes (CNTs) conductometric sensor for detecting nitrite in water. A simple hydrothermal method was employed for preparation of Cu_2O and CNTs for improvement of the sensor sensitivity. Electrochemical impedance spectroscopy (EIS) was used in obtaining the measurement for the detection mechanism due to the finding of electron transport on the electrode because of nitrite addition. Changes in charge transfer resistance before and after nitrite were observed which can be presumed that the higher the content of nitrite, the higher the changes in the charge transfer resistance. Wu et al. then concluded that the proposed sensor was able to detect a broad range concentration of nitrite, and it is highly sensitive (Wu et al. 2018). The performance for potentiometric, amperometic and conductometric sensors is summarized in Table 13.2.

13.4 Application of Electrochemical Sensors

Electrochemical sensors have been used for various applications as illustrated in Fig. 13.4.

13.4.1 Electrochemical Biosensors for Glucose Detection

To this date, electrochemical glucose biosensors have been widely used in glucose detection especially in diabetes diagnosis, providing a better life for diabetic patients. The use of electrochemical glucose sensors is still not well-known, but the previous research has proven that this biosensor is comparable to commercialized glucose sensors. In general, glucose biosensors can be categorized into two categories: enzymatic sensors and non-enzymatic sensors. Enzymatic sensors are depended on the features of GOx, where GOx is unstable, and the enzyme activities are affected by the environment surrounding such as humidity and temperature. While for non-enzymatic glucose sensors, they are based on the direct oxidation of glucose on the surface of an electrode with a transition metal center and excellent electrocatalytic activity (Hovancová et al. 2017). A reliable glucose sensor should have the following properties: selectivity, sensitivity, stability, accuracy, precision, inexpensive and easy fabrication. Glucose sensors should be sensitive to glucose with sensitivity a of 1 μA mM^{-1} cm^{-2} and having precision and accuracy within the error scale of $\pm$ 20% for 95% of the application time (Toghill and Compton 2010).

The use of metallic nanomaterials with different size of surface morphology enhanced the electrocatalytic activity due to the enlargement of active electrode surface area. Aside from structure and morphology of the electrode, the coating substance and pH value of the substance affected the performance of the sensor. Platinum (Pt) and gold (Au) are used as the electrode in detecting glucose in acidic, neutral and alkalic solutions, where glucose is unreactive in acidic solutions. It was also discovered that adsorption of chloride ions toward Pt surface was the weakest in alkaline solution, while it was the greatest in acidic solution (Hovancová et al. 2017; Tian et al. 2014). Nevertheless, there is a drawback for some of the glucose sensors, which chances in adsorbing poisonous compounds such as amino acids, uric acid

Table 13.2 Electrochemical sensor method for various detection of analyte

Analyte	Electrode modification	Detection techniques	Sensitivity, $\mu A\ mM^{-1}\ cm^{-2}$	Detection limit, μM	Linear range, mM	References
Nitrate	Cu electrode	Amperometry	–	4.2	0.1–2.5	Gamboa et al. (2009)
Nitrite	Cu-MOF/rGO/GCEs	Chronoamperometry	43.736	33 nM	3 μM–40 mM	Saraf et al. (2016)
Ammonium	Polymeric membrane-PpCO-SANE-modified Au microelectrode	Potentiometry	– /10 and 12 s	0.04 mM	1 μM–40 mM	Gallardo-Gonzalez et al. (2019)
Nitrate	Chitosan/bentonite nanocomposite-based ISE	Potentiometry	−54.6 mV/decade	–	20 mM–0.8 M	Mendoza et al. (2014)
Ammonium	Natural Zeolite Cli-noptilolite Au Interdigitated microelectrode (IDμE)	Conductometric	2.5 MmS/Mm/10 ± 2 s	0.01	0.01 μM–1 mM	Saiapina et al. (2012)
Nitrite	Cu_2O-CNTs/Au electrode	EIS	–	0.1 Nm–1 mM	0.0188 nM	Wu et al. (2018)
Glucose, uric acid	Cu/SPAuE	Amperometry	–	0.21	1–1800	Kurniawan et al. (2019)
Cysteine	Ag_3CuS_2	Amperometry	0.10–0.11 $\mu A\ mM^{-1}$	0.024–0.036	1–100 μM	Vladislavić et al. (2018)
Nitrobenzene	f–MWCNT/SPCE	Amperometry	0.6685	45	50 nM to 1170 μM	Govindasamy et al. (2016)
Enrofloxacin (ENR)	ENR-PVC, ENR-graphite coated electrode	Potentiometry	–	8.8×10^{-6} mol L^{-1}	1.0×10^{-2} to 1.0×10^{-5} mol L^{-1}	Salama et al. (2018)
L-Cys	MWCNT-FeTsPc	Amperometry	–	1 μM	10 μM to 0.2 mM	Devasenathipathy et al. (2015)

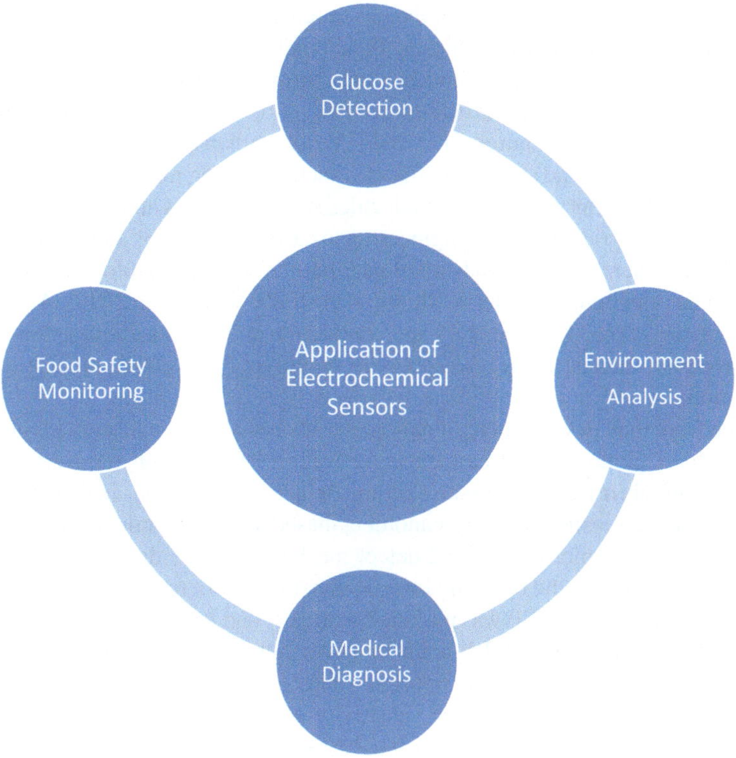

Fig. 13.4 Application of electrochemical sensors

(UA) or ascorbic acids (ACA) are high (Hovancová et al. 2017). These compounds can cling onto the electrode surface, thus preventing the glucose molecules from being adsorbed. Coating materials such as Nafion (Hovancová et al. 2017; Guang-Xian et al. 2015) and polyethylene glycol (Hovancová et al. 2017; Zaidi and Shin 2016) are utilized to minimize adsorption of unwanted toxic chemicals. With this modification, the performance of glucose sensors can be improved and painless as well as frequent blood glucose measurement can be developed.

13.4.2 Electrochemical Biosensors for Medical Diagnosis

Detection of biomolecules such as DNA, proteins, genetic disorders diagnosis and measurement of drug response was based on electrochemical biosensors. Electrochemical methods have been discussed as one of important analytical tool in examining biochemical behavior in the environment, providing parameters of thermodynamic and kinetic depending on their conditions. Carbon-based nanomaterials have gained interest in wound healing, anti-microbial applications and drug delivery, owing to their high surface-to-volume ratio, shape and structural compositions (Shinde et al. 2020; Laurent et al. 2014; Veerapandian and Yun 2013). NPs show potential in detection and staging of diseases such as cancer, making them reliable in detection of pathogens and other biomolecules such as proteins and nucleic acid. NPs are also used as carriers in tissue engineering to carry medications and molecules in a controlled manner. Antibiotic-adsorbed nanoparticles can be employed in scaffolds to overcome infections of bacteria (Shinde et al. 2020; Vieira et al. 2017). Antimicrobial properties owned by silver NPs and conduction and surface conjugation in gold NPs have gain interest from researchers in employing NPs for tissue engineering application.

Aside from that, development of nanomaterials has also been discussed for detection of blood clotting and treatments. Activation of thrombin (converts soluble fibrinogen to fibrin) causes blood clotting, which then can be analyzed through electrochemical sensing. Carbon-nanotubes-based electrochemical strain sensors are implanted into micropillars and detect mechanical force due to blood clotting droplets, at which then the resulting force is transmitted to a resistive nanotube strain sensor (Liu et al. 2018). GCE was modified with rGO and AuNPs to produce a molecularly imprinted polymer (MIP)-aptamer (Apt) hybrid-based thrombin sensor. In comparison with non-molecularly imprinted and non-aptamer molecularly imprinted sensors, MIP-Apt exhibits greater sensitivity and selectivity due to the hybrid recognition of MIP and Apt. The proposed MIP-Apt sensors demonstrated good linear range of 2.5×10^{-9} mg/mL to 1.3×10^{-6} mg/mL with LOD of 1.6×10^{-10} mg/mL (Yang et al. 2019). Another work of Cheng and co-worker modified the metal–organic framework with platinum particles in creating a dual-readout colorimetric/electrochemical aptamer based in detecting thrombin. The thrombin sensor shows low LOD, 0.33 fM and 0.17 pM for electrochemical and colorimetric method, respectively (Cheng et al. 2019).

A novel work of Wang and co-workers demonstrated a biosensor in detecting nucleic acid and residue removal by modifying nanoparticle-based biosensors (NB)- and antarctic thermal sensitive uracil-DNA-glycosylase (ATSU)-supplemented polymerase spiral reaction (PSR;NB-ATSU-PSR). Synthesis of nucleic acid and removal of residue is carried out in single pot; hence, the utilization of a closed-tube process can remove any unwanted results due to residue. Klebsiella pneumoniae is used in demonstrating the applicability of NB-ATSU-PSR assay, and it was observed the amplification products can be detected at as low as 100 fg genomioc DNAs and from ~ 550 colony-forming unit (CFU) in 1 mL of spiked sputum samples (Wang et al. 2019). The use of electrochemical technique has also been employed in tumor biomarker detection. Chang and co-authors designed a homogenous tumor biomarkers electrochemical biosensor with reference to metal–organic frameworks (MOFs). The functionalized MOFs were created by using UIO-66-NH_2 as nanocontainer to load electroactive dyes and sealing them with dsDNA. Two functionalized MOFs (MB@IUO and TMB@UIO) were utilized in detecting let-7a and miRNA-21 simultaneously. The proposed sensor results in detection limit of 3.6 and 8.2 for lt-7a and minRNA-21, lower than reported strategies focusing on single miRNAs detection. The developed sensor has been effectively detecting the target miRNAs spiked in a serum sample, making them as a reliable application in diagnosis of diseases and clinical biomedicine (Chang et al. 2019).

13.4.3 Electrochemical Biosensors for Environmental Analysis

Air pollution has become a serious problem, causing poor air quality that can be harmful for humankind and ecosystem. Transportation and industrial activity sometimes produce pollutants that exceed the limit enforced by the law, therefore it is crucial for air quality to be regularly monitored. Aside from that, water and soil contamination are the result of heavy metals that are bound in fossil fuels and water minerals and being released to the environment. Excess level of heavy metals in water and soil can cause denaturation to both human well-being and environment. Degradation in the environment is possible due to microbial degradation and chemical hydrolysis as tons of pesticides are used in agriculture and horticulture each year (Cosnier 2014). Intermediate products, more toxic than initial products, are generated during this process. With these problems occurring, portable, highly sensitive and low-cost analytical instruments are needed in measuring and detecting these compounds.

Environment friendly electrochemical sensors are commercially available with metal and metal oxide electrochemical sensors design arc applicd. Dctcction of cadmium in real life samples (soil, tea, vegetables, cow liver) was developed by fabricating an ion-selective electrode (ISE) with electro polymerization of 4-vinylpyridine on 2B pencil graphite as ionosphere for Cd^{2+}. The modified electrode shows excellent electrocatalytic activity with a wide range of 4.0–7.5 and detection limit of 2.5×10^{-8} M (Wee Ling et al. 2012). Detection of hydrazine (N_2H_4) was done by an AuNPs modified graphite pencil electrode (GPE). GPE was immersed in the ascorbic solution mixed with gold (III) chloride and undergoes heating for 15 min at 75 °C. AuNPs were attached directly onto GPE, exhibiting quantification limit of 100 nM and limit of detection 42 nM, using square-wave voltammetry as mode of detection (Abdul Aziz and Kawde 2013). Coupling of Ag/FeOOH was bound on GPE for amperometric detection of hydrogen peroxide. FeOOH nanomaterials with broad surface area can give superior support for AgNPs immobilization. After 3 weeks of modification, the proposed electrode was able to display excellent long-term stability by maintaining 90% of the signal and reached LOD of 22.8 mM. With the great attributes shown by the modified electrode, it was used in detecting hydrogen peroxide in disinfectants (Zhang and Zheng 2015).

Rana et al. pre-treated GPE in NaOH solution to charge the surface of GPE for detecting phenol. GPE was capable of self-electro polymerization of phenol on its surface upon charging, making it more sensitive toward phenol. A very low detection limit of 4.17 nM was attained by the pre-charged sensor (Rana and Kawde 2016). Another work by Fartas et al. developed a graphene-decorated gold nanoparticles/chitosan (Gr–Au–Chit/Tyr) nanocomposite-modified screen-printed carbon electrode (SPCE) for detection of phenol. The solution casting method was used to construct the nanocomposites films and observed the proposed sensor performance via differential pulse voltammetry (DPV) and cyclic voltammetry (CV). With sensitivity of 0.624 A/M and detection limit of 0.016 M ($S/N = 3$), the biosensor displays linearity toward phenol concentration within the range of 0.05–15 M. The biosensor demonstrates high repeatability, stability and selectivity within one month of usage (Fartas et al. 2017).

13.4.4 Electrochemical Biosensors for Food Monitoring

Issues on food safety are concerning, and almost every year, there will be cases of food poisoning or existence of illegal additives and ingredients used in the food and beverages production. Still, proper guidelines or enforcement on food safety are lacking, leading to foodborne diseases. The issue with foodborne diseases is that they are affecting both health and economical systems of a country. Proper analytical techniques are needed leading to the use of electrochemical methods to ease detection of contaminants and additives in a large-scale food and beverages production. Like other industries, simple, easy operation and highly sensitive analytical methods are required to analyze while improving the quality and safety in food production.

Excessive number of inedible substances in such as melamine, clenbuterol and Sudan 1 (food dye) affected the safety of the food products. For instance, melamine-contaminated milk leads to kidney failures for infants. MIL-53@XC-72 nanohybrid modified GCE was developed to detect melamine in the dairy products, exhibiting sensitive and selective performance with linear range of 0.04–10 μM and LOD of 0.005 μM ($S/N = 3$) (Zhang et al. 2016). On the other hand, Sudan 1 in chili powder and ketchup was determined by AuNPs/rGO electrode. The modified sensors have proven satisfactory selectivity and sensitivity (Li et al. 2015). Caffeine, an active alkaloid component, is present in coffee-based, teas and in soft drinks. Excessive consumption of caffeine can cause headaches and high heart rate. AuNPs synthesized with chitosan were used in detecting caffeine. The synthesized AuNPS-chitosan reduced the oxalic acid in the caffeine, exhibiting a wide detection range of 5.0 μM to 50.0 Mm and limit of detection of 1.0 μM (Trani et al. 2017). Carbon-based sensors are also applied in food industry, for instance in detecting ascorbic acid (AA). AA acts as an antioxidant in fighting free-radical induced diseases and development of collagen. As graphene is actively used in the electrochemical method, a novel graphene/platinum (Pt) electrochemical sensor was developed by Sun et al. (2011). With the properties owned by graphene, it helps in uniformly distribute Pt nanoparticles thus increases the electrocatalytic activity. The modified graphene/Pt sensor has a good performance with detection limit of 0.15 μM and calibration range in between 0.15 and 34.40 μM.

Electrochemical sensors are used in detecting microbial pathogens such as bacteria, viruses and protozoa that may be present during the production of food products, or sometimes, they are used in the food products itself. A new electrochemical biosensor for detection of Listeria *monocytogenes* (LM) is based on mouse monoclonal antibody adsorbed on self-assembled monolayers (SAM)-modified gold (Au) electrodes. LM in milk can be detected without sample pre-treatment, making it a reliable and simple technique for LM detection with easy-ease operation and high sensitivity at low cost (Cheng et al. 2014). *Salmonella typhimurium* was detected an with iron/gold core/shell nanoparticles (Fe@Au) electrochemical biosensor, conjugated with anti-*Salmonella*. Square-wave anodic stripping voltammetry with the use of CdS nanocrystals determined *S. typhimurium* on screen-printed carbon electrodes. The modified sensor achieved a calibration curve of 1×10^1 and 1×10^6 cells/mL and LOD of 13 cells/mL. Detection of bacteria in dairy products at low concentration is possible with the modified method, and effective for fast (less than 1 h) exposure of *S. typhimurium* in the physical sample (Freitas et al. 2014).

13.5 Conclusion

Electrochemical sensors are becoming an increasingly important analytical instrument due to the growth of demand for sensitive, fast and selective analyte detection. In comparison with spectroscopic and chromatographic devices, electrochemical biosensors can be easily altered in detection of a wide range of analytes while remaining affordable. Aside from that, the sensors can be integrated into portable, sturdy or miniaturized devices according to specific applications. For environmental analysis, electrochemical sensors are capable to track down onsite pollution and meet the environmental requirements. Relevant information at different steps for ease of decision-making process is obtained due to control checking for different points of measurement. Strategies for materials incorporation of nanomaterials varying from noble metals, carbon-based and inexpensive metal oxides for the electrochemical electrode in enhancing the speed, sensitivity and selectivity of the sensors have piqued the researchers' interest. Development of nanotechnology and custom engineering biorecognition components promotes the development of reliable and useful sensors and electrochemical sensors. Combination of multifunctional materials, recognition elements and electrochemical techniques improves the selectivity, sensitivity, stability and reproducibility of the sensors while promoting growth of assays and bioassays sensors. Typically, the fields of electrochemical sensors are still growing and continues to find new areas for various implementations.

Acknowledgements The authors would like to acknowledge Center for Women Advancement and Leadership (CWAL) for the opportunity given and FRGS for the fundamental support.

References

Abdel-Karim R, Reda Y, Abdel-Fattah A (2020) Review—nanostructured materials-based nanosensors. J Electrochem Soc 167(3):037554

Abdul Aziz M, Kawde AN (2013) Gold nanoparticle-modified graphite pencil electrode for the high-sensitivity detection of hydrazine. Talanta 115:214–221

Alam AU, Clyne D, Jin H, Hu NX, Deen MJ (2020) Fully integrated, simple, and low-cost electrochemical sensor array for in situ water quality monitoring. ACS Sens 5(2):412–422

Baracu AM, Dinu Gugoasa LA (2021) Review—recent advances in microfabrication, design and applications of amperometric sensors and biosensors. J Electrochem Soc 168(3):037503

Bard AJ, Faulkner LR (2001) Electrochemical methods: fundamentals and applications. Wiley

Berepiki A, Kent R, MacHado LFM, Dixon N (2020) Development of high-performance whole cell biosensors aided by statistical modeling. ACS Synth Biol 9(3):576–589

Biswas P, Karn AK, Balasubramanian P, Kale PG (2017) Biosensor for detection of dissolved chromium in potable water: a review. Biosens Bioelectron 94:589–604

Bruen D, Delaney C, Florea L, Diamond D (2017) Glucose sensing for diabetes monitoring: recent developments. Sensors (Switzerland) 17(8):1–21

Chang J, Wang X, Wang J, Li H, Li F (2019) Nucleic acid-functionalized metal-organic framework-based homogeneous electrochemical biosensor for simultaneous detection of multiple tumor biomarkers. Anal Chem 91(5):3604–3610

Chaubey A, Malhotra BD (2002) Mediated biosensors. Biosens Bioelectron 17(6–7):441–456

Cheng C, Peng Y, Bai J, Zhang X, Liu Y, Fan X, Ning B, Gao Z (2014) Rapid detection of Listeria monocytogenes in milk by self-assembled electrochemical immunosensor. Sens Actuat B Chem 190:900–906

Cheng T, Li X, Huang P, Wang H, Wang M, Yang W (2019) Colorimetric and electrochemical (dual) thrombin assay based on the use of a platinum nanoparticle modified metal-organic framework (type Fe-MIL-88) acting as a peroxidase mimic. Microchim Acta 186(2):6–13

Cosnier S (2014) Electrochemical nucleic acid aptamer–based biosensors. In: Electrochemical biosensors

Danielsson B (1991) Calorimetric biosensors. Biochem Soc Trans 19(1):26–28

De Carvalho CCCR (2011) Enzymatic and whole cell catalysis: finding new strategies for old processes. Biotechnol Adv 29(1):75–83

Devasenathipathy R, Mani V, Chen SM, Manibalan K, Huang ST (2015) Determination of 4-nitrophenol at iron phthalocyanine decorated graphene nanosheets film modified electrode. Int J Electrochem Sci 10(2):1384–1392

Faridbod F, Gupta VK, Zamani HA (2011) Electrochemical sensors and biosensors. Int J Electrochem 2011:1–2

Fartas FM, Abdullah J, Yusof NA, Sulaiman Y, Saiman MI (2017) Biosensor based on tyrosinase immobilized on graphene-decorated gold nanoparticle/chitosan for phenolic detection in aqueous. Sensors (Switzerland) 17(5)

Freitas M, Viswanathan S, Nouws HPA, Oliveira MBPP, Delerue-Matos C (2014) Iron oxide/gold core/shell nanomagnetic probes and CdS biolabels for amplified electrochemical immunosensing of *Salmonella typhimurium*. Biosens Bioelectron 51:195–200

Gallardo-Gonzalez J, Baraket A, Boudjaoui S, Metzner T, Hauser F, Rößler T, Krause S, Zine N, Streklas A, Alcácer A, Bausells J, Errachid A (2019) A fully integrated passive microfluidic Lab-on-a-Chip for real-time electrochemical detection of ammonium: Sewage applications. Sci Total Environ 653:1223–1230

Gamboa JCM, Peña RC, Paixão TRLC, Bertotti M (2009) A renewable copper electrode as an amperometric flow detector for nitrate determination in mineral water and soft drink samples. Talanta 80(2):581–585

Govindasamy M, Mani V, Chen SM, Subramani B, Devasenathipathy R, Tamilarasan S (2016) Highly sensitive amperometric sensor for nitrobenzene detection using functionalized multiwalled-carbon nanotubes modified screen printed carbon electrode. Int J Electrochem Sci 11(12):10837–10846

Grieshaber D, MacKenzie R, Vörös J, Reimhult E (2008) Electrochemical biosensors—sensor principles and architectures. Sensors 8(3):1400–1458

Guang-Xian Z, Zhang W-X, Sun Y-M, Wei Y-Q, Lei Y, Peng H-P, Liu A-L, Chen Y-Z, Lin X-H (2015) A nonenzymatic amperometric glucose sensor based on three dimensional nano- structure gold electrode. Sens Actuat B Chem 212:72–77

Gugoasa LA, Stefan-van Staden RI, van Staden JF, Coroş M, Pruneanu S (2019) Electrochemical determination of bisphenol A in saliva by a novel three-dimensional (3D) printed gold-reduced graphene oxide (rGO) composite paste electrode. Anal Lett 52(16):2583–2606

Gui Q, Lawson T, Shan S, Yan L, Liu Y (2017) The application of whole cell-based biosensors for use in environmental analysis and in medical diagnostics. Sensors (Switzerland) 17(7):1–17

Gupta VK, Jain AK, Al KM, Bhargava SK, Raisoni JR (2008) Electroanalytical studies on cobalt(II) selective potentiometric sensor based on bridge modified calixarene in poly(vinyl chloride). Electrochim Acta 53(16):5409–5414

Hanrahan G, Patil DG, Wang J (2004) Electrochemical sensors for environmental monitoring: design, development and applications. J Environ Monit 6(8):657–664

Hassan MH, Vyas C, Grieve B (2021) Recent advances in enzymatic and non-enzymatic electrochemical glucose sensing

Hovancová J, Šišoláková I, Oriňaková R, Oriňak A (2017) Nanomaterial-based electrochemical sensors for detection of glucose and insulin. J Solid State Electrochem 21(8):2147–2166

Ju J, Hsieh CM, Tian Y, Kang J, Chia R, Chang H, Bai Y, Xu C, Wang X, Liu Q (2020) Surface enhanced Raman Spectroscopy based biosensor with a microneedle array for minimally invasive in vivo glucose measurements. ACS Sens 5(6):1777–1785

Justino CIL, Freitas AC, Pereira R, Duarte AC, Rocha Santos TAP (2015) Recent developments in recognition elements for chemical sensors and biosensors. TrAC Trends Anal Chem 68:2–17

Karimi-Maleh H, Karimi F, Alizadeh M, Sanati AL (2020) Electrochemical sensors, a bright future in the fabrication of portable kits in analytical systems. Chem Rec 20(7):682–769

Kasemo B (2002) Biol Surf Sci 500:1–22

Kausar ASMZ, Reza AW, Latef TA, Ullah MH, Karim ME (2015) Optical nano antennas: state of the art, scope and challenges as a biosensor along with human exposure to nano-toxicology. Sensors (Switzerland) 15(4):8787–8831

Kormosh ZA, Hunka IP, Bazel YR (2009) A potentiometric sensor for the determination of diclofenac. J Anal Chem 64(8):853–858

Kurniawan A, Kurniawan F, Gunawan F, Chou SH, Wang MJ (2019) Disposable electrochemical sensor based on copper-electrodeposited screen-printed gold electrode and its application in sensing L-Cysteine. In: Electrochimica acta, vol 293. Elsevier Ltd

Kylilis N, Riangrungroj P, Lai HE, Salema V, Fernández LÁ, Stan GBV, Freemont PS, Polizzi KM (2019) Whole-cell biosensor with tunable limit of detection enables low-cost agglutination assays for medical diagnostic applications. ACS Sens 4(2):370–378

Labib M, Sargent EH, Kelley SO (2016) Electrochemical methods for the analysis of clinically relevant biomolecules. Chem Rev 116(16):9001–9090

Laurent S, Saei AA, Behzadi S, Panahifar A, Mahmoudi M (2014) Superparamagnetic iron oxide nanoparticles for delivery of therapeutic agents: opportunities and challenges. Expert Opin Drug Deliv 11(9):1449–1470

Lee CS, Kyu Kim S, Kim M (2009) Ion-sensitive field-effect transistor for biological sensing. Sensors 9(9):7111–7131

Li J, Feng H, Li J, Feng Y, Zhang Y, Jiang J, Qian D (2015) Fabrication of gold nanoparticles-decorated reduced graphene oxide as a high performance electrochemical sensing platform for the detection of toxicant Sudan I. Electrochim Acta 167:226–236

Li F, Yu Z, Han X, Lai RY (2019) Electrochemical aptamer-based sensors for food and water analysis: a review. Anal Chim Acta 1051:1–23

Lim SA, Ahmed MU (2016) Electrochemical immunosensors and their recent nanomaterial-based signal amplification strategies: a review. RSC Adv 6(30):24995–25014

Lima SA, Ahmed MU (2019) Introduction to immunosensors. In: Detection science. Royal Society of Chemistry, pp 1–20

Liu Y, Jiang X, Cao W, Sun J, Gao F (2018) Detection of thrombin based on fluorescence energy transfer between semiconducting polymer dots and BHQ-labelled aptamers. Sensors (Switzerland) 18(2)

Macdonald MA, Andreas HA (2014) Method for equivalent circuit determination for electrochemical impedance spectroscopy data of protein adsorption on solid surfaces. Electrochim Acta 129:290–299

Mendoza MO, Ortega EP, De Fuentes OA, Prokhorov Y, Luna Barcenas JG (2014) Chitosan/bentonite nanocomposite: Preliminary studies of its potentiometric response to nitrate ions in water. In: 2014 IEEE 9th IberoAmerican congress on sensors, IBERSENSOR 2014—conference proceedings, pp 7–10

Menon S, Mathew MR, Sam S, Keerthi K, Kumar KG (2020) Recent advances and challenges in electrochemical biosensors for emerging and re-emerging infectious diseases. J Electroanal Chem 878:114596

Moussa S, Mauzeroll J (2019) Review—microelectrodes: an overview of probe development and bioelectrochemistry applications from 2013 to 2018. J Electrochem Soc 166(6):G25–G38

Naresh V, Lee N (2021) A review on biosensors and recent development of nanostructured materials-enabled biosensors. Sensors (Switzerland) 21(4):1–35

Pachauri V, Ingebrandt S (2016) Biologically sensitive field-effect transistors: from ISFETs to NanoFETs. Essays Biochem 60(1):81–90

Patel BC, Sinha GR, Goel N (2020) Chapter 1. Introduction to sensors. In: Advances in modern sensors. IOP Publishing, pp 1–21

Pisoschi AM (2016) Potentiometric biosensors: concept and analytical applications-an editorial. Biochem Anal Biochem 5(3):19–20

Rana A, Kawde AN (2016) Open-circuit electrochemical polymerization for the sensitive detection of phenols. Electroanalysis 28(4):898–902

Ryu H, Thompson D, Huang Y, Li B, Lei Y (2020) Electrochemical sensors for nitrogen species: a review. Sens Actuators Rep 2(1):100022

Saiapina OY, Dzyadevych SV, Walcarius A, Jaffrezic-Renault N (2012) A novel highly sensitive zeolite-based conductometric microsensor for ammonium determination. Anal Lett 45(11):1467–1484

Salama FM, Attia KA, Abouserie AA, El-Olemy A, Abolmagd E (2018) Potentiometric determination of Enrofloxacin using PVC and coated graphite sensors. Glob Drugs Ther 3(3)

Saraf M, Rajak R, Mobin SM (2016) A fascinating multitasking Cu-MOF/rGO hybrid for high performance supercapacitors and highly sensitive and selective electrochemical nitrite sensors. J Mater Chem A 4(42):16432–16445

Schroeder HW, Cavacini L (2010) Structure and function of immunoglobulins. J Allergy Clin Immunol 125(1):S41–S52

Shen J, Li Y, Gu H, Xia F, Zuo X (2014) Recent development of sandwich assay based on the nanobiotechnologies for proteins, nucleic acids, small molecules, and ions. Chem Rev 114(15):7631–7677

Shinde RB, Veerapandian M, Kaushik A, and Manickam P (2020) State-of-art bio-assay systems and electrochemical approaches for nanotoxicity assessment. Front Bioeng Biotechnol

Simões FR, Xavier MG (2017) Chapter 6—Electrochemical sensors. In: Micro and nano technologies. Elsevier Inc.

Sivarajasekar N, Dharshni RA, Dharshini RD (2019) Thermal biosensors and their applications. Am Int J Res Sci

Skoog DA, Holler FJ, Crouch SR (2016) Principles of instrumental analysis, 7th edn. Cengage Learning

Stradiotto NR, Yamanaka H, Zanoni MVB (2003) Electrochemical sensors: a powerful tool in analytical chemistry. J Braz Chem Soc 14(2):159–173

Sun CL, Lee HH, Yang JM, Wu CC (2011) The simultaneous electrochemical detection of ascorbic acid, dopamine, and uric acid using graphene/size-selected Pt nanocomposites. Biosens Bioelectron 26(8):3450–3455

Tian K, Prestgard M, Tiwari A (2014) A review of recent advances in nonenzymatic glucose sensors. Mater Sci Eng C Mater Biol Appl 41:100–118

Toghill KE, Compton RG (2010) Electrochemical non-enzymatic glucose sensors: a perspective and an evaluation. Int J Electrochem Sci 5(9):1246–1301

Trani A, Petrucci R, Marrosu G, Zane D, Curulli A (2017) Selective electrochemical determination of caffeine at a gold-chitosan nanocomposite sensor: may little change on nanocomposites synthesis affect selectivity? J Electroanal Chem 788(2016):99–106

Tumur M, Kanberoglu GS, Coldur F (2020) A novel potentiometric PVC-membrane cysteamine-selective electrode based on cysteamine-phosphomolybdate ion-pair. In: Current pharmaceutical analysis. Bentham Science Publishers, pp 168–175

Veerapandian M, Yun K (2013) Methylene blue dye coated silver-silica nanoparticles with dual functionality fabricated by injection pump and ultrasonochemistry. Mater Res Bull 48(5):1817–1823

Vieira S, Vial S, Reis RL, Oliveira JM (2017) Nanoparticles for bone tissue engineering. Biotechnol Prog 33(3):590–611

Vladislavić N, Rončević IŠ, Buljac M, Brinić S, Krivić D, Buzuk M (2018) Electroanalytical determination of cysteine using the electrodes based on ternary silver-copper sulfides. Sensors (Switzerland) 18(11)

Wang L, Sipe DM, Xu Y, Lin Q (2008) A MEMS thermal biosensor for metabolic monitoring applications. J Microelectromech Syst 17(2):318–327

Wang M, Ma J, Chang Q, Fan X, Zhang G, Zhang F, Peng W, Li Y (2018) Fabrication of a novel ZnO–CoO/rGO nanocomposite for nonenzymatic detection of glucose and hydrogen peroxide. Ceram Int 44(5):5250–5256

Wang Y, Jiao WW, Wang Y, Sun L, Li JQ, Wang ZM, Xiao J, Shen C, Xu F, Qi H, Wang YH, Guo YJ, Shen AD (2019) Simultaneous nucleic acids detection and elimination of carryover contamination with nanoparticles-based biosensor- and antarctic thermal sensitive uracil-DNA-glycosylase-supplemented polymerase spiral reaction. Front Bioeng Biotechnol 7(December):1–11

Wee Ling JL, Khan A, Saad B, Ab Ghani S (2012) Electro polymerized 4-vinyl pyridine on 2B pencil graphite as ionophore for cadmium (II). Talanta 88:477–483

Willander M, Khun K, Ibupoto ZH (2014) Metal oxide nanosensors using polymeric membranes, enzymes and antibody receptors as ion and molecular recognition elements. Sensors (Switzerland) 14(5):8605–8632

Wu L, Zhang X, Wang M, He L, Zhang Z (2018) Preparation of Cu_2O/CNTs composite and its application as sensing platform for detecting nitrite in water environment. Meas J Int Meas Confed 128(October 2017):189–196

Yang S, Li G, Qu C, Wang G, Wang D (2017) Simple synthesis of ZnO nanoparticles on N-doped reduced graphene oxide for the electrocatalytic sensing of l-cysteine. RSC Adv 7(56):35004–35011

Yang S, Teng Y, Cao Q, Bai C, Fang Z, Xu W (2019) Electrochemical sensor based on molecularly imprinted polymer-aptamer hybrid receptor for voltammetric detection of thrombin. J Electrochem Soc 166(2):B23–B28

Zaidi SA, Shin JH (2016) Recent developments in nanostructure based electrochemical glucose sensors. Talanta 149:30–42

Zhad HR, Lai RY (2015) Comparison of nanostructured silver-modified silver and carbon ultramicroelectrodes for electrochemical detection of nitrate. Anal Chim Acta 892:153–159

Zhang J, Zheng J (2015) An enzyme-free hydrogen peroxide sensor based on Ag/FeOOH nanocomposites. Anal Methods 7(5):1788–1793

Zhang W, Xu G, Liu R, Chen J, Li X, Zhang Y, Zhang Y (2016) Novel MOFs@XC-72-Nafion nanohybrid modified glassy carbon electrode for the sensitive determination of melamine. Electrochim Acta 211:689–696

Zhao W, Chen G (2009) Comparison of active and passive damping methods for application in high power active power filter with LCL-filter. In: 1st international conference on sustainable power generation and supply, SUPERGEN '09

Zhou W, Jimmy Huang PJ, Ding J, Liu J (2014) Aptamer-based biosensors for biomedical diagnostics. Analyst 139(11):2627–2640

Chapter 14
Development of a Wireless Solar Power Transmission for Battery Chargers

Nur Amirah Liyana Abdul Hadi, Noor Hidayah Mohd Yunus, and Mohd Shahrul Mohd Nadzir

Abstract This paper presents a wireless power transmission technology from solar energy to efficiently charge a phone battery. The idea was derived from the issues of the cable supply costs for needs in wired charging as well as the limited non-renewable energy resources for electricity supplies. This paper proposed a solar power wireless charging system for mobile phones which should be able to monitor the presence of solar power displayed on the liquid–crystal display (LCD) I2C. The system is composed of an Arduino Uno as a microcontroller, photovoltaic (PV) solar panel, both primary and secondary copper coils at the transmitter and receiver (transceiver) circuits, LC-tuned circuit and a power-charging unit. The PV solar panel converts solar energy into electrical energy. For wireless power transmission, magnetic resonant coupling based on the Faraday law was utilized at the transceiver coils. The LC-tuned circuit was designed to resonate at 900 ± 15 kHz. The result indicated the electricity supplies from the solar charging corresponded to the solar energy source in terms of time taken, environment and weather per day. The proposed system has been designed to charge a phone battery without a physical connection and the need for a power plug. Solar-based renewable energy for wireless power transmission will be the charging solution of the future in the modern era.

N. A. L. A. Hadi
Electrical Technology Section, Universiti Kuala Lumpur British Malaysian Institute, Batu 8, Jalan Sungai Pusu, 53100 Gombak, Selangor, Malaysia
e-mail: namirah.hadi@s.unikl.edu.my

N. H. M. Yunus (✉)
Advanced Telecommunication Technology, Communication Technology Section, Universiti Kuala Lumpur British Malaysian Institute, Batu 8, Jalan Sungai Pusu, 53100 Gombak, Selangor, Malaysia
e-mail: noorhidayahm@unikl.edu.my

M. S. M. Nadzir
Department of Earth Sciences and Environment, Faculty of Science and Technology, Universiti Kebangsaan Malaysia, 43600 Bangi, Selangor, Malaysia
e-mail: shahrulnadzir@ukm.edu.my

A. Ismail et al. (eds.), *Advancements in Materials Science and Technology Led by Women*, Advanced Structured Materials 165,
https://doi.org/10.1007/978-3-031-21959-7_14

Keywords Solar panel · Arduino Uno · Wireless power · Copper coil · Coils resonance

14.1 Introduction

Several shortcomings in traditional energy sources have forced energy providers to seek renewable energies (Liu and Yin 2019; Yunus et al. 2017a). Solar energy is becoming the choice as a renewable energy source because it is an environmentally friendly energy source without incurring costs. Solar energy can generate electricity without causing environmental damage. Electricity would be generated by the light of an electrode immersed in a conductive fluid (Bisquert et al. 2004; Yun et al. 2022). When photons from the sun strike a solar panel, they knock electrons away from atoms and generate electricity. For wireless charging, this began using the principle of wireless power transmission (WPT) to create a wide range of technologies, including electric smart watches, surgically implanted gadgets, electric vehicles and now cell phone chargers. When an alternating current (AC) travels through a close loop coil, a magnetic field is formed around the coil, according to the general concept of WPT (Barman et al. 2015).

A wireless solar battery charger is a power electronic device that converts solar radiation into electrical energy for the purpose of charging batteries (Dhal et al. 2016; Yunus et al. 2017b). This is accomplished by converting, regulating and conditioning the flow of electrical energy from a source which is a solar panel to charge cell phones according to load requirements. However, the alternative of charging batteries at public charging stations is difficult because these facilities are frequently crowded, limited and require generators as well (Huang et al. 2019). Mobile phone theft is also prevalent at public charging stations. There is a possibility of a fire incident occurring at the charging station due to excessive cable supply load and unprofessionally designed connections. In addition, wire wear and tear will arise as a result of frequent use and incorrect wire handling. As a result, wires must be serviced regularly to maintain transmission capacity and efficiency.

For that reason, this paper aims to introduce the design of a solar power wireless transmission system for a battery charger with voltage values monitoring display. Faraday law has been implemented for the concept of wireless power transmission. There are several research articles related to this project. Researchers (Yilmaz et al. 2018) introduced a charge controller technique for solar PV. This introduces a controller that controls the flow of reverse current, aids in battery protection when the battery voltage increases to some extent, extends the battery life cycle and minimizes system maintenance efficiently. These proposed charge controllers include a shunt as well as a series regulator. In small solar PV arrays, shunt controllers are used. When the battery voltage reaches a predetermined voltage, the series controller opens the PV array circuit from the battery. With a series arrangement, this regulator switch element offers a channel between the battery and the array (Boyea 2021). The work by Aziz et al. (2019) is on revolutionary inductively coupled remote power trade. This creates a safe, efficient and irreplaceable technique for transferring capabilities to remote static equipment or thought-provoking flexible tools. Studies by LokeshReddy et al. (2017) disclosed the use of solar cells to capture and store light energy, generating more than 5 W in both indoor and outdoor settings. The system uses a matching network to maximize power transfer, tune coils and minimize losses by moving power from the battery through both transmission and receiving inductive coupling circuits. Through the use of a female USB type-B connection on the receiving circuit, the system can support any standard charging.

14.2 Methodology

14.2.1 Project Development

Figure 14.1 shows the block diagram of this project. The battery charging is from solar energy and raising the voltage to 5 V is required by the Arduino Uno. At this point, the Arduino Uno has been turned on and linked to the USB port booster. The Arduino Uno will continue to run until the battery runs out. The battery will begin charging automatically when there is adequate sunlight. DC power is then used to track the maximum power point. Following that, the maximum power point tracking checks the output panel and compresses it to the battery voltage. Basically, it calculates the best power that can be generated by the panel to charge the battery and converts the received voltage to the optimum voltage to deliver the largest amperes to the battery.

Figure 14.2 shows the flowchart of this project. Initially, the solar energy goes to the solar panel, then goes into the Arduino. After that, the energy result will be read by the Arduino which serves to protect against overvoltage. The Arduino controller monitors the battery voltage. When it detects a low voltage, it allows the entire flow of current from the panel to enter the battery. When the battery voltage reaches a certain point, the energy flow is reduced to prevent the battery from being overcharged.

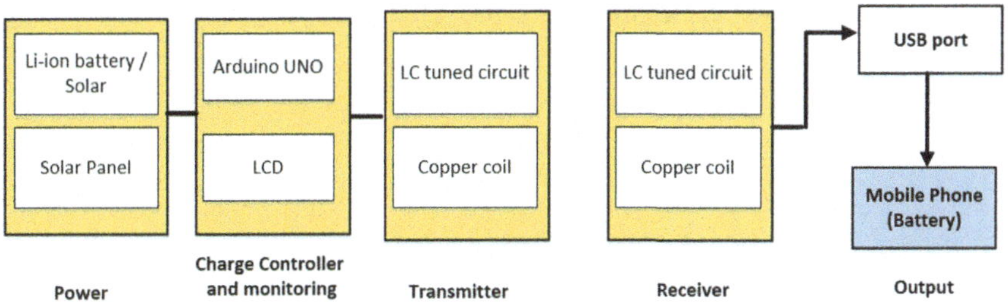

Fig. 14.1 Block diagram of solar power transmission charger

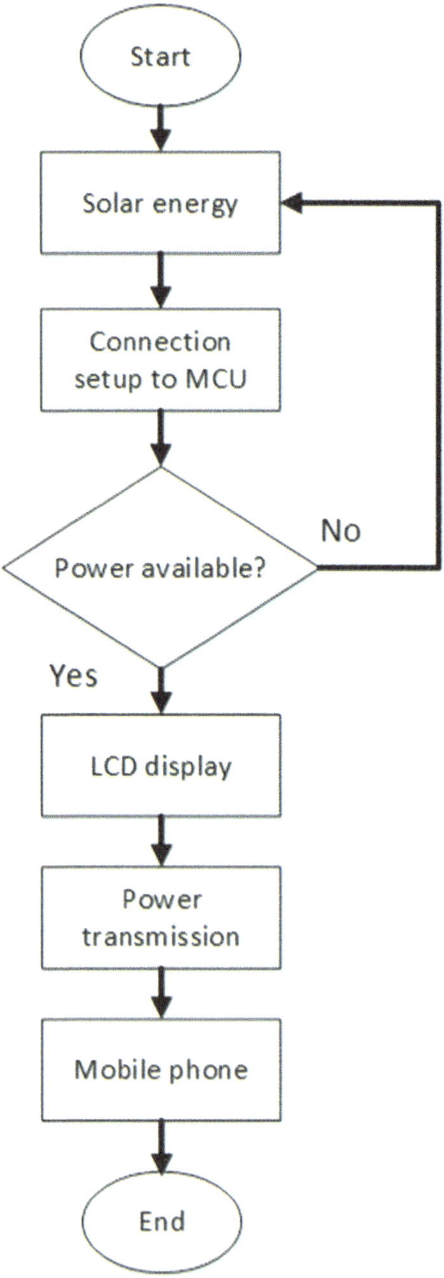

Fig. 14.2 Flowchart of solar power transmission charger

The controller allows energy to flow fully into the battery when the battery voltage decreases due to lack of sunlight or high electrical energy consumption. The transmitter which is the diode inside the solar charge controller, on the other hand, only permits current to flow in one direction from the solar panel to the battery. Then, after the energy is transmitted, the receiver will take over the energy. The secondary coil drives voltage and current to flow so that energy can be supplied to the mobile phones.

This project works to convert solar energy into electricity. Then, the charge controller keeps the charging voltage from overheating and then goes into the primary coil, which acts as a power transmitter. The secondary coil, which is the receiving part, functions as magnetic induction or wireless. Finally, the LED light will illuminate when the receiver part is complete or indicates a voltage ready to be charged to the mobile phone. When the LED is not illuminated, the transmission operation must be repeated until the LED illuminates.

14.2.2 Hardware Development

Figure 14.3 shows the set of Arduino Uno. The Arduino Uno is basically a free operating system. The basic program and function of the operating system are to make the Arduino Uno activated. It is sufficient to connect the module to the Arduino port to switch on the panel, LCD display and battery.

Figure 14.4 shows a set of solar panels. A storage device is required since the hardware implementation process is to store the energy generated by solar cells. A battery is a type of storage device commonly used to store electrical energy generated

Fig. 14.3 Set for Arduino Uno

by solar cells. The charge controller is an essential equipment for replacing a battery with solar PV. As a result, the charge controller must be linked in series between the solar panels and the battery.

Figure 14.5 shows the transmitter and the receiver sides. The transmitter side is comprised of 4 capacitors, a 10 kΩ resistor, an IC L7805CV, a 1000 μF capacitor and an LED and a primary coil. The receiver side only has a secondary coil and a USB charger for the mobile phone.

Fig. 14.4 Solar panel with battery

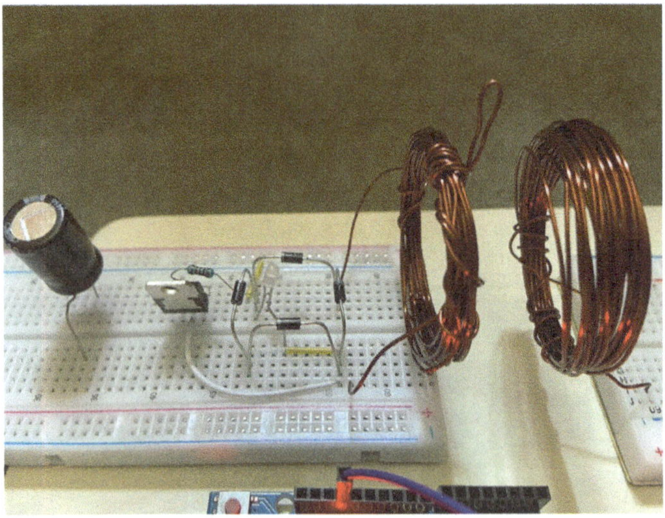

Fig. 14.5 Transmitter and receiver end sides

14.3 Results and Discussion

Figures 14.6 and 14.7 show a prototype in which the Arduino is powered by a 9 V battery and charged by the solar panel, respectively. Basically, the Arduino Uno is linked to a LCD and a solar panel to display voltage readings. LCD and solar panel are ensured to be connected to the correct ports; GND to GND, VCC to 5 V, SDA to A4 and SCL to A5 are the connections for the LCD port. For the solar panel connection, it is punctured at GND to GND and VCC to A2.

Figure 14.8 shows the graph of solar charging at the time taken on the day. This data collection is based on the values and conditions collected on average over a month according to the time of the presence of sunlight in Malaysia starting from 8.00 AM to 6.00 PM. The most obvious is the increase in voltage value as a result of the presence of sunlight.

Initially, at around 8.00 AM, it was found that the solar panel only had 2.45 V for full charging at this time. At 10.00 AM, it was found that the voltage taken was higher than at 8.00 AM which was 3.98 V for a fully solar-charged battery. The fastest time for a fully charged solar is at 2.00 PM. Finally, it was checked at about 6.00 PM that 3.76 V data had been collected. This means it is the best time to charge the battery, and it indicates that the battery has been fully charged at noon. However, it dropped to 3 V in the evening.

Fig. 14.6 Prototype of solar power transmission charger

Fig. 14.7 Prototype when charging a mobile phone

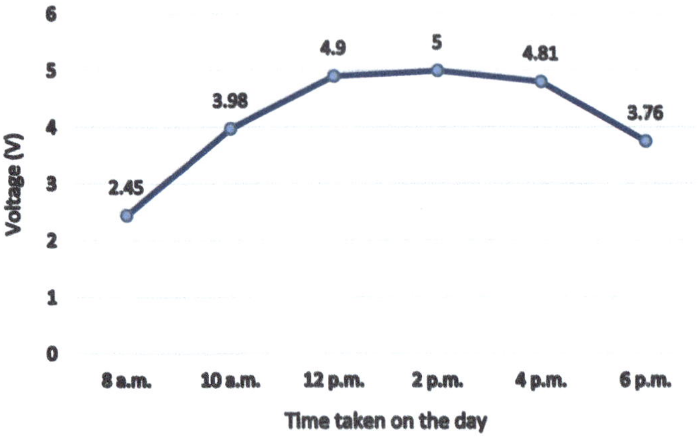

Fig. 14.8 Graph of solar charging at the time taken on the day

14.4 Conclusion

In this paper, the development of a wireless solar power battery charger based on magnetic resonance coupling is presented. The implementation by experimenting and troubleshooting with the battery charging unit was carried out several times to demonstrate the wireless charging operation. The project has achieved a functional battery charger unit, which captures solar sources through the PV solar panel and display the received voltage values. However, extending more distance during wireless charging is unsuccessful to produce sufficient power.

The improvement in hardware design in terms of copper coil turns can be implemented in order to charge the battery efficiently. Future work should be carried out to research the electromagnetic field of wireless power transmission and adapt the remote wireless charging capabilities.

Acknowledgements The authors wish to thank Universiti Kuala Lumpur British Malaysian Institute and Center for Research and Innovation, CoRI of Universiti Kuala Lumpur, Malaysia. We also like to thank UKM for the support given to the success of this project.

References

Aziz AF, Romlie MF, Baharudin Z (2019) Review of inductively coupled power transfer for electric vehicle charging. IET Pow Electron 12(14):3611–3623

Barman SD, Reza AW, Kumar N et al (2015) Wireless powering by magnetic resonant coupling: recent trends in wireless power transfer system and its applications. Renew Sust Energ Rev 51:1525–1552

Bisquert J, Zaban A, Greenshtein M et al (2004) Determination of rate constants for charge transfer and the distribution of semiconductor and electrolyte electronic energy levels in dye-sensitized solar cells by open-circuit photovoltage decay method. J Am Chem Soc 126(41):13550–13559

Boyea S (2021) Source independent power converter for cell phone charging, doctoral dissertation, Worcester Polytechnic Institute

Dhal SB, Agarwal A, Agarwal K (2016) Solar powered mobile power bank systems. Am J Electr Electron Eng 4(5):148–151

Huang P, Ma Z, Xiao L et al (2019) Geographic information system-assisted optimal design of renewable powered electric vehicle charging stations in high-density cities. Appl Energy 255(2):113855

Liu J, Yin Y (2019) An integrated method for sustainable energy storing node optimization selection in China. Energy Convers Manage 199

LokeshReddy M, Kumar PP, Chandra SA et al (2017) Comparative study on charge controller techniques for solar PV system. Energy Procedia 17:1070–1077

Yilmaz U, Kircay A, Borekci S (2018) PV system fuzzy logic MPPT method and PI control as a charge controller. Renew Sust Energ Rev 81:994–1001

Yun C, Dinh TD, Hwang S (2022) Chemical electrification at solid/liquid/air interface by surface dipole of self-assembled monolayer and harvesting energy of moving water. J Colloid Interf Sci 615:59–68

Yunus NHM, Sampe J, Yunas J et al (2017b) MEMS based RF energy harvester for battery-less remote control: a review. AJASSP 14(2):316–324

Yunus NHM, Sampe J, Yunas J et al (2017a) Parameter design of microstrip patch antenna operating at dual microwave-band for RF energy harvester application. In: 2017a IEEE regional symposium on micro and nanoelectronics (RSM), pp 92–95

Chapter 15
Short-Term Study on the Potential of Oil Palm Frond Biochar for Acid Sulphate Soil Amelioration

Amelia Md Som, Abdul Addahary Abdul Yahya, Padmini Karananidi, and Robert Thomas Bachmann

Abstract The potential of oil palm frond biochar to ameliorate acid sulphate soil was evaluated. In Malaysia, acid sulphate soil can be found mainly in coastal regions where waterlogged soil was drained for agriculture purposes. The soil has a very low pH value and high dissolved aluminium concentration which are both detrimental for plant growth. In this study, biochar was produced from sundried, 3 × 3 cm oil palm frond by allothermal pyrolysis at 345 °C for 45 min. Acid sulphate soil was obtained from Kampung Sungai Raya (Negeri Sembilan), characterized and incubated with three oil palm frond biochar application rates (20, 40 and 60 g/kg). The changes on acid sulphate soil pH and electrical conductivity (EC) value were monitored for 40 days. Oil palm frond biochar application rate had a significant effect ($p < 0.05$) on acid sulphate soil pH and EC. While incubation period had no significant effect ($p > 0.05$) on EC, the pH at oil palm frond biochar application rates of 40 and 60 g/kg was significantly greater than the control and 20 g/kg ($p < 0.05$). Our study clearly demonstrates that acid sulphate soil from Kampug Sungai Raya can benefit from the addition of low-temperature oil palm frond biochar at moderate dosage rates of 20–60 kg per kg soil.

Keywords Acid sulphate soil · OPF biochar

A. M. Som (✉) · A. A. A. Yahya · P. Karananidi · R. T. Bachmann
Green Chemistry and Sustainable Technology Cluster, Universiti Kuala Lumpur Malaysian Institute of Chemical and Bioengineering Technology, Bandar Vendor Taboh Naning, 78000 Alor Gajah, Melaka, Malaysia
e-mail: ameliamds@unikl.edu.my

A. A. A. Yahya
e-mail: addahary.yahya@s.unikl.edu.my

P. Karananidi
e-mail: padmini.karananidi@s.unikl.edu.my

R. T. Bachmann
e-mail: bachmann@unikl.edu.my

A. Ismail et al. (eds.), *Advancements in Materials Science and Technology Led by Women*, Advanced Structured Materials 165,
https://doi.org/10.1007/978-3-031-21959-7_15

15.1 Introduction

Acid sulphate soil is the result from oxidation of sulphidic material in the soil which occurs when water logged soil such as peat and coastal alluvium are dried or drained due to natural changes or by man intervention for agriculture or other purposes (Fanning et al. 2017). It is characterized by low pH (< 4) and high amount of soluble aluminium (Dent 1986). Acid sulphate soils contain pyrite (FeS_2), the main oxidisable sulphidic material that forms from microbial decomposition of organic matter under anaerobic conditions. Once the sulfidic material is exposed to oxygen, it will oxidize and produce sulphuric acid according to Eq. (15.1).

$$4FeS_2 + 15O_2 + 14H_2O \rightarrow 4Fe(OH)_3 + 8SO_4^{2-} + 16H^+ \tag{15.1}$$

The presence of a sulphuric horizon above the sulphidic material can be verified by the presence of 0.05% of water-soluble sulphate and jarosite ($KFe_3(SO_4)_2(OH)_6$) (Soil Survey Staff 2014). Jarosite is a straw-yellow mineral formed as a result of pyrite oxidation. The acidic nature initiates dissolution of clay minerals releasing Fe and Al (Shamshuddin et al. 2014). Figure 15.1 illustrates aluminium speciation at different pH by using the Visual Minteq software (USEPA 1991). It shows that at pH < 4, most of aluminium is present as Al^{3+}. Accumulation of aluminium in the root tissue will prevent elongation and division of cells resulting in a stunted and deformed root system. Soluble aluminium may also react with soluble phosphate and forms insoluble compounds such as $AlPO_4$ and hence reduces the availability of phosphate for plant intake (Rorison 1973).

Globally, there are about 14–24 million ha of acid sulphate soil (Fanning et al. 2017), and about 7.5 million ha are found mainly in Southeast Asia (Kawalec 1972). In Malaysia, acid sulphate soils are present in the plains of the west coast states of Peninsular Malaysia, east coast of Kelantan (Shamshuddin et al. 2004a) and Sarawak (Teng 2005). Undisturbed acid sulphate soil is stable and environmentally friendly; however, concerns over food security have resulted in an increased utilization of marginalized soils for agricultural purpose. Once disturbed, sulphidic material will

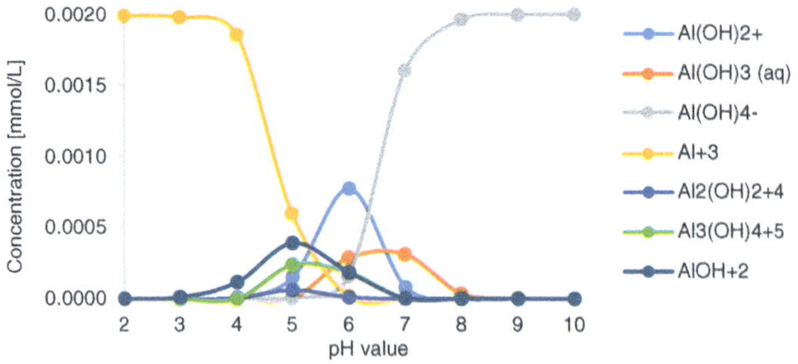

Fig. 15.1 Speciation of aluminium in water as a function of pH

start to oxidize producing a sulphuric horizon. Three courses of action have been proposed in dealing with acid sulphate soil: avoiding disturbance, minimizing disturbance or rehabilitation (CSIRO Land and Water 2008). Application of lime ($Ca(OH)_2$, CaO or $CaCO_3$) is the conventional rehabilitation method. However, applying up 60 t of lime per ha can be an economic burden for most farmers. Studies on alternative ameliorants able to substitute lime are currently receiving considerable amount of attention in the scientific community. For example, application of organic matter will result in the depletion of oxygen due to its decomposition by aerobic bacteria which in turn slows down the pyrite oxidation (Michael et al. 2015). A number of studies have been carried out using accessible organic waste materials such as wheat straw (Jayalath et al. 2016), leafy materials (Castellano et al. 2015) peat, palm oil mill sludge, rice straw, sawdust and poultry dungs (Shamshuddin et al. 2004a) to improve crop growth on acidic soil. The use of biochar, a carbonaceous porous material produced by heating of biomass in presence of limited amount of oxygen, has also been proposed (Manickam et al. 2015). Biochar generally has a good adsorption capacity and its ability to adsorb aluminium (Qian and Chen 2013) has the potential to reduce Al toxicity to crops. Biochar in combination with compost has significantly increased bioavailability of phosphorus (Qayyum et al. 2015). Application of 5 wt.% rice husk biochar was able to increase the pH by 0.8 pH units after 95 days (Manickam et al. 2015). Similarly, the soil pH increased by up to 0.7 pH units upon application 40 t/ha of oil palm empty fruit bunch (EFB) biochar after 118 days (Abu Bakar et al. 2015). Both studies also reported an increase in crop yield by 20–100%. In general, rice husk and EFB biochars have shown better ameliorating effects in comparison to poultry litter biochar (Fraser et al. 2012; Hass et al. 2012). Amelioration effects of biochar should be investigated on a "char by char" basis due to its varying properties (Jeffery et al. 2015). Even though the basic properties such as alkaline pH are similar, its surface areas and the presence of specific functional groups on its surface varies considerably based on its precursor and its pyrolysis temperature. Beside rice husk and EFB, there are several other locally accessible biomass types such as oil palm fronds that can be used for conversion into biochar.

Malaysia is the second largest producer of palm oil in the world, and by end of December 2021, the total area planted by oil palm was 5,737,731 ha (MPOB 2022). The best planting density is 143 trees planted per hectare of land (FAO 2017) while the fronds have to be pruned twice yearly. In total, 82.5 kg of fronds/(palm year) are produced (Chan et al. 1981), and this has generated masses of agricultural waste. Studies have been conducted on the potential use of oil palm fronds for ruminants (Wan Zahari et al. 2003), adsorption of heavy metals (Som et al. 2012), carbon source in biorefineries (Tan et al. 2016), fermentation substrate (Che Maail et al. 2014) and bioethanol production (Abdullah et al. 2015). The potential use of oil palm frond for acid sulphate soil amelioration has not been explored so far.

In this study, we produced biochar from oil palm fronds and evaluated the effect of biochar application rate and incubation time on acidic sulphate soil in terms of pH, electrical conductivity and sulphate content in laboratory scale experiments.

Fig. 15.2 Sampling site. Kampung Sungai Raya, 71,150 Linggi, Negeri Sembilan, Malaysia. 2° 25′ 48.0″ N and 101° 57′ 29.5″E

15.2 Methodology

15.2.1 Soil Location

The acidic soil was collected from 20 cm top layer at Kampung Sungai Raya, 71,150 Linggi, Negeri Sembilan, Malaysia which is located at 2° 25′ 48.0″ N and 101° 57′ 29.5″ E (Fig. 15.2). The soil was spread on plastic sheet and air-dried for 2 weeks before the experiments commenced.

15.2.2 Oil Palm Frond Collection

The stems of oil palm fronds were chosen as feedstock to produce the biochar. Fresh oil palm fronds were collected from an oil palm plantation at Kampong Cherana Putih, Alor Gajah, Melaka (Nizam 2016).

15.2.3 Preparation of Oil Palm Frond Biochar

Fresh oil palm fronds were cut into 3 cm × 3 cm dimensions. The oil palm fronds were sundried for three days to reduce its water content. The sundried oil palm fronds were spread as a single layer on the aluminium foil (Fig. 15.3) and wrapped with aluminium foil. A second layer of aluminium foil was wrapped around the oil palm fronds in order to reduce the presence of oxygen during pyrolysis. The pyrolysis process parameters were 345 °C for 45 min, which were chosen based on previous studies to produce biochar with nearest to neutral pH value and minimum hydrophobicity (Wan Zahari et al. 2003; Abdullah et al. 2015).

(a) (b) (c)

Fig. 15.3 **a** Freshly cut oil palm frond, **b** sundried oil palm frond and **c** oil palm frond biochar

Mix	Biochar Application
A	0
B	20 g/kg
C	40 g/kg
D	60 g/kg

Fig. 15.4 Soil design mix in the containers

15.2.4 Preparation of Soil Sample and Incubation

Approximately, 14 kg of soil was used for this project. The soil was air-dried and sieved to eliminate pebbles and other foreign objects. The soil was weighed and 1000 g transferred into plastic containers. The containers were perforated with a small hole to allow gaseous exchange. Oil palm frond biochar was added to each of the container according to experimental design in Fig. 15.4 and mixed thoroughly. The samples were incubated for 40 days at room temperature on a work bench, and samples were collected every 5 days for pH and conductivity determination. All experiments were carried out in triplicate.

15.2.5 pH and Conductivity Measurements

For pH and conductivity measurements, an in-house method base on BS1377: Part 3:1990 was followed. In brief, 30 ± 0.1 g of soil was weighed and placed in a 100 mL beaker. 75 mL of distilled water was added into the beaker and the suspension stirred for a few minutes. The beakers were then covered with a cover glass and allowed to stand for at least 8 h. The suspension pH was measured using a Mettler Toledo pH 5 Easy meter. The same sample was used for electrical conductivity (EC) measurement with a Mettler Toledo 5 Easy conductivity meter.

15.2.6 *Particle Size Distribution*

The particle size distribution was determined using the Malvern Mastersizer 2000 particle size analyser. The analyser was set at obscuration of 5%, 1700 rpm stir rates, and stability was reached at 3 min.

15.2.7 *Water-Soluble Sulphate*

Water-soluble sulphate was determined by using BS 1377: Part 3:1990. A 2:1 water extract was obtained from soil samples dried at 75 °C overnight. The sulphate in the extract was precipitated by using $BaCl_2$.

15.2.8 *Statistical Study*

The significant effect of application rate and incubation time was analysed by using analysis of variance (ANOVA) in Minitab 17.

15.3 Results and Discussion

15.3.1 *Characterization of Soil and Oil Palm Frond Biochar*

The acid sulphate soil was collected from the 20 cm top layer in what was suspected to be a sulphuric horizon. The soil initial pH was 3.7 that indicates the presence of sulphidic materials. The soil sulphuric horizon was further confirmed with the soil water-soluble sulphate content of 0.07% (Soil Survey Staff 2014). According to our particle size analaysis, the soil consists of 60% silt, 30% of very fine and fine sand and 10% of clay (Fig. 15.5). This classifies the soil as silt loam under soil texture triangle (Cruz 1983). The low amount of clay further indicates the presence of sulphuric acid from oxidation of pyrite that caused disintegration of clay minerals (Shamshuddin et al. 2004b).

The oil palm frond biochar used in this study has a pH value of 7.6. Other physico-chemical properties of the oil palm frond biochar have been discussed elsewhere (Som et al. 2012).

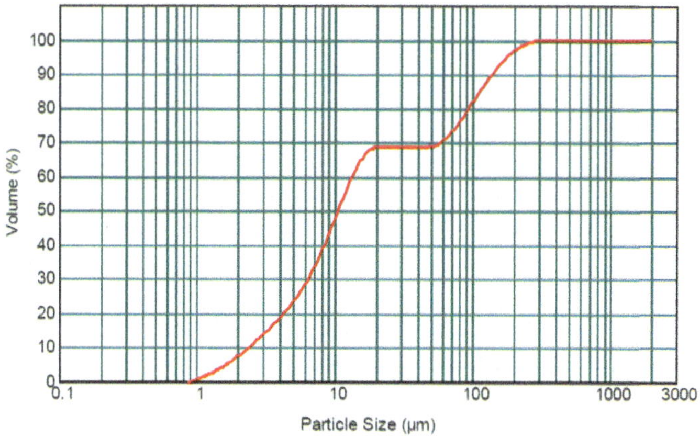

Fig. 15.5 Particle size distribution of acid sulphate soil

15.3.2 Effect on Soil pH

The soil pH value increased upon incremental addition of 20, 40 and 60 g/kg oil palm frond biochar to 4.13, 4.31 and 4.56, respectively. Oil palm frond biochar's alkaline nature may have caused the initial rise of soil pH value while a minor dilution effect caused by the substitution of acid sulphate soil with biochar also contributed to the observed increase. Soluble organic and inorganic alkalis present on the biochar have been reported to contribute for the short-term acid amelioration (Silber et al. 2010). It was proposed that cations (K^+, Na^+, Mg^{2+} and Si^{2+}) presence in the feedstock such as oil palm fronds will form carbonates during pyrolysis process and become the main alkaline components in the biochar (Brewer et al. 2012). Correlation studies have also shown that soil acidity is more strongly correlated to biochar alkalinity value ($R^2 = 0.95$) rather than its pH value ($R^2 = 0.46$) (Yuan and Xu 2011). The alkalinity neutralizes the acidity presence in the soil. Nevertheless, in this study, pH decreased after 5 days for 40 and 60 g/kg amendment and on the 10 days for 20 g/kg application. Soluble alkalis present in the biochar may have been fully utilized to neutralise the acidity presence. Nevertheless, the pH values were still above those of the control (0 g/kg). It was anticipated that acid sulphate soil will experience a drop in pH when incubated under moist condition at room temperature due to pyrite oxidation (Soil Survey Staff 2014). However, in this study, control samples only recorded a reduction of 0.02 pH unit, whereas larger reductions were temporarily observed in soil with oil palm frond biochar (Fig. 15.6). The pH decrease was more prominent with higher increment of biochar application. After 5 days, the pH value of soil amended with oil palm frond biochar started to climb due to onset of soil compaction caused by watering. In comparison, the control sample fluctuated steadily at 0.02 pH unit. Low-temperature oil palm frond biochar has functional groups such as $-COO^-$ present on its surface (Som et al. 2012) that would contributes to its alkalinity (Yuan and Xu 2011). It has been identified that surface organic functional

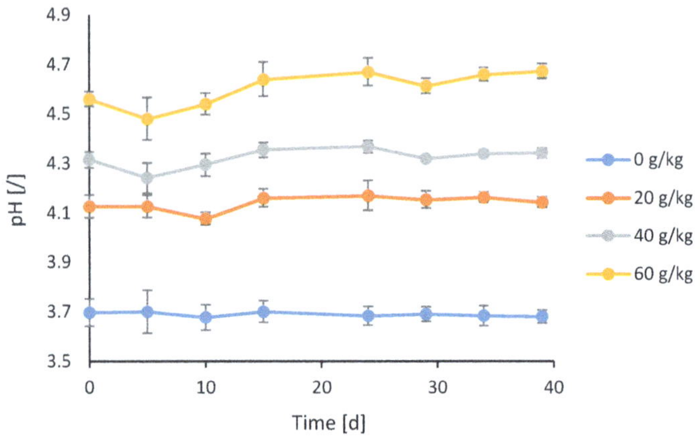

Fig. 15.6 Changes in pH value upon application of 20, 40 and 60 g/kg of oil palm frond biochar and changes during incubation. 0 g/kg are control sample. Standard deviation of means is marked as error bars ($n = 3$)

groups on the biochar contribute to long-term soil CEC and pH buffering capacity (Mao et al. 2012). Negatively charge phenolic, carboxyl and hydroxyl group will bind with H^+ ions from the soil solution (Chintala et al. 2014). The pH fluctuated until it reached 4.14, 4.34 and 4.67 pH unit at the end of incubation period for 20, 40 and 60 g/kg, respectively. One-way ANOVA showed that there are significant differences ($p < 0.05$) on the effect of oil palm frond biochar application rate on pH value, whereas there are no significant differences ($p > 0.05$) on the incubation period for 20 g/kg application but significant difference for 40 and 60 g/kg ($p < 0.05$).

15.3.3 Effect on Soil Electrical Conductivity

Soil EC indicates the presence of soluble positively (Na^+, Ca^{+2}, K^+ and Mg^{+2}) or negatively charged (Cl^-, SO_4^{-2}, CO_3^{-2}, HCO_3^-, NO_3^{-2} and PO_4^{-3}) ions. These ions are important indicators for soil health (Laishram et al. 2012). The original soil tested had an EC of 450 μS/cm. The relationship between the apparent soil electrical conductivity and average clay content in specific soil profile has been described as

$$\text{EC25} = 2.735 + 1.044\,\text{TW} \tag{15.2}$$

whereby EC25 is the electro-conductivity at 25 °C (10 × μS/cm), and TW is the relative weight fraction of clay in the sample (w/w%) (Domsch and Giebel 2001).

According to Eq. (15.2), 10 w/w % clay should have an EC25 value of 132 μS/cm. However, EC value detected in this study was three times greater. It may have been caused by the presence of water-soluble sulphates and may also indicate seeping of

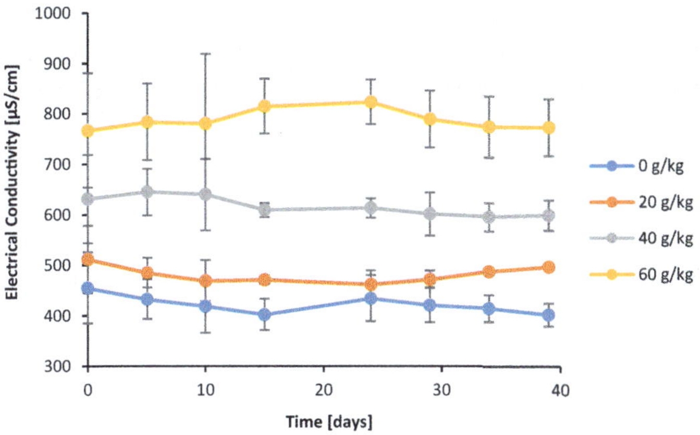

Fig. 15.7 EC value upon application of 20, 40 and 60 g/Kg of oil palm frond biochar and changes during incubation. 0 g/Kg is control sample. Standard deviation of means is marked as error bars ($n = 3$)

salinity to the soil. EC value of 450 μS/cm indicates sodicity of soil (FAO 1999). The area where the soil was collected is sporadically flooded with seawater during high tide that may have caused the soil sodicity to occur.

Figure 15.7 proves that incorporation of oil palm frond biochar increased the EC. The EC values increased by nearly 100 units with every 10 g increment of oil palm frond biochar application rates. Oil palm frond biochar has been reported to contain a high amount of base cations K^+, Ca^{2+}, Mg^{2+} and Na^+ (Som et al. 2012) which upon release to soil water contributed to EC. The electrostatic forces that bind cations onto the biochar surface may not be strong enough, especially in the presence of high H^+ concentration causing their release into soil pore water (Chintala et al. 2014). Furthermore, incorporation of biochar has already been found to increase exchangeable base cations of the soil (Yuan and Xu 2011). Application rate and incubation time was reported to have a significant effect on EC. In our study, one-way ANOVA shows that there are significant differences ($p < 0.05$) on the effect of oil palm frond biochar application rate on EC value, but there are no significant differences ($p > 0.05$) on the incubation period.

15.4 Conclusion

Oil palm frond biochar was able to increase ASS pH value from 3.7 to a maximum of 4.7 by incremental application of 20, 40 and 60 g/kg. One-way ANOVA shows that the application rate has a significant difference ($p < 0.05$) towards pH value. The pH value fluctuated throughout the incubation period. Incubation period shows significant difference ($p < 0.05$) for 40 and 60 g/kg but no significant difference for 20 g/kg ($p > 0.05$). Acid sulphate soil EC value also increased upon incremental application of oil palm frond biochar with significant difference ($p < 0.05$). Incubation period shows no significant difference ($p > 0.05$) in the EC value for each application rate. The presence of sulphuric horizon was ascertained from the presence of 0.07% of water-soluble sulphate and low pH value of 3.7. Our study shows that oil palm frond biochar application can assist with the amelioration of acid sulphate soil. Future study must focus on integration effect of oil palm frond biochar, acid sulphate soil, plant growth, fertilizer and other additives.

Acknowledgements The writers would like express appreciation to UniKL MICET for providing facilities to conduct this study

References

Abdullah S, Shirai Y, Bahrin E, Hassan M (2015) Fresh oil palm frond juice as a renewable, non-food, non-cellulosic and complete medium for direct bioethanol production. Ind Crops Prod 63:357–361

Abu Bakar R, Abdul Razak Z, Hajar S, Jalili B, Chin L, Poh C (2015) Influence of oil palm empty fruit bunch biochar on floodwater pH and yield components of rice cultivated on acid sulphate soil under rice intensification practices. Plant Prod Sci 18:491–500

Brewer C, Hu Y, Schmidt-Rohr K, Loynachan T, Laird D, Brown R (2012) Extent of pyrolysis impacts on fast pyrolysis biochar properties. J Environ Qual 41:1115–1122

Castellano M, Mueller K, Olk D, Sawyer J, Six J (2015) Integrating plant litter quality, soil organic matter stabilization, and the carbon saturation concept. Glob Chang Biol 21:3200–3209

Chan K, Watson I, Kim L (1981) Use of oil-palm waste material for increased production. In: Proceedings of conference on soil sciences and agricultural development in Malaysia. Malaysian Soil Science Society, Kuala Lumpur, pp 213–241

Che Maail C, Ariffin H, Hassan M, Shah U, Shirai Y (2014) Oil palm frond juice as future fermentation substrate: a feasibility study. Biomed Res Int 2014:8

Chintala R, Mollinedo J, Schumacher T, Malo D, Julson J (2014) Effect of biochar on chemical properties of acidic soil. Arch Agron Soil Sci 60:393–404

Cruz C (1983) Soil quality considerations in the selection of sites for aquaculture. In: Fishpond engineering: a technical manual for small-and medium-scale coastal fish farms, Southeast Asia

CSIRO Land and Water (2008) Soil Management in the agricultural development areas

Dent D (1986) Acid sulphate soils: a baseline for research and development. In: Publication 39 international institute for land reclamation and improvement, Wageningen

Domsch H, Giebel A (2001) Soil electrical conductivity and texture. Landtechnik, pp 318–319

Fanning D, Rabenhorst M, Fitzpatrick R (2017) Historical developement in understanding of acid sulfate soils. Geoderma 308:191–206

FAO (1999) Soil salinity assessment. Book, p 165

FAO (2017), Palm oil. In: The plantarion. http://www.fao.org/docrep/006/t0309e/T0309E03.htm

Fraser M, Baldwin D, Rees G, Silvester E, Whitworth K (2012) Rehabilitation options for inland waterways impacted by sulfidic sediments—field trials in a south-eastern Australian wetland. J Environ Manage 102:71–78

Hass A, Gonzalez J, Lima I, Godwin H, Halvorson J, Boyer D (2012) Chicken manure biochar as liming and nutrient source for acid appalachian soil. J Environ Qual 41:1096

Jayalath N, Mosley L, Fitzpatrick W, Marschner P (2016) Addition of organic matter influences pH changes in reduced and oxidised acid sulfate soils. Geoderma 262:125–132

Jeffery S, Meinders M, Stoof C, Bezemer T, van de Voorde T, Mommer L, van Groenigen J (2015) Biochar application does not improve the soil hydrological function of a sandy soil. Geoderma 251–252:47–54

Kawalec A (1972) World distribution of acid sulphate soils. References and map. In: Dost H (ed) Acid sulphate soils proceedings of the international symposium. International institute for land reclamation and improvement, Wageningen, p 293

Laishram J, Saxena K, Maikhuri R, Rao K (2012) Soil quality and soil health: a review. Int J Ecol Environ Sci 38:19–37

Manickam T, Cornelissen G, Bachmann R, Ibrahim I, Mulder J, Hale S (2015) Biochar application in Malaysian sandy and acid sulfate soils: soil amelioration effects and improved crop production over two cropping seasons. Sustainability 7:16756–16770

Mao J-D, Johnson R, Lehmann J, Olk D, Neves E, Thompson M, Schmidt-Rohr K (2012) Abundant and stable char residues in soils: implications for soil fertility and carbon sequestration. Environ Sci Technol 46:9571

Michael P, Fitzpatrick R, Reid R (2015) The role of organic matter in ameliorationg acid sulfate soils with sulfuric horizons. Geoderma 255–256:42–49

MPOB (2017) Oil Palm Estates, January–December 2016. In: The economic development division. http://bepi.mpob.gov.my/index.php/en/statistics/area.html

MPOB (2022) Economics and industry development division. https://bepi.mpob.gov.my/index.php/en/. Accessed 17 July 2022

Nizam F (2016) Comparison on the basic characteristic of oil palm frond biochars and the potential of biochar as source of potassium for oil palm cultivation in peat soil. Thesis. Universiti Kuala Lumpur

Qayyum M, Ashraf I, Abid M, Steffens D (2015) Effect of biochar, lime, and compost application on phosphorus adsorption in a ferralsol. J Plant Nutr Soil Sci 178:576

Qian L, Chen B (2013) Dual role of biochars as adsorbents for aluminum: the effects of oxygen-containing organic components and the scattering of silicate particles. Environ Sci Technol 47:8759–8768

Rorison I (1973) The effect of soil acidity on the nutrient uptake and physiology of plants. In: Proceedings of the international symposium on acid sulfate soils. Wageningen, The Netherlands, p 293

Shamshuddin J, Muhrizal S, Fauziah I, Van Ranst E (2004a) A laboratory study of pyrite oxidation in acid sulfate soils. Commun Soil Sci Plant Anal 35:117–129

Shamshuddin J, Muhrizal S, Fauziah I, Husni M (2004b) Effects of adding organic materials to an acid sulfate soil on the growth of cocoa (*Theobroma cacao L.*) seedlings. Sci Total Environ 323:33–45

Shamshuddin J, Elisa A, Shazana M, Fauziah C, Panhwar Q, Naher U (2014) Properties and management of acid sulfate soils in southeast Asia for sustainable cultivation of rice, oil palm, and cocoa, 1st edn. Elsevier Inc

Silber A, Levkovitch I, Graber E (2010) pH-dependent mineral release and surface properties of cornstraw biochar: agronomic implications. Environ Sci Technol 44:9318–9323

Soil Survey Staff (2014) Keys to soil taxonomy. Soil Conserv Serv 12:410. https://doi.org/10.1109/TIP.2005.854494

Som A, Wang Z, Al-Tabbaa A (2012) Palm frond biochar production and characterisation. Earth Environ Sci Trans R Soc Edinburgh 103:39–50

Tan J, Jahim J, Harun S, Wu T, Mumtaz T (2016) Utilization of oil palm fronds as a sustainable carbon source in biorefineries. Int J Hydrogen Energy 41:4896–4906

Teng C (2005) The characteristics and soil-forming processes in acid sulfate soil in Sarawak. Soil Management Division, Department of Agriculture, Sarawak, Malaysia

USEPA (1991) MINTEQA, A geochemical assessment database and test cases for environmental systems:Ver.3.0 user's manual. EPA/600/3–91/-21. USEPA, Athens

Wan Zahari M, Abu Hassan O, Wong H, Liang J (2003) Utilization of oil palm frond—based diets for beef and dairy production in Malaysia. Asian-Australasian J Anim Sci 16:625–634
Yuan J, Xu R (2011) The amelioration effects of low temperature biochar generated from nine crop residues on an acidic Ultisol. Soil Use Manag 27:110–115

Chapter 16
The Stress Analysis of the Jack Rod Crankshaft on a Single-Cylinder Engine: A Study Using the Finite Element Method

Eida Nadirah Roslin, Ikhsan Wafiy Ishak, Mohd Zaki Bahrom, and Hasan Muhamad Abid Hasan

Abstract The crankshaft is an important component in an engine that converts the reciprocating displacement of the piston into a rotary motion. The maximum possibility of crankshaft failure on the crankpin occurs because the load of pistons and connecting rods are indirectly induced onto the crankshaft. Therefore, crankshaft failure will occur due to the decrease in fatigue strength. The main purpose of this research is to investigate the stress performance on single-cylinder engine crankshaft by setting up differing heights of the crankpin. By using finite element analysis, the analytical data on the stress and the effect of the crankshaft in the occurrence of geometry changes has shown that the high stress value is the main factor that contributes towards the reduction of the crankshaft lifecycle in the engine. A comprehensive study on the material types is recommended for future improvement on the study.

Keywords Crankshaft · Single-cylinder engine · Jack rod · Stress analysis

16.1 Introduction

The main function of the crankshaft is to convert the mechanical energy on the conversion of the piston driven by pneumatic energy to move back and forth and it is due to pressure changes that occur as a result of the combustion reaction. The crankshaft is also an engine component that involves an internal combustion process for high

E. N. Roslin · I. W. Ishak · M. Z. Bahrom (✉) · H. M. A. Hasan
Automotive Engineering Section, Universiti Kuala Lumpur Malaysia France Institute, Section 14, Jalan Teras Jernang, 43650 Bandar Baru Bangi, Selangor Darul Ehsan, Malaysia
e-mail: zakibahrom@unikl.edu.my

E. N. Roslin
e-mail: eidanadirah@unikl.edu.my

I. W. Ishak
e-mail: ikhsan.ishak@s.unikl.edu.my

H. M. A. Hasan
e-mail: hasanmuhamad@unikl.edu.my

A. Ismail et al. (eds.), *Advancements in Materials Science and Technology Led by Women*, Advanced Structured Materials 165,
https://doi.org/10.1007/978-3-031-21959-7_16

volume production and it is usually used in vehicle engines. In engine operation, the reciprocating motion of the piston is a linear motion and is converted to rotary motion by passing through the crankshaft (Mathapati and Dhamejani 2015). The maximum gas pressure on the piston will release the maximum force on the crankpin in the plane of the crankshaft which will cause only the bending of the shaft to occur (Joshi and Patel 2017). In addition, the crankpin and crank web of the crankshaft will experience a bending moment. The maximum possibility of crankshaft failure on the crankpin occurs because the load of pistons and connecting rods are indirectly induced on the crankshaft. Therefore, crankshaft failure will occur due to a decrease in fatigue strength.

The pressure on the shaft is subjected to a force variable; however, in general, it should be divided into two positions. First, at the maximum bending position, it is likely to allow failure in the central region of the crank or at both ends ((Amit and Jaydeepsinh 2014; Durga Prasad et al. 2018)). The pressure in this position is a maximum pressure but it is a small fraction of the maximum. Furthermore, there is a possibility of a curved failure on the wrench that makes the interfacing bar to be affected by the shear in the maximum bending position (Aatish et al. 2020). Crankshafts are generally subject to torsional stresses and bending stresses arising from the weight of the crankshaft itself or the possibility of misalignment between journal bearings or the weight of components on the crankshaft (Kumar et al. 2014). The reliability and service life of the engine are influenced by the crankshaft having stiffness and intensity or not. This is because the design on the crankshaft and its load is quite complex which uses classical mechanical methods to analyse structures that often have limited characteristics (Yingkui and Zhibo 2011; Ding and Li 2011).

The finite element analysis (FEA) by using the ANSYS software is to analyse the stresses, models and harmonics on the crankshaft. Hence, the results of the stress analysis on the crankshaft found the values of the maximum stress point, maximum deformation and critical points areas on the crankshaft (Mathapati and Dhamejani 2015). The force that has been applied through the crankshaft will make it very stressed on the crank pin area with the web crank and the junction of the journal with the web crank. Therefore, it will provide a response on the fillet radius area to reduce stress and this stress has a high level of achievement on material strength and fatigue life (Karthick et al. 2021; Thriveni and Chandraiah 2013; Metkar et al. 2013). The location at the failure critical on the crankshaft is in the critical fillet area that connects the crankpin with the web crank and the web crank with the shaft (Mathapati and Dhamejani 2015; Sorte and Sheikh 2013; Pandiyan et al. 2018). The reduction in crankshaft weight will provide cost reduction on the crankshaft as well as improvements upon engine performance in the internal combustion engine (Karthick et al. 2021). Maximum deformation often occurs in the mid-crankpin region and maximum stresses appear in the area between the crank cheeks and the journal but the design against the crankshaft is safe or strong because the von Mises stress value is less than the yield stress value of the material (Joshi and Patel 2017; Amit and Jaydeepsinh 2014). The edge of the main journal is an area of high stresses (Solanki and Dodiya 2014; Singh et al. 2014).

The performance single-cylinder engine crankshaft has various heights on the crankshaft in other terms it is called as "Jack rod". In this paper, the values emphasized in crankpin heights start from 1.15, 2, 3 and to 4 mm. The main purpose of this paper is to observe the change in resistance resulting from these differences made in crankpin height and whether it provides variable changes. In addition, observations in the changes that occur at each height difference on the crankpin involve a variable of force applied to the crankshaft whether it allows the occurrence of bending moments and in the area most affected. Thus, the crankshaft performance for the 135 cc 4-stroke motorcycle model is made the reference in this paper. Hence, the analysis is performed after performing the modelling process on the crankshaft. In addition, static analysis was chosen in order to determine von Mises stress, strain and deformation. A force variable will be applied to each crankpin height difference to observe how the stress builds up and the resistance that will result from each crankpin height difference on the crankshaft. Thus, the results released by ANSYS software will be the main output of this research.

16.2 Methodology

The research is focused to determine the stress on the crankpin height difference on the performance of a single-cylinder crankshaft. Therefore, CATIA V5 software and ANSYS workbench are used in this research. Performance single-cylinder crankshaft on the Yamaha 135 LC model is used as references. Measurements on the performance of the crankshaft must be made accurately and without errors to ensure that every result that comes out is a reliable source. Furthermore, the single cylinder performance crankshaft was modelled with four different crankpin heights, namely 1.15, 2, 3 and 4 mm. Therefore, there are four crankshafts to be modelled in this software. Once the design is completed, the performance single cylinder of this crankshaft will be simulated using ANSYS. All data will be collected and analysed.

16.2.1 Analytical Calculation

The design calculation of a single-cylinder petrol engine crankshaft with specific two-wheeler Yamaha 135LC regular model is as per description. The dimension of the engine is given in Table 16.1.

- Moment on pin:

$$M = \frac{P_{\mathrm{p}}}{2} \times \frac{l_{\mathrm{c}}}{2} = \frac{20000}{2} \times \frac{45}{2} = 225 \times 10^{3}\,\mathrm{Nmm} \tag{16.1}$$

Table 16.1 Yamaha 135LC engine specification

Engine type	4-stroke, 1 cylinder. Liquid-cooled
Valve system	SOHC, 4-valve
Cylinder bore	54 mm
Stroke	58.7 mm
Displacement	134.4 cm^3
Compressor ratio	10.9:1
Maximum power	8.45 kW/8500 r/min
Maximum torque	11.65 Nm/5500 r/min

- Section module of crankpin:

$$Z = \frac{\pi}{32} \times (d_c)^3 = \frac{\pi}{32} \times \left(27.8^3\right) = 2109.82\,\text{mm}^3. \tag{16.2}$$

- Torque obtained at a maximum power of Yamaha 135LC engine:

$$P = \frac{2\pi NT}{60} \text{ where } 8.45 \times 10^3 = \frac{2 \times \pi \times 8500 \times T}{60}. \tag{16.3}$$

$$T = \frac{8.45 \times 10^3 \times 60}{2 \times \pi \times 8500} = \frac{507000}{53407.0751} = 9.4931 \times 10^3\,\text{Nmm} \tag{16.4}$$

- Von Mises stress:

$$\sigma_{\text{von}} = \frac{M_{\text{eq}}}{Z} \tag{16.5}$$

where

$$M_{\text{eq}} = \text{equivalent bending moment}$$

So, the Equivalent bending moment

$$M_{\text{ev}} = \sqrt{(k_b \times M)^2 + \frac{3}{4}(k_c \times T)^2} \tag{16.6}$$

where

$$k_b = \text{Combined shock and fatigue for bending} = 1$$
$$k_t = \text{Combined shock and fatigue for bending} = 1$$
$$M_{\max} = \text{Bending moment}$$
$$= \sqrt{\left(1 \times 225 \times 10^3\right)^2 + \frac{3}{4}(1 \times 9.4931 \times 10^3)^2} = 225.15 \times 10^3\,\text{Nmm}$$

Now,

$$\sigma_{\mathrm{von}} = 106.72\,\mathrm{MPa}$$

- Stain

$$\varepsilon = \frac{\sigma_{\mathrm{von}}}{\varepsilon} = \frac{106.72}{21000} = 0.0005082 \quad (16.7)$$

Analytical calculation at force 40,000 N

Description	Value
Moment on pin, M	450×10^3 Nmm
Section module of crankpin, Z	2109.82 mm^3
Torque at maximum power, P	9.4931×10^3 Nmm
Von Mises stress, σ_{von}	213.32 MPa
Strain, ε	0.001524

Analytical calculation at force 80,000 N

Description	Value
Moment on pin, M	900×10^3 Nmm
Section module of crankpin, Z	2109.82 mm^3
Torque at maximum power, P	9.4931×10^3 Nmm
Von Mises stress, σ_{von}	426.59 MPa
Strain, ε	0.002031

Analytical calculation at force 100,000 N

Description	Value
Moment on pin, M	1125×10^3 Nmm
Section module of crankpin, Z	2109.82 mm^3
Torque at maximum power, P	9.4931×10^3 Nmm
Von Mises stress, σ_{von}	533.24 MPa
Strain, ε	0.002539

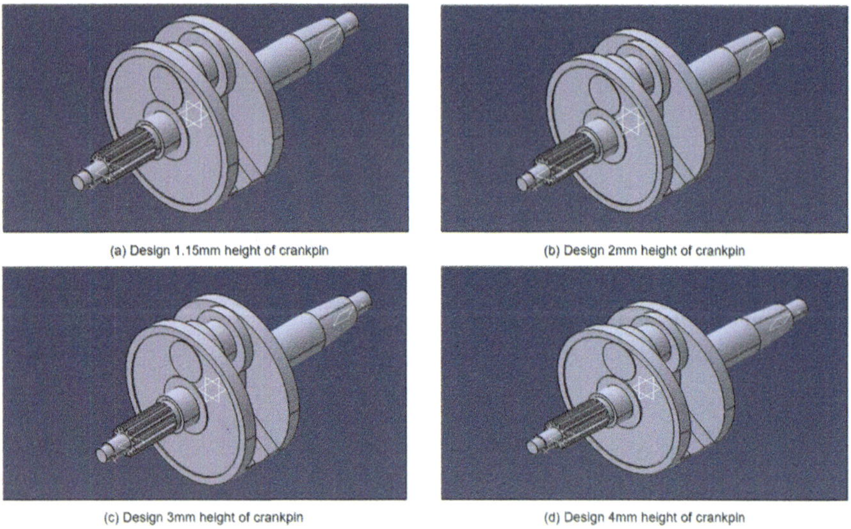

Fig. 16.1 Models of single-cylinder engine crankshaft

16.2.2 Models of Single-Cylinder Engine Crankshaft

Four differences crankpin height of single-cylinder performance crankshaft model have been modelled in the CATIA V5 software and exported into ANSYS workbench software which is saved as STP file format to perform further analysis. The models are shown in Fig. 16.1.

16.2.3 Material Properties of Crankshaft

The material properties of the crankshaft are given in Table 16.2.

Table 16.2 Material properties of forged steel crankshaft

Material type	Forged steel
Density (kg/m^3)	:
1.15 mm	8395
2 mm	8420
3 mm	8444
4 mm	8469
Young's modulus (GPa)	210
Poisson's ratio	0.3
Yield strength (MPa)	540
Ultimate tensile strength (MPa)	845

16.2.4 Crankshaft Meshing

The finite element model for the crankshaft geometry will be meshed with tetrahedral elements. Therefore, mesh refinement will be done on the crankpin fillet area and journal fillet in order to obtain a fine mesh on the fillet area. Elements of tetrahedral shapes are used to connect complex geometries that have been imported into the ANSYS workbench software. Mesh optimization is carried out until results from FEA and analytical solutions are obtained and similar to each other. Table 16.3 gives the details of the crankshaft meshing and Fig. 16.2 shows the meshing of the single-cylinder engine crankshaft.

Table 16.3 Details of crankshaft meshing

Height of crankpin (mm)	Type of meshing	Sizing (mm)	Nodes	Elements
1.15	Tetrahedral shapes	1.0	334,847	196,513
2	Tetrahedral shapes	1.0	334,976	196,561
3	Tetrahedral shapes	1.0	338,660	199,109
4	Tetrahedral shapes	1.0	335,912	197,180

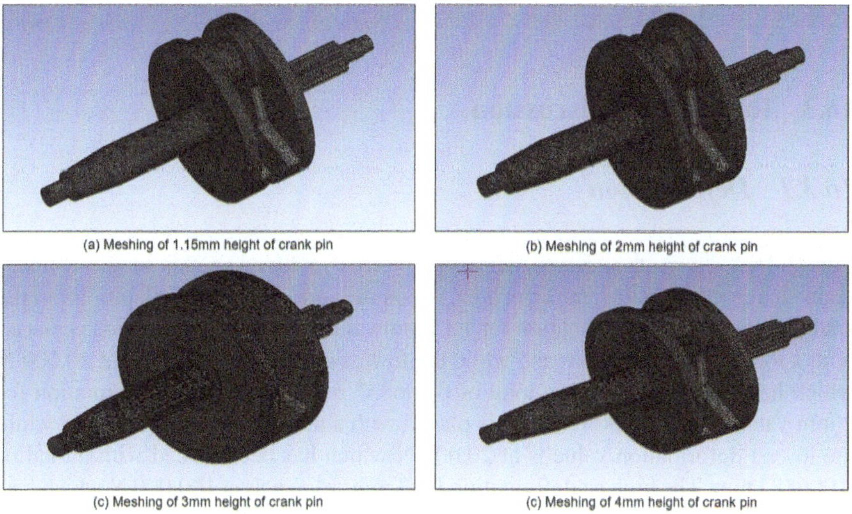

Fig. 16.2 Meshing of single-cylinder engine crankshaft

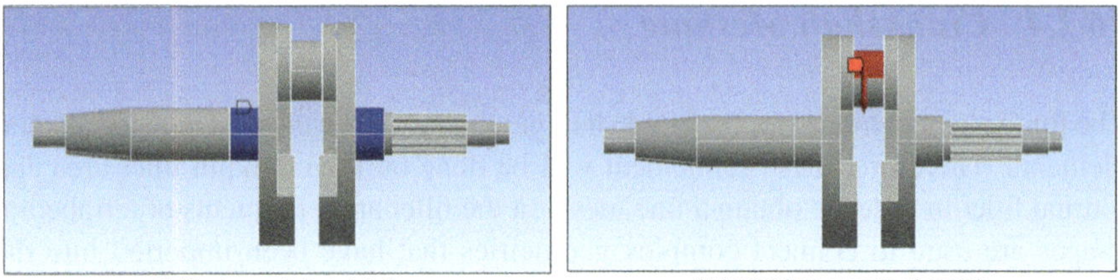

(a) Fixed support of crankshaft (b) Applied force at centre of crankpin

Fig. 16.3 Applied load and boundary conditions

16.2.5 Apply Load and Boundary Conditions

The component is charged with a load to observe whether the crankshaft is in the maximum bending moment position or is in the dead centre. In this study, the force will be variable with the gas load on the piston obtained is 20000, 40,000, 60,000, 80,000 and 100,000 N, which is applied at the centre of the crankpin. Therefore, load data was taken from the crankshaft journal. The boundary condition is based on the supporting condition of the crankshaft, which is that crankshaft journals are fixed on both sides of the crank web. Figure 16.3 shows the applied load and boundary conditions.

16.3 Results and Discussion

16.3.1 Deformation

Figure 16.4 shows the critical deformation area involved in this single-cylinder crankshaft located in the crank pin area, i.e. on the surface of the crank pin. From the analysis, the highest deformation for 1.15 mm value is at 100,000 N which is placed with a total of 0.034184 mm and while the lowest deformation value is at 20,000 N which has been placed with a total of 0.006837 mm. The highest deformation for 2 mm value is at 100,000 N which is placed with a total of 0.034408 mm and while the lowest deformation value is at 20,000 N which has been placed with a total of 0.006882 mm. The highest deformation for 3 mm value was at 100,000 N which was placed with a total of 0.034743 mm and while the lowest deformation value was at 20,000 N which was placed with a total of 0.006949 mm. The highest deformation for 4 mm value is at 100,000 N which is placed with a total of 0.0350990 mm and while the lowest deformation value is at 20,000 N which has been placed with a total of 0.0070199 mm. Hence, the higher the force applied, the higher the deformation value.

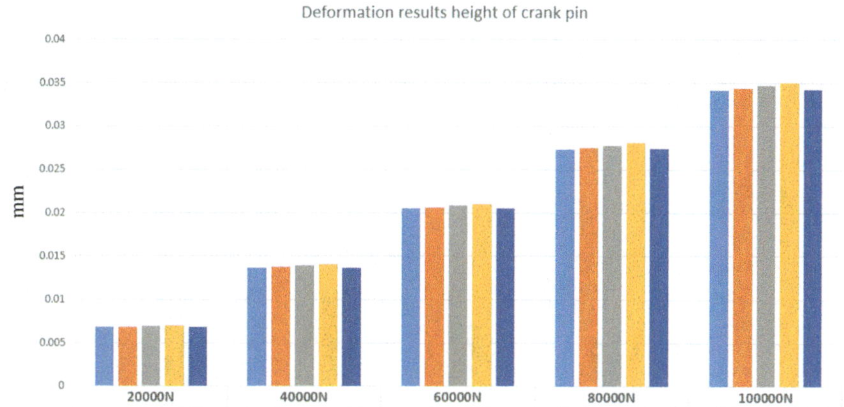

	20000N	40000N	60000N	80000N	100000N
1.15mm	0.0068368	0.013674	0.02051	0.027347	0.034184
2mm	0.0068815	0.013763	0.020645	0.027526	0.034408
3mm	0.0069486	0.013897	0.020846	0.027794	0.034743
4mm	0.0070199	0.01404	0.02106	0.028079	0.035099
Standard	0.0068507	0.013701	0.020552	0.027403	0.034253

1.15mm 2mm 3mm 4mm Standard

Fig. 16.4 Graphs and data for deformation on height of crank pin

16.3.2 Strain

Based on the strain results obtained, there are two results from the strain, namely the analytical result and the computational result. These results analytical results and software results which produce different data outputs. Figure 16.5 shows that the critical strain area involved in this single-cylinder engine crankshaft is located in the area between the fillet crank web and the shaft. The strain value of the highest strain analytical result is at 100,000 N which is placed with a total of 0.0005082 and while the lowest strain value is at 20,000 N which has been placed with a total of 0.0025390. The highest strain for 1.15 mm value on the result software is at 100,000 N which is placed with a total of 0.0046062 and while the lowest strain value is at 20,000 N which has been placed with a total of 0.0009212. The highest strain value for 2 mm on the result software is at 100,000 N which is placed with a total of 0.0040417 and while the lowest strain value is at 20,000 N which has been placed with a total of 0.0008083. The highest deformation for 3 mm value was at 100,000 N which was placed with a total of 0.034743 mm and while the lowest deformation value was at 20,000 N which was placed with a total of 0.006949 mm. The highest strain for 4 mm value on the result software is at 100,000 N which is placed with a total of 0.0048726 and while the lowest strain value is at 20,000 N which has been placed with a total of 0.0009745. Hence, the higher the force applied, the higher the strain value.

Referring to Fig. 16.5, the higher the value of force applied resulted higher value of strain obtained. The crankshaft is also expanded based on the amount of force that has been applied. The value of stress and strain will increase together if it is

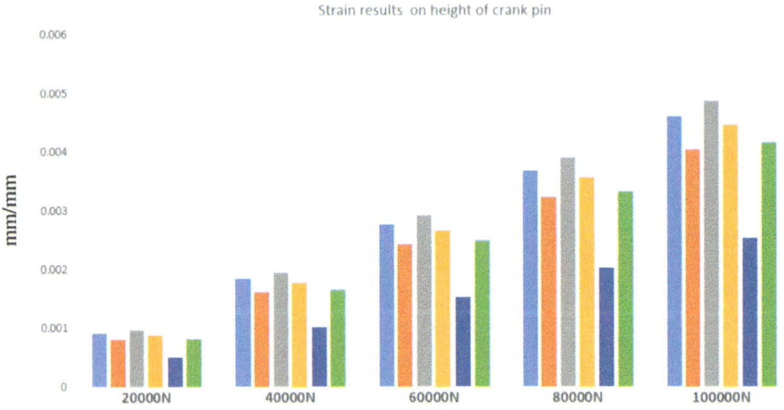

	20000N	40000N	60000N	80000N	100000N
1.15mm	0.00092123	0.0018425	0.0027637	0.0036849	0.0046062
2mm	0.00080834	0.0016167	0.002425	0.0032334	0.0040417
3mm	0.00097451	0.001949	0.0029235	0.003898	0.0048726
4mm	0.00089452	0.001789	0.0026836	0.0035781	0.0044726
Analytical result	0.0005082	0.0010158	0.001524	0.002031	0.002539
Standard	0.00083147	0.0016629	0.0024944	0.0033259	0.0041573

Fig. 16.5 Graphs and data for strain on height of crank pin

in the elastic limit phase. Moreover, the percentage rate difference values at all the crank pin heights have shown that the strain rate percentage values at 3 mm crankpin heights are the highest compared to the crank pin heights at 1.15, 2 and 4 mm. This can be translated because the value of stress experienced at the 3 mm crankpin height is high which has been influenced by the location and height of the crank pin.

16.3.3 Stress

Figure 16.6 shows that the critical stress area involved in this single-cylinder engine crankshaft is located in the area between the fillet crank web and the crank pin. The highest stress value on the analytical result is at 100,000 N which is placed with a total of 533.24 MPa and while the lowest stress value is at 20,000 N which has been placed with a total of 106.72 MPa. The highest stress for 1.15 mm value on the result software is at 100,000 N which is placed with a total of 563.47 MPa and while the lowest stress value is at 20,000 N which has been placed with a total of 112.69 MPa. The highest stress for 2 mm value on the result software is at 100,000 N which is placed with a total of 575.49 MPa and while the lowest stress value is at 20,000 N which has been placed with a total of 115.10 MPa. The highest stress for 3 mm value on the software result is at 100,000 N which is placed with a total of 684.91 MPa and while the lowest stress value is at 20,000 N which has been placed with a total of 136.98 MPa. The highest stress for 4 mm value on the result software is at 100,000 N

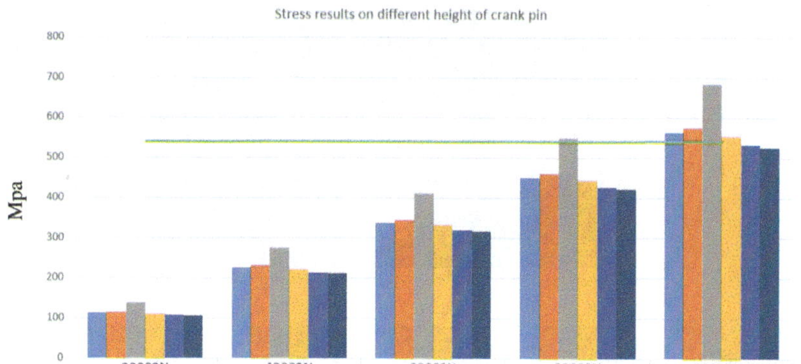

	20000N	40000N	60000N	80000N	100000N
1.15mm	112.69	225.39	338.08	450.77	563.47
2mm	115.1	230.2	345.3	460.39	575.49
3mm	136.98	273.96	410.94	547.93	684.91
4mm	111.12	222.25	333.37	444.5	555.62
Analytical result	106.72	213.32	319.95	426.59	533.24
Standard	105.2	210.4	315.59	420.79	525.99
Yield strength	540	540	540	540	540

1.15mm 2mm 3mm 4mm Analytical result Standard Yield strength

Fig. 16.6 Graphs and data for stress on height of crank pin

which is placed with a total of 555.62 MPa and while the lowest stress value is at 20,000 N which has been placed with a total of 111.12 MPa.

This shows the crankshaft at 100,000 N for all crankshaft heights, i.e., 1.15, 2, 3 and 4 mm in phase to fracture or cracking because the stress value obtained has exceeded the yield strength value of 540 MPa. However, for a height of 3 mm, the phase to fracture or crack is experienced earlier at a force value of 80000 N because the resulting stress value exceeds the yield stress value.

The critical stress area for 3 mm height of crank involved in this single-cylinder engine crankshaft is located in the area between the fillet crank web and the crank pin. The higher the value of stress placed, the higher the value of stress obtained. As the stress of the crankshaft is increased, this will decrease fatigue life of the crankshaft. The critical strain area involved in the crankshaft of this single-cylinder engine is located in the area between the fillet crank web and the shaft. The higher the value of force applied, the higher the value of strain obtained. This is because the higher the force that will make the crankshaft to expand according to the amount of force that has been applied. Thus at the critical deformation area for 3 mm height of crankpin involved in this single-cylinder crankshaft located in the crank pin area, i.e., on the surface of the crank pin. Normal deformation may occur due to the maximum load of the crankshaft that needs to be taken for confirmation of work with von Mises stresses.

The stress at 3 mm height of crankpin is higher than the height of other crankpin location. This means fatigue life at 3 mm height of crankpin will decrease. This is because the location of the 3 mm height of the crankpin is not practical because it will put pressure on the web crank to accommodate the crankpin when it experiences any

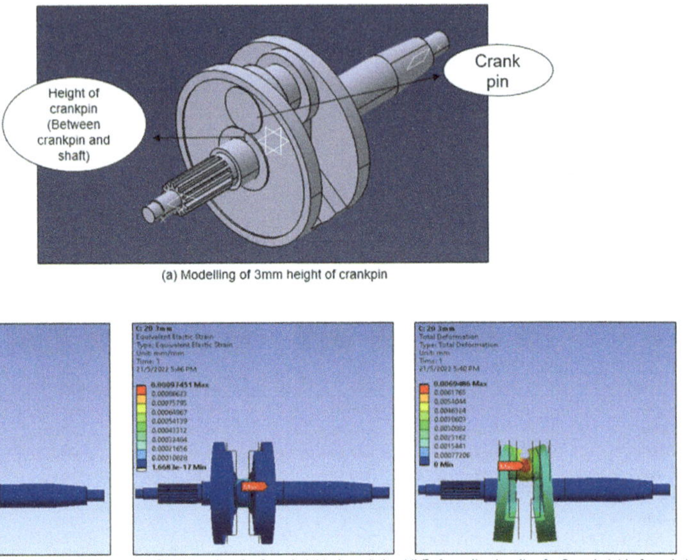

(a) Modelling of 3mm height of crankpin

(b) Stress location for 3mm height of crankpin (c) Strain location for 3mm height of crankpin (d) Deformation location for 3mm height of crankpin

Fig. 16.7 Results and detail for 3 mm height of crankpin

pressure. However, the deformation value for 3 mm height of crankpin is lower than the deformation value at 4 mm. This is because the geometry shape at the crankpin height at 3 mm is lower than 4 mm. If it is getting higher, the crankpin position will be at the end of the web crank which will cause the deformation value to increase. In addition, for the Yamaha 135 LC, crankshaft model at 3 mm height of crankpin is not practiced in the field of modification and it is difficult to find in the market. After performing this simulation analysis, it is proven that this 3 mm height of crankpin has a high stress value compared to the crankpin height at 1.15, 2 and 4 mm. Figure 16.7 shows the results and detail for 3 mm height of crankpin.

16.4 Conclusion

As part of this study, the main purpose is about stress effect on the performance of a single-cylinder four-stroke cycle engine. There are four different crank pin heights on the crankshaft starting from 1.15, 2, 3 and 4 mm from similar engines which were studied in this research. The single-cylinder engine crankshaft models with different crank pin heights, i.e. 1.15, 2, 3 and 4 mm were created by CATIA software. Then, the model that has been designed in CATIA software has been imported into ANSYS software. Thus, finite element analysis using ANSYS software and analytical methods has been performed. The maximum stress results in all differences in crankshaft height in the area between the crank pin and the crank web where the edge of the crank pin is the main area for maximum stress in the crankshaft. Mesh refinement is performed on the fillet crank pin and fillet shaft, so that a fine mesh is obtained on the fillet area, which is generally a critical location on the crankshaft. The failure in the crankshaft is initiated at the crank pin area and strain is initiated at the area between the crank web and the shaft. The results of comparison at all crankpin heights with different force values show the effect of stress on the crankshaft. Finally, the difference between the analytical result and the software result is dependent on the height of the crank pin. This is because the data output produced by the analytical result is the same even though the geometry on the crankshaft has changed. Therefore, this can be summarized that the software has produced data based on geometry and parameters, while the analytic result is not sensitive with the changes that occur in the crankshaft geometry.

Acknowledgements We would like to thank Universiti Kuala Lumpur for the supports and assistants towards the completion of this research.

References

Aatish C, Abhishek M, Devendra R et al (2020) Design and Analysis of Crankshaft using Forged Steel & Composite Materials. Int J Res Appl Sci Eng Tech 8(5):1107–1112

Amit S, Jaydeepsinh D (2014) Design and Stress Analysis of Crankshaft for Single Cylinder 4-Stroke Diesel Engine. Int J Res Appl Sci Eng Tech 2(5):320–324

Ding Y, Li X (2011) Crankshaft Strength Analysis of a Diesel Engine using Finite Element Method. 2011 Asia-Pacific Power and Energy Engineering Conference. IEEE, p 1

Durga Prasad K, Jagath Narayana K, Kiranmayee N (2018) Design and Stress Analysis of Crankshaft for Single Cylinder 4 Stroke Diesel Engine. Int J Eng Res Tech 7(11):154–159

Joshi JJ, Patel DM (2017) Design and Failure Analysis of Single Cylinder Petrol Engine Crankshaft using ANSYS Software. Int J Eng Sci Comput 7(4):10549–10555

Karthick L, Mallireddy N, Yogaraja J et al (2021) Modelling and Analysis of an EN8 crankshaft material in comparison with Forged steel crankshaft. Mater Today 47(17):6168–6172

Kumar MS, Ragunathan S, Suresh M (2014) Analysis of Crankpin Failure in a Single Cylinder Engine. Int J Mech Eng Robot Res 3(4):260

Mathapati NC, Dhamejani CL (2015) FEA of A Crankshaft in Crank-pin Web Fillet Region for Improving Fatigue Life. Int J Inno Eng Res Tech 2(6):1–8

Metkar RM, Sunnapwar VK, Hiwase SD et al (2013) Evaluation of FEM based fracture mechanics technique to estimate life of an automotive forged steel crankshaft of a single cylinder diesel engine. Procedia Eng 51:567–572

Pandiyan A, Arunkumar G, Patel A et al (2018) Design and optimization of crankshaft for single cylinder 4-Stroke spark ignition engine using coupled steady-state thermal structural analysis. Int J Mech Eng Techn 9(7):135–145

Singh AK, Praveen KS, Tripathi AK et al (2014) FEA of the crankshafts design by using Ansys workbench for nickel chrome steel and structural steel. Int J Sci Eng Res 5(4):1249–1253

Solanki A, Dodiya J (2014) Design and stress analysis of crankshaft for single cylinder 4-stroke diesel engine. Int J Res Appl Sci Eng Techn 2(5):320–324

Sorte SM, Sheikh SM (2013) Stress Analysis and Design Optimization of Crankpin. Int J Sci Mod Eng 1(4):18–20

Thriveni K, Chandraiah BJ (2013) Modeling and analysis of the crankshaft using ANSYS software. Int J Comp Eng Res 3(5):84–89

Yingkui G, Zhibo Z (2011) Strength Analysis of Diesel Engine Crankshaft based on PRO/E and ANSYS. 3rd International Conference on Measuring Technology and Mechatronics Automation, vol 3. IEEE, p 362

Chapter 17
Antimicrobial Peptides, An Alternative Antimicrobial Agent Against Multi-drug-Resistant Microbes: Source, Application, and Potential

Nur Husna Hafiza Lyana Ramzah, Tong Woei Yenn, Wing-Hin Lee, Ching-Yee Loo, Wen-Nee Tan, and Leong Chean Ring

Abstract Pathogenic microorganisms are often associated with infectious diseases in humans, plants, or animals. Over the past 100 years, antibiotics have been often used to combat the infections caused by these pathogenic microorganisms. The frequent and overuse of antibiotics has led to the rapid emergence of multi-drug-resistant microorganisms. There has been a worldwide quest for a new antibacterial agent. In recent years, antimicrobial peptides (AMPs), produced naturally by bacteria, insects, amphibians, and mammals, have gained attention as an alternative antimicrobial agent against multi-drug resistant microbes. Contrary to conventional antibiotics, which typically work by targeting a specific high-affinity antimicrobial target and may lead to the development of resistance in microorganisms, AMPs exert various antimicrobial activities that may offer a strategy to stop bacterial resistance. In addition, AMPs have a wide range of applications as they are effective in

N. H. H. L. Ramzah · L. C. Ring (✉)
Universiti Kuala Lumpur, Malaysian Institute of Chemical and Engineering Technology, Lot 1988, Kawasan Perindustrian Bandar Vendor, Taboh Naning, 78000 Alor Gajah, Melaka, Malaysia
e-mail: crleong@unikl.edu.my

N. H. H. L. Ramzah
e-mail: husna.ramzah29@s.unikl.edu.my

T. W. Yenn
Institute of Medical Science Technology, Universiti Kuala Lumpur, A1-1, Jalan TKS 1, Taman Kajang Sentral, 43000 Kajang, Selangor, Malaysia
e-mail: wytong@unikl.edu.my

W.-H. Lee · C.-Y. Loo
Faculty of Pharmacy and Health Sciences, Universiti Kuala Lumpur Royal College of Medicine Perak, Ipoh, Perak, Malaysia
e-mail: whlee@unikl.edu.my

C.-Y. Loo
e-mail: cyloo@unikl.edu.my

W.-N. Tan
School of Distance Education, Universiti Sains Malaysia, 11800 Gelugor, Pulau Pinang, Malaysia
e-mail: tanwn@usm.my

A. Ismail et al. (eds.), *Advancements in Materials Science and Technology Led by Women*, Advanced Structured Materials 165,
https://doi.org/10.1007/978-3-031-21959-7_17

eradicating microorganisms and are non-toxic or harmful to humans and the environment. This paper reviews the sources, application, and potential of AMPs against multi-drug-resistant microorganisms and their risks to humankind.

Keywords Antimicrobial peptides · Multi-drug-resistant microorganisms · Source · Applications · Potential

17.1 Introduction

In this new era of medicine, antibiotics are not only prescribed to treat infectious diseases, but they have also emerged as therapeutic agents for various medical procedures including chemotherapy for cancer treatment, organ transplantation, hip replacement surgery, and many other practices (Prestinaci et al. 2015). Patients who undergo these treatments are more likely to develop a high risk of morbidity and mortality as they are more susceptible to bacteria-associated infections. Nevertheless, in the past few decades, bacteria-associated infections have developed resistance toward every antibiotic introduced on the market (Prestinaci et al. 2015). Frequent and widely use of antibiotics has caused bacteria to build resistance against them, resulting in poor efficiency of the drugs (Holt 2018). The World Health Organization (WHO) mentioned that the world is moving toward a post-antibiotic era where infectious diseases are possibly fatal if no action is taken to eliminate multi-drug-resistant microorganisms. In 2016, the World Bank reported that multi-drug-resistant microorganisms would cause a financial burden on low-income and middle-income countries (World Health Organization ROFS-EA 2018; Naeemmudeen et al. 2021). As infections caused by multi-drug-resistant microorganisms increase, death and loss of productivity worldwide would also surge. As stated in the review by Adebisi et al. in 2021, it is estimated that by 2050, multi-drug-resistant microorganisms could result in 10 million deaths globally, in which people are dying more due to drug-resistant infections rather than cancer (Adebisi et al. 2021; Akpan et al. 2020). Additionally, during the past three decades, fewer new antibiotics have received approval and been introduced into the market. Therefore, there is an urgent need to develop novel antimicrobial agents to overcome the issues of infections caused by multi-drug-resistant microorganisms.

Antimicrobial peptides (AMPs) are generally known as small biological molecules (<10 kDa) that possess inhibitory activity against bacteria, fungi, protozoa, and some viruses. AMPs are produced naturally by certain plants, animals, and microorganisms as defense mechanisms against the invasion and attack of other microorganisms (Maria-Neto et al. 2015). The first AMP—Gramicidin, an antibacterial compound produced from a soil-based *Bacillus* strain, was discovered more than 80 years ago (Rima et al. 2021). Since then, great interest has been generated among the scientific community due to their excellent property of almost no or limited resistance development (Rima et al. 2021; Yu et al. 2018). 3240 AMPs have been discovered and reported to the antimicrobial peptide database (APD3). AMPs can be another option for available antibiotics that develop resistance rather quickly (Maria-Neto et al. 2015; Rima et al. 2021; Forde and Devocelle 2015). Thus, this paper focuses on the source, application, and potential of AMPs moving forward to fight against multi-drug-resistant microorganisms in this globalized era.

17.2 Source, Mechanism of Action, Application, and Potential of Antimicrobial Peptides

17.2.1 The Source of AMP and Diversity

AMPs can be found in prokaryotes (microorganisms such as bacteria) and eukaryotes (fungi, plants, and animals) (Bagley 2014; Berglund et al. 2015; Rai et al. 2016). As of 2021, there are 3324 AMPs recorded in the APD from 6 life kingdoms; 391 from bacteria, 5 from archaea, 8 from protists, 22 from fungi, 364 from plants, and 2446 from animals (AMP Database Search 2017). AMPs can be classified according to their biological sources; bacterial AMPs (bacteriocins), plant AMPs, and animal AMPs (insect AMPs, amphibian AMPs, fish AMPs, reptile AMPs, mammal AMPs, and others, depending on the source family names) (AMP Database Search 2017).

Figure 17.1 shows the main sources of AMPs and some of their examples.

17.2.1.1 Bacterial AMPs

To minimize the usage of antibiotics and consequently infections caused by multi-drug-resistant microorganisms, peptides, or small proteins from bacteria (bacteriocins) have been developed to eradicate other bacteria (Jamali et al. 2019; Martinez

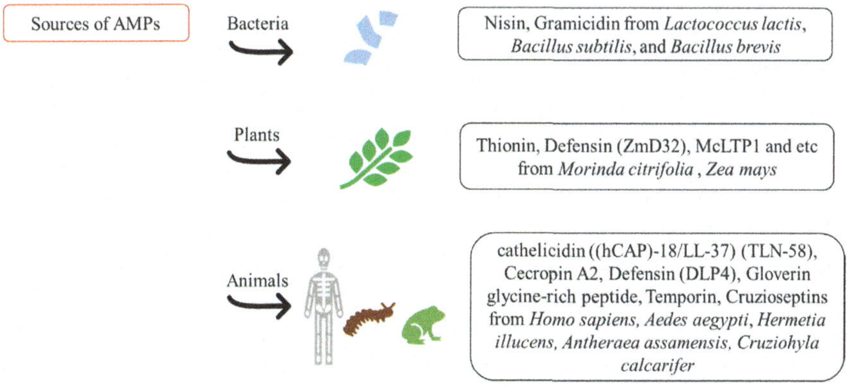

Fig. 17.1 AMPs and their main sources

et al. 2013). In a study by Lazdunski in 1988, bacteriocins were first characterized in Gram-negative bacteria and applied to ensure food safety (Jamali et al. 2019; Lazdunski 1988). Bacteriocins are beneficial as natural AMPs in foods where risks of food safety and infections associated with bacteria in humans and livestock can be lowered (Jamali et al. 2019). In 2018, isolation of *Bacillus cereus* from Tunisian dairy products produced an AMP when grown in casein through the production of extracellular enzymes (Ouertani et al. 2018). This AMP was active against some *Bacillus* and *Listeria monocytogenes* species. Ouertani et al. stated that this production of new AMP from contaminants like *B. cereus* can further garner researchers' interest in developing eco-friendly preservatives in dairy products from specific proteolytic enzymes, thus reducing the usage of chemical preservatives. Another AMP produced from bacteria was developed by Zhu et al. in (2021), a human oral actinomyces-sourced defensin–actinomycesin. Actinomycesin was present in the defensins from actinomyces found in the human oral cavity, ruminant rumen, and dental plaque, which it showed great adaption to many bacteria species in said human body parts. Actinomycesin inhibited bacterial peptidoglycan synthesis and imposed low toxicity on human and mammalian cells in the heart and nervous system. This AMP was also active against *Streptococcus pneumoniae*-infected mice and *MRSA*-induced infection mice.

17.2.1.2 Plant AMPs

In plants, AMPs serve as protection against bacterial and fungal infection, as plants are lacked of adaptive immunity (Kumar et al. 2018). Plant AMPs can be classified based on their sequence similarity, cysteine motifs presence, and tertiary structures (Hammami et al. 2009; Li et al. 2021). The classification of these AMPs includes thionin, plant defensin, hevein-like peptides, a-hairpinin, lipid transfer protein, snakins, and cyclotide family (Li et al. 2021). AMPs derived from plants can be detected in their leaves, flowers, seed, and tubers (Kumar et al. 2018; Jenssen et al. 2006). In 2017, an AMP that belonged to lipid transfer protein was isolated from *Morinda citrifolia L.* or noni seeds and identified as $McLTP_1$ $_($Souza et al. 2018$_)$. This AMP exhibited antibacterial properties as it can reduce *Staphylococcus aureus* and *Staphylococcus epidermis* planktonic growth, with maximal inhibition up to 50% and 98%, respectively. Biofilm formation of *S. aureus* strains tested was also successfully inhibited by $McLTP_1$. In the in vivo experiments, $McLTP_1$-treated mice were still alive after three days before or after cecal ligation and puncture (CLP). $McLTP_1$ also reduced body weight loss, fever, leukocytosis, organ damage, and inflammatory serum cytokines level for sepsis-induced animals. Based on the study conducted by Kerenga et al. (2019), they have studied the relationship between the charge of an AMP and its activity in media with elevated concentrations by using plant defensins. At certain salt concentrations, it was observed that AMPs experienced a loss of activity, thus limiting their clinical applications. This leads to poor electrostatic interactions between the cationic AMPs and anionic surfaces of microbial cells when salt is present. Kerenga and his colleagues have discovered ZmD32, a defensin from *Zea mays* or corn with charge of + 10.1 and pH 7 to be antifungal and antibacterial. ZmD32 was active against fungal species in media with high concentration of salt and retained activity against *Candida albicans* biofilm, Gram-negative and Gram-positive bacteria.

17.2.1.3 Animal AMPs

Invertebrates and vertebrates require host defense peptides (HDPs) as their innate immune systems (Wang 2014). Like plants, invertebrates such as insects and crustaceans do not have an adaptive immune system. Hence, their primary defense mechanism is the innate defense system. As for vertebrates, their innate immune systems help kill microbes in the early stages and assist the adaptive immune system in being active against infections.

Human host defense peptides protect against infections caused by microbes and can be detected in different stages of human growth. For instance, cathelicidin LL-37 can be found in the skin of newborn infants while human beta-defensin 2 is detected in the elderly human bodies (Gschwandtner et al. 2014; Huan et al. 2020). Some major families of human AMPs include defensins, histatins, cathelicidins, dermicin, hepcidins, AMPs derived from known proteins, antimicrobial proteins, and many more (Wang 2014). In 2016, palmoplantar pustulosis (PPP) vesicle was found to contain AMP human cathelicidin (hCAP)-18/LL-37 (Murakami et al. 2017). hCAP-18/LL-37 was depleted using an LL-37 antibody affinity column after lesional PPP vesicles were recovered from PPP patients. A designed recombinant hCAP-18 peptide was produced and treated with the depleted PPP vesicle fluid, in which an extra form was later identified as TLN-58. TLN-58 has shown antibacterial action against *S. aureus*, *S. epidermidis*, and *Group A Streptococcus*, similar to LL-37. On the other hand, dermcidin is detected in human sweat, unlike human defensins and cathelicidins found in inflamed and injured skin (Wang 2014; Rieg et al. 2004). It is observed that the dermcidin level found in healthy people is similar to that of infected patients (Wang 2014; Rieg et al. 2014), and it might be associated with cancer and atherosclerosis (Wang 2014; Schittek 2012; Ghosh et al. 2012).

AMPs derived from insects can be classified into three main groups according to amino acid sequence and structures; cecropins which have linear peptides with a-helix but no cysteine residues, defensins with 6–8 conserved cysteine residues, 3 or 4 disulfide bridges, and 3 domains consisting in a flexible amino-terminal loop and lastly, peptides with proline and/or glycine residues (Makarova et al. 2018; Wu et al. 2018). In 2017, a study conducted by Zheng et al. have discovered cecropin A2 peptide from mosquito *Aedes aegypti* which was active against *Pseudomonas aeruginosa* and other Gram-negative bacterial biofilm, with minimum inhibitory concentration (MIC) recorded to be ranging from 2 to 64 μg/mL (Sahoo et al. 2021; Zheng et al. 2017). Another study in 2015, a defensin-like peptide known as DLP4 was isolated from black soldier fly (*Hermetia illucens)* larvae (Sahoo et al. 2021; Park et al. 2015). This AMP was effective in eradicating *S. aureus*, *S. epidermis*, *Escherichia coli,* and *Bacillus subtilis* biofilm formation with MIC of 0.02–1.17 μM. One of the examples of AMP that has glycine residues is detected in 2018, known as gloverin glycine-rich peptide (Sahoo et al. 2021; Nayak et al. 2018). This AMP was derived from muga silkworm (*Antheraea assamensis*) and could eliminate bacterial biofilm formation of *E. coli* and *Enterobacter cloacae* with MIC of 16.24 and 18.56 μg/mL.

AMPs from the frog or *anuran* species have been studied and developed over the past few years, for their benefits in lowering risks of antimicrobial resistance. According to Guangshun Wang, based on the APD, 99.9% of the AMPs derived from frogs are less than 50 amino acids long and have an average length of 24 residues and a net charge of + 2.5 (Wang 2020; Mangoni and Casciaro 2020). Amphibian peptides such as temporins, brevinins, and esculentins are correlated through multiple length-dependent relationships; when the length of the peptide increases, the net charge increases and hydrophobic content decreases (Wang 2020). Peptides like glycine, leucine, lysine, and proline, on the other hand, are linearly correlated with the peptide length (Wang 2020). One example of the *anuran* AMPs discovered is a novel temporin peptide, Temporin-PF (TPF), extracted from ranid frog *Pelophylax fukienensis* skin (Zai et al. 2021). TPF showed excellent activity against Gram-positive bacteria but not against Gram-negative bacteria. The aggregation pattern of this AMP was also studied by conducting "salting out" assays and MD simulation to ensure the peptide conformation. When aromaticity was added to the AMP, its bioactivities were not improved and binding affinity to the lipid membrane of microbes decreased. In 2020, in a study by Cuesta et al., two novel cruzioseptins peptides, namely cruzioseptin-16 and -17 were produced from leaf frog *Cruziohyla calcarifer* through solid-phase peptide synthesis (Cuesta et al. 2021). These amphibian AMPs were 21–23 residues long, alpha-helical cationic peptides, and active against *E. coli, S. aureus,* and *C. albicans.* AMPs, their sources, and target microorganisms are summarized in Table 17.1.

17.2.2 AMP Mechanisms of Action

AMPs are advantageous compared to conventional antibiotics as they are more active against resistance-developing bacteria due to their electrostatic interactions with the bacterial cell membrane (Boparai and Sharma 2019; Hollmann et al. 2018; Pfalzgraff et al. 2018). As more AMPs are recently being discovered, it is vital to study and comprehend the mechanism of action of these peptides to ensure continuous improvement in their performance against multi-drug resistant microbes. There are five main processes of interaction between AMPs and their hosts; (1) initial contact with target cells through biochemical or biophysical relationships such as electrostatic or hydrophobic interactions, (2) adjusting their structures in the target membrane (conformation into a helix or barrel structure), (3) aggregation until an active stoichiometric level is reached, (4) disrupting the target membrane by permeabilization or depolarization, or causing other direct/indirect, temporary or permanent irregularity in its function, and (5) obtaining access to the targets via the membrane or from within the cell (Moravej et al. 2018; Koprivnjak and Peschel 2011). AMPs

Table 17.1 AMPs from various sources and their targeting microorganisms

Sources	AMP	Targeting organisms	References
Bacteria	AMP derived from *B. cereus* RC6 strain	*B. cereus* ATCC 11,778 (BC45), *B. thuringiensis* USDA HD22, and *L. monocytogenes* DISTAM MACa1	Ouertani et al. (2018)
	Defensin (Actinomycesin)	*S. pneumoniae* and *MRSA*	Zhu et al. (2021)
Plants	McLTP1	*S. aureus* and *S. epidermis*	Souza et al. (2018)
	Defensin (ZmD32)	*C. albicans*, Gram-negative, and Gram-positive bacteria	Kerenga et al. (2019)
Animals	Design recombinant of human cathelicidin ((hCAP)-18/LL-37) (TLN-58)	*S. aureus, S. epidermidis*, and *Group A-Streptococcus*	Murakami et al. (2017)
	Cecropin A2	*P. aeruginosa* and Gram-negative bacteria	Sahoo et al. (2021); Zheng et al. (2017)
	Defensin (DLP4)	*S. aureus, S. epidermis, E. coli,* and *B. subtilis*	Sahoo et al. (2021), Park et al. (2015)
	Gloverin glycine-rich peptide	*E. coli* and *E. cloacae*	Sahoo et al. (2021), Nayak et al. (2018)
	Temporin (Temporin-PF (TPF))	Gram-positive bacteria	Zai et al. (2021)
	Cruzioseptins (cruzioseptin-16 and -17)	*E. coli, S. aureus,* and *C. albicans*	Cuesta et al. (2021)

can be divided into two mechanisms of action: direct killing (membrane acting and non-membrane acting) and immune modulation (Kumar et al. 2018; Boparai and Sharma 2019) as in Figure. Peptides acting via membrane acting mechanism disrupt the cell membrane of microorganisms, while non-membrane peptides can traverse the membrane without causing damage (Boparai and Sharma 2019; Hancock and Patrzykat 2002). AMPs acting via immune modulation mechanism can generate an immune response to attract other immune cells and regulate inflammation in case of infections (Kumar et al. 2018) (Fig. 17.2).

17.2.2.1 Direct Killing: Membrane Targeting Mechanism of Action

After initial electrostatic and hydrophobic interactions between bacteria and AMPs, AMPs will aggregate on the bacterial membrane at a particular concentration (Kumar et al. 2018; Epand et al. 2016; Andersson et al. 2016). This process stage is best represented in models: toroidal pore model, barrel-stave pore model, and carpet-like

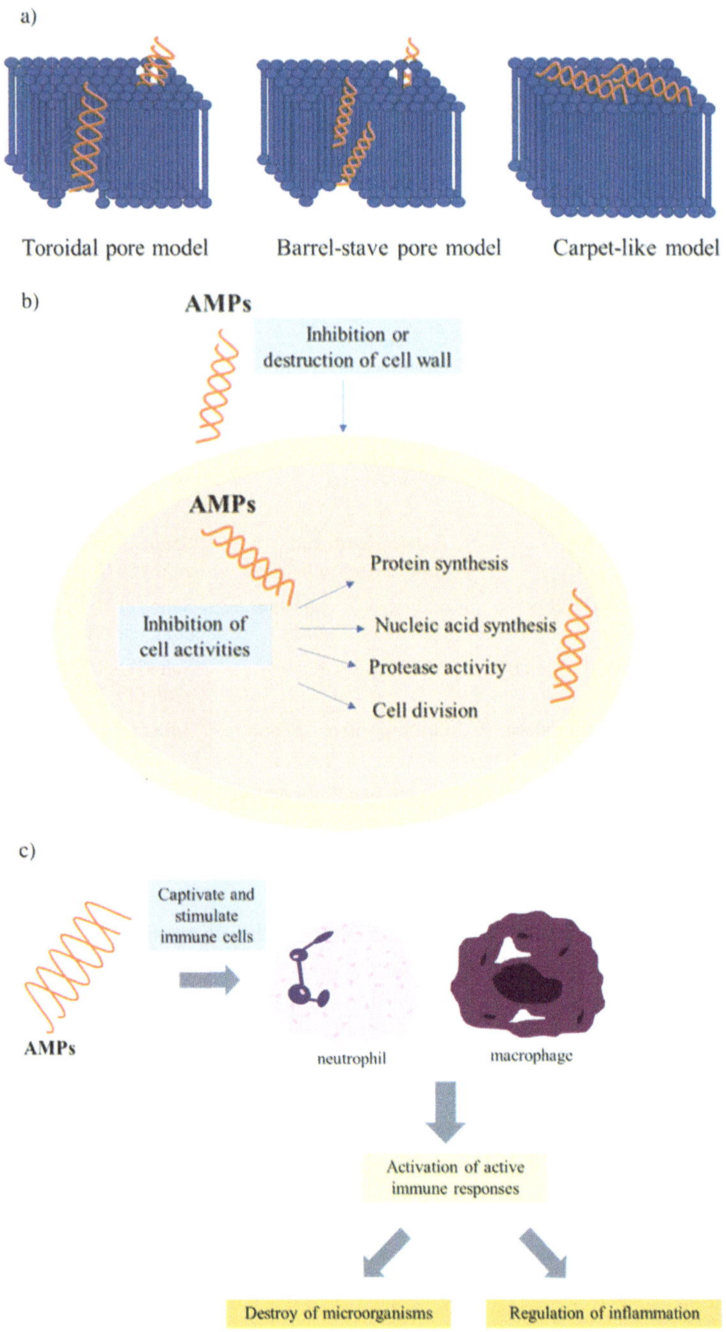

Fig. 17.2 Mechanism action of AMPs: **a** Direct killing mechanism via membrane-targeting, **b** direct killing mechanism via non-membrane targeting, and **c** immune modulation mechanism

model (Huan et al. 2020). Two major categories of models are transmembrane pore model and non-pore model, where toroidal pore and barrel-stave pore models belong to the first category (Kumar et al. 2018).

Toroidal pore model or wormhole model indicates that AMPs are inserted perpendicularly in the lipid bilayer; however, specific peptide–peptide interactions are absent (Kumar et al. 2018; WC, 2010). Hence, the peptides cause the lipid bilayer to curve and pores are formed by the peptides as well as the phospholipid head group. The major difference between the toroidal pore model and barrel-stave model is the bilayer net arrangement, where the hydrophobic and hydrophilic arrangements of lipids in toroidal pore model are interrupted unlike in the latter model. This condition allows the interaction between the lipid tail and lipid head. Some peptides may enter the cytoplasm via inner cytoplasmic leaflet when the pores disintegrate, allowing them to target intracellular components. Some AMPs that exhibit the toroidal pore model are magainin 2, lacticin Q, and arenicin (Huan et al. 2020).

For the barrel-stave pore model, the AMPs first oriented parallel to the membrane and subsequently entered the lipid bilayer perpendicularly (Kumar et al. 2018; Ehrenstein and Lecar 1977). Peptide-peptide interactions are hence present. In the barrel-stave model, multimers of AMPs penetrate the cell membrane and produce channels that allow cytoplasmic outflow (Huan et al. 2020). It is also reported that AMPs in this pore model can disrupt cell membranes and cause cell death (Huan et al. 2020; Lohner and Prossnigg 2009). The amphipathic structure of peptide plays an important role in forming pores for this model; the hydrophobic part will interact with lipids while hydrophilic residues will form lumen of the channels (Kumar et al. 2018; Brogden 2005; Breukink and Kruijff 1999). AMPs such as alamethicin (Kumar et al. 2018; WC W, 2010), pardaxin (Kumar et al. 2018; Rapaport and Shai 1991; Shai et al. 1990), and protegrins (Kumar et al. 2018; Brogden 2005) are shown to form these model channels.

AMPs may also function without generating pores in the membrane, which is presented by the carpet model (Kumar et al. 2018). In this model, AMPs are arranged parallel to the cell membrane, forming a "carpet" such that; the hydrophilic end is in contact with the solution while the hydrophobic end is in contact with the phospholipid bilayer (Huan et al. 2020). The carpet on the membrane surface is formed when the AMPs are attached to the lipid bilayer and a certain concentration threshold is achieved (Kumar et al. 2018). As interactions on the membrane surface are disturbed, membrane integrity is then destroyed in a detergent-like manner by forming micelles (Kumar et al. 2018; Huan et al. 2020; Oren and Shai 1998). Membrane bilayer structure made up of micelles is known as the detergent-like model (Kumar et al. 2018). In the carpet-like model, there is no need for peptide–peptide interactions between the membrane-bound peptide monomers also for the peptide to transverse into the hydrophobic core to create transmembrane channels or specific peptide structures (Kumar et al. 2018; Yeaman and Yount 2003). Some of the examples of AMPs that act like the model are cecropin (Kumar et al. 2018; Sitaram and Nagaraj 1999), indolicidin (Kumar et al. 2018; Rozek et al. 2000), aurein 1.2 (Kumar et al. 2018; Fernandez et al. 2012), and LL-37 (Kumar et al. 2018; Shai 2002).

17.2.2.2 Direct Killing: Non-membrane Targeting Mechanism of Action

After entering the cytoplasm via direct penetration or endocytosis, AMPs will identify their surroundings and later act on the target cells (Huan et al. 2020). Inhibition of AMPs on the target cell can be categorized into protein biosynthesis, nucleic acid biosynthesis, protease activity, and cell division (Huan et al. 2020).

Transcription, translation, and assembly into functional peptides are affected by the AMPs where they disrupt related enzymes and effector molecules (Huan et al. 2020). Based on a study by Mardirossian et al., 2 AMPs called Bac7 1–35 and Tur1A were observed to exhibit this mechanism of action (Huan et al. 2020; Mardirossian et al. 2014). To inhibit protein translation in *E. coli* and *Thermus thermophilus*, Bac7 135 targets ribosomes while Tur1A inhibits the transitions from the early phase to extension phase. Nucleic acid biosynthesis is inhibited as key enzymes are affected or nucleic acid molecules are degraded by AMPs. This is proven by TFP1-1TC24, an AMP derived from tongues, which gets into the cells after cell membrane disruption and later causes DNA and RNA degradation (Huan et al. 2020; He et al. 2017).

Metabolic activities of microorganisms can be disrupted when protease activity is inhibited (Huan et al. 2020). It was observed that histatin 5 peptide effectively inhibits proteases discharged by the host and bacteria. Another example of AMP that acts in the same manner is cathelicidin-BF peptide, derived from the venom of *Bungarus fasciatus* (Huan et al. 2020; Shu et al. 2019). This AMP could inhibit thrombin-induced platelet aggregation and block receptor 4, activated by protease.

For inhibition of cell division in microbes, one example of AMPs that exhibit said mechanism of action is APP(GLARALTRLLRQLTRQLTRA) (Huan et al. 2020; Li et al. 2016). *C. albicans* were destroyed due to the ability of this AMP to penetrate the cells efficiently, had strong DNA-binding affinity, and induced S-phase arrest in the intracellular environment. AMPs can also target differences in cell wall composition like lipopolysaccharide, lipid A, and mannoproteins (Huan et al. 2020). Teichoic acid and lipoteichoic acid in the cell wall of microbes can also be the target of AMPs, in which AMPs with low toxicity are developed to exhibit this action.

17.2.2.3 Immune Modulation Mechanism of Action

Besides the direct killing mechanism of action via the two methods, AMPs also display various immunomodulatory activities (Hilchie et al. 2013). AMPs can captivate and stimulate immune cells to further destroy the microorganisms and/or regulate inflammation-induced (Kumar et al. 2018; Hilchie et al. 2013; J. Afacan N, T.Y. Yeung A, M. Pena O, E.W. Hancock R, 2012; Mader and Hoskin 2006). AMPs are the first defense mechanism to come in contact with intruding microorganisms as they are generated by immune cells like neutrophils and macrophages (Kumar et al. 2018; Jenssen et al. 2006). Some of the immune responses activated by AMPs are (1) white blood cells activation, attraction, and differentiation, (2) angiogenesis stimulation, (3) lowering expression of proinflammatory chemokines to reduce inflammation, and (4) controlling expression of chemokines and reactive oxygen/nitrogen species

(Hilchie et al. 2013; Afacan et al. 2012; Nijnik and Hancock 2009; Lai and Gallo 2009; Hancock et al. 2012). AMPs recorded to exhibit immunomodulatory activities in vivo according to Hilchie et al. are HNP-1 (Hilchie et al. 2013; Welling et al. 1998; Tani et al. 2000) and IDR-1 (Hilchie et al. 2013; Scott et al. 2007). Defensin HNP-1 produced three disulfide bridges that connect Cys1-Cys6, Cys2-Cys4, and Cys3-Cys5. It was active in both in vitro and in vivo immunomodulatory activities. In in vivo activity, even at 0.4 ng of concentration, HNP-1 could protect mice-induced *Klebsiella pneumoniae* and *S. aureus* (Hilchie et al. 2013; Welling et al. 1998). Despite being a less effective AMP, HNP-1 is biologically active due to its immune-cell function manipulation, direct antimicrobial activities, or both (Hilchie et al. 2013). Nevertheless, HNP-1 tends to lose its antibacterial properties because of several conditions; monovalent and divalent cations, glycosaminoglycans, and serum (Hilchie et al. 2013); thus, this proves that HNP-1 is mostly biologically active due to its immunomodulatory activities. IDR-1, on the other hand, was inactive in in vitro activity but could protect mice-induced Gram-negative and Gram-positive bacteria; methicillin-resistant *S. aureus*, vancomycin-resistant *Enterococcus,* and *Salmonella typhimurium* (Hilchie et al. 2013; Scott et al. 2007).

17.2.3 The Applications of AMP

As diverse as the sources and mechanism of action of the AMPs, there are various applications of AMPs to this date, including in medicine and healthcare settings, food industry, agriculture, aquaculture, and cosmetics industry. Applications of AMPs are summarized in Fig. 17.3.

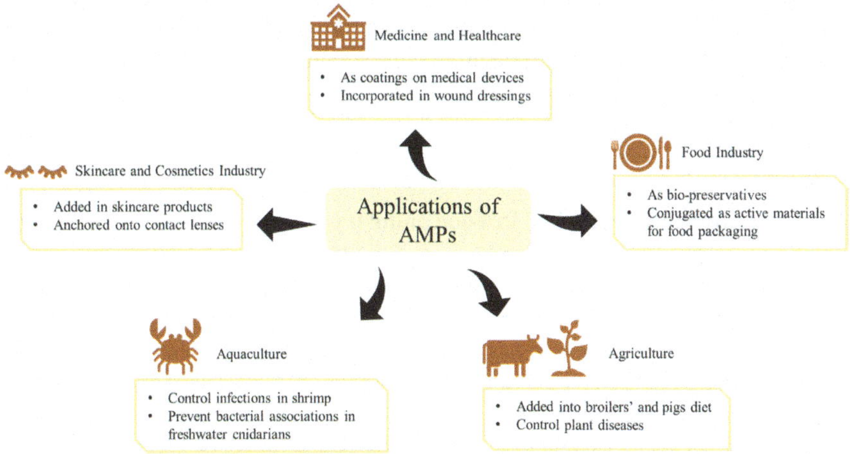

Fig. 17.3 Applications of AMPs in various industries

17.2.3.1 Application of AMP in Medicine and Healthcare Settings

In healthcare settings, biofilms caused by microorganisms can be formed on the surfaces of devices such as catheters, orthopedic implants, ventilators, heart valves, stents, shunts, and contact lenses (Kazemzadeh-Narbat et al. 2021). Medical devices such as implants can be coated with AMPs due to their efficacy even at low concentrations and functionalized surfaces of these devices can release AMPs locally in situ. In this case, polymers are commonly used in conjugating AMPs to an immobilized and functionalized polymer, where AMPs can be specifically released to coat the implant surface. AMPs are susceptible to proteolytic enzymes as they can hydrolyze peptide bonds and alter the structure, which eventually reduces the antimicrobial activity of the AMPs. Thus, liposomes can be used to encapsulate AMPs to protect their functions and properties. Another coating type of biomedical device used was surface tethering on titanium surface in 2016 and 2017 (Nie et al. 2016, 2017; Riool et al. 2017). Bacitracin peptide was coupled onto medical devices with titanium material and could reduce *S. aureus* adhesion in vitro and implant and tissue colonization by the bacteria in vivo.

AMPs conjugated with polymers are also used in wound treatment. To promote wound healing, dressing with profound properties must be used (Lin et al. 2019). Natural polymer materials are used in wound healing due to their good biocompatibility and biodegradability (Lin et al. 2019; Asghari et al. 2016). A study by Lin et al. in 2019 has conjugated AMP Tet213 onto a natural composite substrate consisting of alginate (ALG), hyaluronic acid (HA), and collagen (COL) through chemicals which are (1-ethyl-3-(3-dimethylaminopropyl)-carbodiimide (EDC) and Nhydroxysuccinimide (NHS) cross-linking method (Lin et al. 2019). Amide bonds between the AMP and the polymers resulted in fine-tuning of the wound dressing properties such as porosity, swelling profile, tensile strength, elongation at break, and biodegradability in vitro. The AMP-polymers composite effectively treated rats infected with *E. coli* and *S. aureus*, where these bacteria were killed in the infected wounds. As release of Tet213 was in good and sustainable manner, rapid healing of the wounds was observed. The wound dressing was also able to regulate inflammation, improved the deposition of collagen also enhance angiogenesis.

17.2.3.2 Application of AMP in Food Industry

Conventionally, the food industry utilizes chemicals to preserve food products to extend their shelf lives. However, it is noted that using chemical preservatives poses several health threats to humans and affects the quality and safety of the food itself. Hence, natural preservatives have been developed over the past few years which include the usage of AMPs. AMPs can preserve food without affecting its quality and safety, consequently extending its shelf-life (Rai et al. 2016; Wang et al. 2016; Gálvez et al. 2014; Song et al. 2014).

Bacteriocin synthesized by lactic acid bacteria (LAB) is widely used as a biopreservative in the food industry (Rai et al. 2016). According to Galvez et al., the advantages of this AMP are; food-containing bacteriocin is safe to consume as protease enzymes rapidly digest bacteriocin in gastrointestinal tract, also active against pathogens (Elayaraja et al. 2014), non-toxic for eukaryotic cells, efficient in eradicating food-spoiling microbes as well as good resistance toward pH and heat (Rai et al. 2016; Gálvez et al. 2014). There are three main methods used to incorporate the usage of bacteriocin in the food industry: (1) production of bacteriocin through bacteria inoculation in food, (2) addition of purified or semi-purified bacteriocin as preservatives, and (3) use of a fermented product with bacteriocin-producing strain in food process (Jones et al. 2005; Chen and Hoover 2003; Müller-Auffermann et al. 2015).

Nisin is one example of bacteriocin used in food preservation as it is active against Gram-positive bacteria even though it shows little or no effect of antimicrobial activity on Gram-negative bacteria, yeasts, or fungi (Müller-Auffermann et al. 2015; Gharsallaoui et al. 2016). Nisin acts on the cell membranes of the microorganisms by binding to their anionic phospholipids (including lipid II) and inserted to the cell membranes. Biosynthetic processes in Gram-positive bacteria are halted which eventually kills them. However, nisin is ineffective on Gram-negative bacteria as it is impermeable to their cell walls (Müller-Auffermann et al. 2015; Han 2020). In dairy products such as cheese, nisin is used to fight against *Lactobacillus* spp, *Listeria monocytogenes*, and *S. aureus,* which cause flavor change and gas production in the cheese (Rai et al. 2016; Upendra et al. 2016). As cheese of different forms is considered processed food, it is prone to spoilage by *Clostridium* spp. due to its high pH and a very low redox potential (Rai et al. 2016; D'Amato and Sinigaglia 2010; Suganthi et al. 2012). In another study in 2013, Zohri et al. have discovered that nisin loaded with chitosan/alginate nanoparticles could preserve ultrafiltered (UF) Feta cheese (Rai et al. 2016; Zohri et al. 2013). It had greater antibacterial strength and better features compared to free nisin. Nisin-loaded nanoparticles had decreased up to fivefold of *S. aureus* and sevenfold of *L. monocytogenes* populations. There were also improvements in the sensory acceptance and physicochemical properties of the tested cheese treated with nisin-encapsulated nanoparticles.

Nisin is also used to preserve canned food products which are potentially spoiled by heat-resistant microorganisms (Rai et al. 2016). Other than pasteurization treatments, nisin is effective in eliminating thermophilic spores that are produced by the microbes. It was proven that the addition of nisin into canned vegetables at room temperature can extend their shelf-lives (Rai et al. 2016; Suganthi et al. 2012). Nisin, nisin-pediocin, or enterocin AS-48 addition into cooked or canned vegetables can hinder growth of endospore and enterotoxins production during storage, also improving thermal treatments against the endospores (Gálvez et al. 2014, 2008; Thomas et al. 2000; López Cabo et al. 2009; Abriouel et al. 2010). In 2007, a study was conducted to test enterocin AS-48 against *Bacillus cereus* LWL1, in which the endospore-forming bacteria was inhibited in all vegetable products tested for up to 30 days (Gálvez et al. 2014; Grande et al. 2007). Enterocin AS-48 with concentration up to 50 μg/mL were also able to exhibit antimicrobial activity against *Bacillus* species and *Paenibacillus* species isolated from purees. In 2010, enterocin EJ97 derived from *Enterococcus faecalis* EJ97 was tested to inhibit *G. stearothermophilus* cells in canned foods for 30 days at 45 °C (Gálvez et al. 2014; Viedma et al. 2009). The microbe was inactivated at a concentration of 7 μg/g. Concurrent application of EJ97 with optimum concentration and heat treatments of 90 and 95 °C on the endospores have increased the antimicrobial effects against the vegetative cells and spores of *G. stearothermophilus*.

In food packaging, AMPs are also used to ensure food quality is preserved and safe for consumption, with extended shelf-life. Consumers have been demanding for better food packaging to provide food safety, prolonged shelf-life, easy handling, and protection from physical hazard (Liu et al. 2021). In context of this, active materials are used for food packaging to destroy microorganisms or inhibit their growth on the food surface or in the surrounding (Agrillo et al. 2019). Polymers loaded with antimicrobial agents or bioactive compounds are chosen due to their excellent physicochemical properties, efficiency, resistance, and long-lasting. Among the wide range of antimicrobial polymers, plastics exhibit great benefits as they are low in cost, easy to use, and abundantly available. Hence, the incorporation of AMPs via covalent or physical binding into these polymers has been studied extensively (Agrillo et al. 2019; Sobczak et al. 2013; Irkin and Esmer 2015).

In 2019, polyethylene terephthalate (PET) material was pre-activated via cold oxygen plasma technique, before being functionalized with 1018K6 peptide to test its efficacy as antimicrobial packaging (Agrillo et al. 2019). When tested using mozzarella cheese as the food sample, aerobic plate count (APC), yeast, and mold counts were decreased after 24 h. 1018K6 was seen to possess antimicrobial properties even after conjugation with PET. This antimicrobial polymer was also active against *Listeria* biofilms.

Meat is a perishable food, thus prone to microbial spoilage and lipid oxidation, due to its high contents of fatty acids, nutrients, and vitamins (Liu et al. 2021). These microbes can affect its physical and biological properties such as color, flavor, texture, odor, and pH imbalance, leading to its deterioration. A study by Gogliettino et al. in 2020 has discovered that PET functionalized with MTP1, a mitochondrial-targeting peptide, reduced microbes causing spoilage, and extended the shelf-life of ricotta cheese and buffalo meat (Gogliettino et al. 2020). MTP1 characterized from a study (Gogliettino et al. 2020; Palmieri et al. 2016) has been functionalized with PET to create an active food packaging to retain the nutritional values and prolong the shelf-life of the food. Another study in the same year was conducted to develop an AMP derived from a fish (*turbot viscera hydrolysate*) (Bi et al. 2020). Several peptides were identified by ultraperformance liquid chromatography-quadrupole time-of-flight mass spectroscopy, and their properties were studied. Among the peptides, Sm-A1 was very active against Gram-positive and Gram-negative bacteria by disrupting the integrity of their cell membrane. To enhance its antibacterial activity and efficiency in eradicating biofilm formation, Sm-A1 was incorporated into hydroxyl-rich polyvinyl alcohol (PVA)/chitosan hydrogel. This AMP-hydrogel was able to preserve salmon muscle from contamination caused by microorganisms and texture deterioration.

17.2.3.3 Application of AMP in Agriculture

AMPs derived from insects are stable against heat and do not threaten eukaryotic cells (Józefiak and Engberg 2017). These AMPs can be used as single peptides, complex of different AMPs, or active fraction of insect proteins in livestock nutrition. Insect AMPs are mostly derived from their larval stage (Józefiak and Engberg 2017). Genes encoding the peptides are developed after a septic injury and their expression goes on for 3 days. In adult insects, AMPs are produced by the fat body cells and various epithelia, which are then secreted into the hemolymph at concentration up to 0.5 mM. It is mentioned that the AMPs' concentration is higher than what is required to kill most microorganisms in vitro (Park et al. 2015; Józefiak and Engberg 2017; Bulet and Stocklin 2005; Choi et al. 2012). Also, termicin, AMP that has antifungal properties is found in hemocyte granules and salivary glands of termite *Pseudacanthotermes spiniger* (Isoptera).

When synthetic cecropin is added in broilers diets in a dose-dependent manner, it can decrease aerobic bacterial counts in their jejunal and caecal digesta and increase intestinal villus height in their duodenum (Józefiak and Engberg 2017; Wen and He 2012). Like in human foods, nisin is also used in animal feed as an antibacterial additive, where consumption of nisin helps to reduce *Bacteroides* and *Enterobacteriaceae* in ileal digesta of broilers (Józefiak and Engberg 2017; Józefiak et al. 2013; Kierończyk et al. 2016). An AMP derived from bovine milk known as lactoferrin has been used as dietary supplement for pigs. Lactoferrin stimulates active peptides of the animals' gastric digestion; it reaches the lower gastrointestinal tract and applies its effects. In pigs, lactoferrin improves their growth, reduces diarrhea, lessens pathogens counts such as *E. coli*, increases number of beneficial bacteria as *Lactobacilli* and *Bifidobacteria* (Józefiak and Engberg 2017; Yoon ct al. 2012; Xiao et al. 2015), and increases height of the villus and depth of crypt in their jejunum and ileum (Józefiak and Engberg 2017; Xiao et al. 2015).

To accommodate for global food security and prevention of crops loss due to plant pathogens, research has been made in agriculture to develop non-toxic and non-polluting treatments (Datta et al. 2015). A study was conducted in 2015 to investigate the application of AMP derived from an active fragment of Dengue virus fusion peptide to control plant disease (Datta et al. 2015). VG16KRKP, an analogue of VG16, was active against Gram-negative E. coli, plant pathogens X. oryzae and X. campestris, and human fungal pathogens *C. albicans* and *C. grubii*. Disease progression in rice and cabbage was studied via application of peptide-treated pathogens, which effectively prevented the disease.

17.2.3.4 Application of AMP in Aquaculture

Marine invertebrates include arthropods, mollusks, and cnidarians (Destoumieux-Garzón et al. 2016). Microbial communities in the aquatic environment can cause several effects to the marine invertebrates such as infectious diseases; temperature-dependent vibriosis; and polymicrobial diseases (Destoumieux-Garzón et al. 2016;

Ben-Haim et al. 2003; Fukui et al. 2010; Petton et al. 2015; Barneah et al. 2007). In marine arthropods, extreme cultural practices can lead to diseases associated with penaeid shrimp. For instance, an outbreak of shrimp disease in 2010 was caused by *Vibrio parahaemolyticus*, which obtained a virulent plasmid containing bacterial toxin. This toxin was observed to be as toxic as *Bacillus* Cry toxin that can kill insects (Destoumieux-Garzón et al. 2016; Lee et al. 2015). While for mollusks or shellfish, exploitation and ecosystem health are the main factors of the emergence of infections. Oysters are mainly infected by *vibrio* strains and have affected the USA, Japan, Australia, and Western Europe for many decades.

Marine invertebrates can survive in the complex marine environment even though they do not possess immunoglobulin, complement, and immune system components (Wu et al. 2021). This is most probably due to their immune defenses (Wu et al. 2021). AMPs derived from the bodies of marine invertebrates are among the best immune effectors of these organisms (Destoumieux-Garzón et al. 2016). They are cationic, hydrophobic and able to target specific cell walls and membranes of the pathogenic microorganisms (Destoumieux-Garzón et al. 2016; Schmitt et al. 2016). AMPs in marine invertebrates control infections in their hosts through RNA interference, and this was shown in the eradication of bacteria in shrimp hemolymph by Type I crustins from *Marsupenaeus japonicus* (Destoumieux-Garzón et al. 2016; Yang et al. 2015; Liu et al. 2015). AMPs are also active against microbiota I marine invertebrates through gene silencing, in which ALFs are active against microbial communities present in shrimp (Destoumieux-Garzón et al. 2016; Ponprateep et al. 2012; Wang et al. 2014). Other than that, an AMP called armicins help to protect freshwater cnidarians from bacterial associations (Destoumieux-Garzón et al. 2016; Franzenburg et al. 2013).

17.2.3.5 Application of AMPs in Skincare and Cosmetics Industry

AMPs are naturally produced by human skin and further application of AMPs can treat skin diseases besides eliminating skin pathogens to improve skin health (Rahnamaeian and Vilcinskas 2015). AMPs are often formulated in ointments, lotions, shampoos, creams, or wound dressings to fight against pathogens on the skin and prevent the complexities of anti-infectives. *C. albicans* can result in skin infections and atopic eczema, where neutrophils are accumulated in the inflammatory responses and enzymatic proteins are secreted. High-molecular-weight-kininogen (HK) is present and causes antifungal AMPs to protect human skin against pathogens. Therapeutic application to eliminate yeast-causing skin infections includes formulations of antifungal and anti-inflammatory agents such as topical azole-type antifungal agents. *Malassezia furfur* a lipophilic yeast can cause recalcitrant skin conditions, and cathelicidin peptide is active against the pathogen (Rahnamaeian and Vilcinskas 2015; López-García et al. 2006).

Besides protecting human skin from diseases, AMPs are also used to eliminate bacterial adhesion in contact lenses (Salvagni et al. 2020). Eye microbial infection, such as severe corneal infection, is often associated with contact lens usage, which may lead to vision loss. Conjugation of two cationic AMPs; melittin and protamine (linked in a single molecule) with contact lenses were able to immobilize the AMP melamine and found to be active against bacteria (Salvagni et al. 2020; Yasir et al. 2018; Chen et al. 2009; Kampshoff et al. 2019). Also, a shorter peptide KNKRKR-RRRRRGGRRRR named as mel4 was observed to possess similar antibacterial properties that can be functionalized into commercial contact lenses (Salvagni et al. 2020; Dutta et al. 2018). Up to 80% of bacterial adhesion caused by *P. aeruginosa* and *S. aureus* was reduced when melimine was used in contact lenses. As melimine is not cytotoxic, contact lenses anchored with this AMP were safe when tested on rabbits and humans (Salvagni et al. 2020; Dutta et al. 2016, 2014). They did not cause ocular irritation and had the same properties of wettability, surface deposits, and lens fitting on rabbits, compared to contact lenses without the AMP. The antimicrobial activity of the contact lenses remained even after being used (Salvagni et al. 2020; Dutta et al. 2014). In 2020, a study by Salvagni et al., a short and an ultrashort AMPs; LKKLLK-LLKKLLKL (LK) and IRIRIRIR (IR) were anchored onto hydrogel contact lenses using 1,4-butanediol diglycidyl ether as a linker at room temperature and 7.4 of pH (Salvagni et al. 2020). 100% and 98% reduction of bacterial adhesion of S. aureus and P. aeruginosa were recorded when these peptides were introduced into the contact lenses. Also, it was observed that optical properties of the AMP-anchored contact lenses, such as transmittance and refractive index were not altered. All in all, conjugating AMPs onto the contact lenses could reduce corneal infections caused by bacteria without changing their properties or purposes.

17.3 The Clinical Potential of Antimicrobial Peptides

Among many factors that contribute toward antimicrobial resistance, main factor causing it to be a serious global issue is the misuse of antibiotics and unavailability of effective antibiotics in the market (Alanis 2005; Browne et al. 2020). Naturally occurring class of AMPs are being researched globally to limit the spread of pathogens that are resistance toward AMPs. Some available antibiotics in therapeutic applications that contain AMPs include bacitracin, dalbavancin, daptomycin, and many more (Browne et al. 2020). In a study to treat skin infections, bacitracin peptide was admitted through 3 routes: topical, ophthalmic, and intramuscular. The mechanism of action of this AMP is by inhibiting cell wall synthesis, and it was able to kill Gram-positive bacteria (Browne et al. 2020; Nguyen et al. 2021). AMP called dalbavancin, derived from teicoplanin derivative, has a similar target organism and mechanism of action like bacitracin (Browne et al. 2020; Chen et al. 2007; Malabarba and Goldstein 2005). Dalbavancin was also used to treat skin infections and administered intravenously.

It was observed that many developed AMPs have encountered failure throughout different stages of clinical trials (Browne et al. 2020). This is due to the complicated nature of human diseases, unlike in animal models or in vitro (Browne et al. 2020; Bart van der Worp et al. 2010). Other than that, AMPs produced do not possess improved activity than current antibiotics available in the market, and thus, hesitation is present to fulfill the need for a particular indication in clinical trials. The toxicity level of these AMPs toward hosts is also to be considered for developing AMPs with safe and better properties. Based on much research to improve and enhance the efficacy of AMPs, chemical modifications of AMPs show definite possibilities (Browne et al. 2020). In clinical trials, many compounds undergo chemical modification to enhance their ability in drug delivery. Digital libraries and modeling software are also seen to facilitate the development of better AMPs (Browne et al. 2020).

17.4 Conclusion

AMPs exhibit various advantages over conventional antibiotics, including the potency, broad-spectrum activity, diversity of sources found in nature, absence of quick development of resistance, low accumulation in tissue, and rapid killing activity. Nevertheless, most of the AMPs with amphipathic structure can incorporate into the anionic cell wall and phospholipid membranes of microorganisms and induce membrane lysis. Thus, they can be exploited as alternative antimicrobial therapeutic agents for microbial infections and have gained significant interest in recent years. More AMPs are being discovered over the years, and these AMPs are applied in medicine and healthcare settings, food industry, agriculture, aquaculture, and cosmetics industry. This review provides a comprehensive insight into the source, application, and potential of AMPs. In healthcare settings, AMPs can be used to coat medical devices due to their efficacy and functionalized surfaces of the devices using polymers can release AMPs locally in situ. AMP-polymers composite Tet213-immobilized ALG/HA/COL dressing effectively killed *E. coli* and *S. aureus* in infected wounds (Lin et al. 2019). In the food industry, chemical preservatives are substituted with natural preservatives that use AMPs without affecting food quality and safety, also extending its shelf-life. In the cosmetics industry, AMPs formulated in ointments, lotions, shampoos, creams, or wound dressings are active against pathogens on the skin. AMPs also eliminate bacterial adhesion in contact lenses that may lead to corneal infections. Indeed, AMPs have shown great potential applications in various industries. On these bases, it is also important to improve the metabolic and chemical stability of these AMPs. The development of novel strategies and formulation is critical to improve the delivery of these AMPs as well as their bioavailability and activity.

Acknowledgements This research was supported by the Malaysia Ministry of Higher Education under the Fundamental Research Grant scheme (FRGS/1/2020/STG01/UNIKL/02/1).

References

Abriouel H, Lucas R, Ben ON et al (2010) Potential applications of the cyclic peptide enterocin AS-48 in the preservation of vegetable foods and beverages. Probiotics Antimicrob Proteins 2:77–89

Adebisi YA, Jumoke AA, Melody O et al (2021) COVID-19 and antimicrobial resistance: a review. Infect Dis Res Treat 14:1–9

J Afacan N, TY Yeung A, M Pena O, EW Hancock R (2012) Therapeutic potential of host defense peptides in antibiotic-resistant infections. Curr Pharm Des 18:807–819

Agrillo B, Balestrieri M, Gogliettino M et al (2019) Functionalized polymeric materials with bio-derived antimicrobial peptides for "active" packaging. Int J Mol Sci 20:1–13

Akpan MR, Isemin NU, Udoh AE, Ashiru-Oredope D (2020) Implementation of antimicrobial stewardship programmes in African countries: a systematic literature review. J Glob Antimicrob Resist 22:317–324

Alanis AJ (2005) Resistance to antibiotics: are we in the post-antibiotic era? Arch Med Res 36:697–705

Andersson DI, Hughes D, Kubicek-Sutherland JZ (2016) Mechanisms and consequences of bacterial resistance to antimicrobial peptides. Drug Resist Updat 26:43–57

Asghari F, Samiei M, Adibkia K et al (2016) Biodegradable and biocompatible polymers for tissue engineering application: a review. Artif Cells Nanomed Biotechnol 45:185–192

Bagley CP (2014) Potential role of synthetic antimicrobial peptides in animal health to combat growing concerns of antibiotic resistance -a review. Wyno Acad J Agric Sci 2:19–28

Barneah O, Ben-Dov E, Kramarsky-Winter E, Kushmaro A (2007) Characterization of black band disease in Red Sea stony corals. Environ Microbiol 9:1995–2006

Bart van der Worp H, Howells DW, Sena ES et al (2010) Can animal models of disease reliably inform human studies? PLOS Med 7:e1000245

Ben-Haim Y, Thompson FL, Thompson CC et al (2003) Vibrio coralliilyticus sp. nov., a temperature-dependent pathogen of the coral *Pocillopora damicornis*. Int J Syst Evol Microbiol 53:309–315

Berglund NA, Piggot TJ, Jefferies D et al (2015) Interaction of the antimicrobial peptide polymyxin B1 with both membranes of E. coli: a molecular dynamics study. PLoS Comput Biol 11:1–18

Bi J, Tian C, Jiang J et al (2020) Antibacterial activity and potential application in food packaging of peptides derived from turbot viscera hydrolysate. J Agric Food Chem 68:9968–9977

Boparai JK, Sharma PK (2019) Mini review on antimicrobial peptides, sources, mechanism and recent applications. Protein Pept Lett 27:4–16

Breukink E, De Kruijff B (1999) The lantibiotic nisin, a special case or not? Biochim Biophys Acta - Biomembr 1462:223–234

Brogden KA (2005) Antimicrobial peptides: pore formers or metabolic inhibitors in bacteria? Nat Rev Microbiol 3:238–250

Browne K, Chakraborty S, Chen R et al (2020) A new era of antibiotics: the clinical potential of antimicrobial peptides. Int J Mol Sci 21:1–23

Bulet P, Stocklin R (2005) Insect antimicrobial peptides: structures, properties and gene regulation. Protein Pept Lett 12:3–11

Chen H, Hoover DG (2003) Bacteriocins and their food applications. Compr Rev Food Sci Food Saf. https://doi.org/10.1111/j.1541-4337.2003.tb00016.x

Chen AY, Zervos MJ, Vazquez JA (2007) Dalbavancin: a novel antimicrobial. Int J Clin Pract 61:853–863

Chen R, Cole N, Willcox MDP et al (2009) Synthesis, characterization and in vitro activity of a surface-attached antimicrobial cationic peptide. Biofouling 25:517–524

Choi WH, Yun JH, Chu JP, Chu KB (2012) Antibacterial effect of extracts of Hermetia illucens (Diptera: Stratiomyidae) larvae against Gram-negative bacteria. Entomol Res 42:219–226

Cuesta SA, Reinoso C, Morales F et al (2021) Novel antimicrobial cruzioseptin peptides extracted from the splendid leaf frog, Cruziohyla calcarifer. Amino Acids 53:853–868

D'Amato D, Sinigaglia M (2010) Antimicrobial agents of microbial origin: Nisin. Appl Altern Food-Preserv Technol Enhanc Food Saf Stab. https://doi.org/10.2174/978160805096311001010083

Datta A, Ghosh A, Airoldi C et al (2015) Antimicrobial peptides: Insights into membrane permeabilization, lipopolysaccharide fragmentation and application in plant disease control. Sci Rep 5:1–15

Destoumieux-Garzón D, Rosa RD, Schmitt P et al (2016) Antimicrobial peptides in marine invertebrate health and disease. Philos Trans R Soc B Biol Sci. https://doi.org/10.1098/rstb.2015.0300

Dutta D, Ozkan J, Willcox MDP (2014) Biocompatibility of antimicrobial melimine lenses: rabbit and human studies. Optom vis Sci 91:570–581

Dutta D, Vijay AK, Kumar N, Willcox MDP (2016) Melimine-coated antimicrobial contact lenses reduce microbial keratitis in an animal model. Invest Ophthalmol vis Sci 57:5616–5624

Dutta D, Kamphuis B, Ozcelik B et al (2018) Development of silicone hydrogel antimicrobial contact lenses with mel4 peptide coating. Optom vis Sci 95:937–946

Ehrenstein G, Lecar H (1977) Electrically gated ionic channels in lipid bilayers. Q Rev Biophys 10:1–34

Elayaraja S, Annamalai N, Mayavu P, Balasubramanian T (2014) Production, purification and characterization of bacteriocin from Lactobacillus murinus AU06 and its broad antibacterial spectrum. Asian Pac J Trop Biomed 4:S305–S311

Epand RM, Walker C, Epand RF, Magarvey NA (2016) Molecular mechanisms of membrane targeting antibiotics. Biochim Biophys Acta Biomembr 1858:980–987

Fernandez DI, Le Brun AP, Whitwell TC et al (2012) The antimicrobial peptide aurein 1.2 disrupts model membranes via the carpet mechanism. Phys Chem Chem Phys 14:15739–15751

Forde E, Devocelle M (2015) Pro-moieties of antimicrobial peptide prodrugs. Molecules 20:1210–1227

Franzenburg S, Walter J, Künzel S et al (2013) Distinct antimicrobial peptide expression determines host species-specific bacterial associations. Proc Natl Acad Sci USA 110:E3730–E3738

Fukui Y, Saitoh SI, Sawabe T (2010) Environmental determinants correlated to *Vibrio harveyi*-mediated death of marine gastropods. Environ Microbiol 12:124–133

Gálvez A, López RL, Abriouel H et al (2008) Application of bacteriocins in the control of foodborne pathogenic and spoilage bacteria. Crit Rev Biotechnol 28:125–152

Gálvez A, José Grande Burgos M, López RL, Pérez Pulido R (2014) Food biopreservation. Springer, New York

Gharsallaoui A, Oulahal N, Joly C, Degraeve P (2016) Nisin as a food preservative: part 1: physicochemical properties, antimicrobial activity, and main uses. Crit Rev Food Sci Nutr 56:1262–1274

Ghosh R, Maji UK, Bhattacharya R, Sinha AK (2012) The role of dermcidin isoform 2: a two-faceted atherosclerotic risk factor for coronary artery disease and the effect of acetyl salicylic acid on it. Thrombosis 2012:1–9

Gogliettino M, Balestrieri M, Ambrosio RL et al (2020) Extending the shelf-life of meat and dairy products via PET-modified packaging activated with the antimicrobial peptide MTP1. Front Microbiol 10:1–11

Grande MJ, López RL, Abriouel H et al (2007) Treatment of vegetable sauces with enterocin AS-48 alone or in combination with phenolic compounds to inhibit proliferation of Staphylococcus aureus. J Food Prot 70:405–411

Gschwandtner M, Zhong S, Tschachler A et al (2014) Fetal human keratinocytes produce large amounts of antimicrobial peptides: involvement of histone-methylation processes. J Invest Dermatol 134:2192–2201

Hammami R, Ben Hamida J, Vergoten G, Fliss I (2009) PhytAMP: a database dedicated to antimicrobial plant peptides. Nucleic Acids Res 37:963–968

Han J (2020) What is nisin preparation (E234) in food? Uses, safety, side effects. In: FoodAdditives.net. https://foodadditives.net/preservatives/nisin/. Accessed 24 Jan 2022

Hancock REW, Patrzykat A (2002) Clinical development of cationic antimicrobial peptides: from natural to novel antibiotics. Curr Drug Targets Infect Disord 2:79–83
Hancock REW, Nijnik A, Philpott DJ (2012) Modulating immunity as a therapy for bacterial infections. Nat Rev Microbiol 10:243–254
He SW, Zhang J, Li NQ et al (2017) A TFPI-1 peptide that induces degradation of bacterial nucleic acids, and inhibits bacterial and viral infection in half-smooth tongue sole, Cynoglossus semilaevis. Fish Shellfish Immunol 60:466–473
Hilchie AL, Wuerth K, Hancock REW (2013) Immune modulation by multifaceted cationic host defense (antimicrobial) peptides. Nat Chem Biol 9:761–768
Hollmann A, Martinez M, Maturana P et al (2018) Antimicrobial peptides: Interaction with model and biological membranes and synergism with chemical antibiotics. Front Chem 6:1–13
Holt E (2018) Anti-microbial resistance on the rise. In: New straits times press Bhd. https://www.nst.com.my/opinion/columnists/2018/12/438444/anti-microbial-resistance-rise. Accessed 12 Jan 2022
Huan Y, Kong Q, Mou H, Yi H (2020) Antimicrobial peptides: classification, design, application and research progress in multiple fields. Front Microbiol 11:1–21
Irkin R, Esmer OK (2015) Novel food packaging systems with natural antimicrobial agents. J Food Sci Technol 52:6095–6111
Jamali H, Krylova K, Dozois CM (2019) The 100 top-cited scientific papers focused on the topic of bacteriocins. Int J Pept Res Ther 25:933–939
Jenssen H, Hamill P, Hancock REW (2006) Peptide antimicrobial agents. Clin Microbiol Rev 19:491–511. https://doi.org/10.1128/CMR.00056-05
Jones E, Salin V, Williams GW (2005) Nisin and the market for commercial bacteriocins. Texas
Józefiak A, Engberg RM (2017) Insect proteins as a potential source of antimicrobial peptides in livestock production. A Review. J Anim Feed Sci 26:87–99
Józefiak D, Kierończyk B, Juśkiewicz J et al (2013) Dietary nisin modulates the gastrointestinal microbial ecology and enhances growth performance of the broiler chickens. PLoS ONE 8:e85347
Kampshoff F, Willcox MDP, Dutta D (2019) A pilot study of the synergy between two antimicrobial peptides and two common antibiotics. Antibiot 8:60
Kazemzadeh-Narbat M, Cheng H, Chabok R et al (2021) Strategies for antimicrobial peptide coatings on medical devices: a review and regulatory science perspective. Crit Rev Biotechnol 41:94–120
Kerenga BK, McKenna JA, Harvey PJ et al (2019) Salt-tolerant antifungal and antibacterial activities of the corn defensin ZmD32. Front Microbiol 10:1–13
Kierończyk B, Pruszyńska-Oszmałek E, Świątkiewicz S et al (2016) The nisin improves broiler chicken growth performance and interacts with salinomycin in terms of gastrointestinal tract microbiota composition. J Anim Feed Sci 25:309–316
Koprivnjak T, Peschel A (2011) Bacterial resistance mechanisms against host defense peptides. Cell Mol Life Sci 68:2243–2254
Kumar P, Kizhakkedathu JN, Straus SK (2018) Antimicrobial peptides: diversity, mechanism of action and strategies to improve the activity and biocompatibility in vivo. Biomolecules. https://doi.org/10.3390/biom8010004
Lai Y, Gallo RL (2009) AMPed up immunity: how antimicrobial peptides have multiple roles in immune defense. Trends Immunol 30:131–141
Lazdunski CJ (1988) Pore-forming colicins: synthesis, extracellular release, mode of action, immunity. Biochimie 70:1291–1296
Lee CT, Chen IT, Yang YT, Ko TP, Huang YT, Huang JY, Huang MF, Lin SJ, Chen CY, Lin SS, Lightner DV(2015) The opportunistic marine pathogen Vibrio parahaemolyticus becomes virulent by acquiring a plasmid that expresses a deadly toxin. Proc Natl Acad Sci 112:E5445–E5445
Li L, Sun J, Xia S et al (2016) Mechanism of antifungal activity of antimicrobial peptide APP, a cell-penetrating peptide derivative, against Candida albicans: intracellular DNA binding and cell cycle arrest. Appl Microbiol Biotechnol 100:3245–3253

Li J, Hu S, Jian W et al (2021) Plant antimicrobial peptides: structures, functions, and applications. Bot Stud. https://doi.org/10.1186/s40529-021-00312-x
Lin Z, Wu T, Wang W et al (2019) Biofunctions of antimicrobial peptide-conjugated alginate/hyaluronic acid/collagen wound dressings promote wound healing of a mixed-bacteria-infected wound. Int J Biol Macromol 140:330–342
Liu N, Lan JF, Sun JJ et al (2015) A novel crustin from Marsupenaeus japonicus promotes hemocyte phagocytosis. Dev Comp Immunol 49:313–322
Liu Y, Sameen DE, Ahmed S et al (2021) Antimicrobial peptides and their application in food packaging. Trends Food Sci Technol 112:471–483
Lohner K, Prossnigg F (2009) Biological activity and structural aspects of PGLa interaction with membrane mimetic systems. Biochim Biophys Acta Biomembr 1788:1656–1666
López Cabo M, Torres B, Rodríguez Herrera JJ et al (2009) Application of nisin and pediocin against resistance and germination of Bacillus spores in sous vide products. J Food Prot 72:515–523
López-García B, Lee PHA, Gallo RL (2006) Expression and potential function of cathelicidin antimicrobial peptides in dermatophytosis and tinea versicolor. J Antimicrob Chemother 57:877–882
Mader JS, Hoskin DW (2006) Cationic antimicrobial peptides as novel cytotoxic agents for cancer treatment. Expert Opin Investig Drugs 15:933–946
Makarova O, Johnston P, Rodriguez-Rojas A et al (2018) Genomics of experimental adaptation of Staphylococcus aureus to a natural combination of insect antimicrobial peptides. Sci Rep 8:1–8
Malabarba A, Goldstein BP (2005) Origin, structure, and activity in vitro and in vivo of dalbavancin. J Antimicrob Chemother 55:ii15–ii20
Mangoni ML, Casciaro B (2020) Development of antimicrobial peptides from amphibians. Antibiotics 9:1–4
Mardirossian M, Grzela R, Giglione C et al (2014) The host antimicrobial peptide Bac71-35 binds to bacterial ribosomal proteins and inhibits protein synthesis. Chem Biol 21:1639–1647
Maria-Neto S, De Almeida KC, Macedo MLR, Franco OL (2015) Understanding bacterial resistance to antimicrobial peptides: From the surface to deep inside. Biochim Biophys Acta - Biomembr 1848:3078–3088
Martinez FAC, Balciunas EM, Converti A et al (2013) Bacteriocin production by Bifidobacterium spp: a review. Biotechnol Adv 31:482–488. https://doi.org/10.1016/j.biotechadv.2013.01.010
Moravej H, Moravej Z, Yazdanparast M et al (2018) Antimicrobial peptides: features, action, and their resistance mechanisms in bacteria. Microb Drug Resist 24:747–767
Müller-Auffermann K, Grijalva F, Jacob F, Hutzler M (2015) Nisin and its usage in breweries: a review and discussion. J Inst Brew 121:309–319
Murakami M, Kameda K, Tsumoto H et al (2017) TLN-58, an additional hCAP18 processing form, found in the lesion vesicle of palmoplantar pustulosis in the skin. J Invest Dermatol 137:322–331
Naeemmudeen NM, Mohd Ghazali NAN, Bahari H et al (2021) Trends in antimicrobial resistance in Malaysia. Med J Malaysia 76:698–705
Nayak T, Mandal SM, Neog K, Ghosh AK (2018) Characterization of a gloverin-like antimicrobial peptide isolated from muga silkworm, Antheraea assamensis. Int J Pept Res Ther 24:337–346
Nguyen R, Khanna NR, Safadi AO, Sun Y (2021) Bacitracin topical. StatPearls, Florida
Nie B, Ao H, Zhou J et al (2016) Biofunctionalization of titanium with bacitracin immobilization shows potential for anti-bacteria, osteogenesis and reduction of macrophage inflammation. Colloids Surf B Biointerfaces 145:728–739
Nie B, Ao H, Long T et al (2017) Immobilizing bacitracin on titanium for prophylaxis of infections and for improving osteoinductivity: an in vivo study. Colloids Surf B Biointerfaces 150:183–191
Nijnik A, Hancock R (2009) Host defence peptides: antimicrobial and immunomodulatory activity and potential applications for tackling antibiotic-resistant infections. Emerg Health Threats J 2:1–7
Oren Z, Shai Y (1998) Mode of action of linear amphipathic α-helical antimicrobial peptides. Biopolymers 47:451–463

Ouertani A, Chaabouni I, Mosbah A et al (2018) Two new secreted proteases generate a casein-derived antimicrobial peptide in Bacillus cereus food born isolate leading to bacterial competition in milk. Front Microbiol 9:1–12

Palmieri G, Balestrieri M, Proroga YTR et al (2016) New antimicrobial peptides against foodborne pathogens: from in silico design to experimental evidence. Food Chem 211:546–554

Park SI, Kim JW, Yoe SM (2015) Purification and characterization of a novel antibacterial peptide from black soldier fly (Hermetia illucens) larvae. Dev Comp Immunol 52:98–106

Petton B, Bruto M, James A et al (2015) Crassostrea gigas mortality in France: the usual suspect, a herpes virus, may not be the killer in this polymicrobial opportunistic disease. Front Microbiol 6:686

Pfalzgraff A, Brandenburg K, Weindl G (2018) Antimicrobial peptides and their therapeutic potential for bacterial skin infections and wounds. Front Pharmacol 9:1–23

Ponprateep S, Tharntada S, Somboonwiwat K, Tassanakajon A (2012) Gene silencing reveals a crucial role for anti-lipopolysaccharide factors from Penaeus monodon in the protection against microbial infections. Fish Shellfish Immunol 32:26–34

Prestinaci F, Pezzotti P, Pantosti A (2015) Antimicrobial resistance: a global multifaceted phenomenon. Pathog Glob Health 109:309–318

Rahnamaeian M, Vilcinskas A (2015) Short antimicrobial peptides as cosmetic ingredients to deter dermatological pathogens. Appl Microbiol Biotechnol 99:8847–8855

Rai M, Pandit R, Gaikwad S, Kövics G (2016) Antimicrobial peptides as natural bio-preservative to enhance the shelf-life of food. J Food Sci Technol 53:3381–3394

Rapaport D, Shai Y (1991) Interaction of fluorescently labeled pardaxin and its analogues with lipid bilayers. J Biol Chem 266:23769–23775

Rieg S, Garbe C, Sauer B et al (2004) Dermcidin is constitutively produced by eccrine sweat glands and is not induced in epidermal cells under inflammatory skin conditions. Br J Dermatol 151:534–539

Rieg S, Saborowski V, Kern WV et al (2014) Expression of the sweat-derived innate defence antimicrobial peptide dermcidin is not impaired in Staphylococcus aureus colonization or recurrent skin infections. Clin Exp Dermatol 39:209–212

Rima M, Rima M, Fajloun Z et al (2021) Antimicrobial peptides: a potent alternative to antibiotics. Antibiotics 10:1 15

Riool M, de Breij A, Drijfhout JW, et al (2017) Antimicrobial peptides in biomedical device manufacturing. Front Chem AUG:1–13

Rozek A, Friedrich CL, Hancock REW (2000) Structure of the bovine antimicrobial peptide indolicidin bound to dodecylphosphocholine and sodium dodecyl sulfate micelles. Biochemistry 39:15765–15774

Sahoo A, Swain SS, Behera A et al (2021) Antimicrobial peptides derived from insects offer a novel therapeutic option to combat biofilm: a review. Front Microbiol. https://doi.org/10.3389/fmicb.2021.661195

Salvagni E, García C, Manresa À et al (2020) Short and ultrashort antimicrobial peptides anchored onto soft commercial contact lenses inhibit bacterial adhesion. Colloids Surf B Biointerfaces 196:111283

Schittek B (2012) The multiple facets of dermcidin in cell survival and host defense. J Innate Immun 4:349–360

Schmitt P, Rosa RD, Destoumieux-Garzón D (2016) An intimate link between antimicrobial peptide sequence diversity and binding to essential components of bacterial membranes. Biochim Biophys Acta Biomembr 1858:958–970

Scott MG, Dullaghan E, Mookherjee N et al (2007) An anti-infective peptide that selectively modulates the innate immune response. Nat Biotechnol 25:465–472

Shai Y (2002) Mode of action of membrane active antimicrobial peptides. Biopolym Pept Sci Sect 66:236–248

Shai Y, Bach D, Yanovsky A (1990) Channel formation properties of synthetic pardaxin and analogues. J Biol Chem 265:20202–20209

Shu G, Chen Y, Liu T et al (2019) Antimicrobial peptide cathelicidin-bf inhibits platelet aggregation by blocking protease-activated receptor 4. Int J Pept Res Ther 25:349–358
Sitaram N, Nagaraj R (1999) Interaction of antimicrobial peptides with biological and model membranes: structural and charge requirements for activity. Biochim Biophys Acta Biomembr 1462:29–54
Sobczak M, Debek C, Oledzka E, Kozłowski R (2013) Polymeric systems of antimicrobial peptides-strategies and potential applications. Molecules 18:14122–14137
Song DF, Zhu MY, Gu Q (2014) Purification and characterization of plantaricin ZJ5, a new bacteriocin produced by Lactobacillus plantarum ZJ5. PLoS ONE 9:1–8. https://doi.org/10.1371/journal.pone.0105549
Souza AA, Costa AS, Campos DCO et al (2018) Lipid transfer protein isolated from noni seeds displays antibacterial activity in vitro and improves survival in lethal sepsis induced by CLP in mice. Biochimie 149:9–17
Suganthi V, Selvarajan E, Subathradevi C, Mohanasrinivasan V (2012) Lantibiotic nisin: natural preservative from lactococcus lactis. Int Res J Pharm 3:13–19
Tani K, Murphy WJ, Chertov O et al (2000) Defensins act as potent adjuvant that promote cellular and humoral immune response in mice to a lymphhoma idiotype and carrier antigents. Int Immunol 12:691–700
Thomas LV, Clarkson MR, Delves-Broughton J (2000) Natural food antimicrobial systems. CRC Press LLC, Boca Raton
University of Nebraska Medical Center (2017) AMP database search. In: Univ. Nebraska Med. Cent. https://aps.unmc.edu/database/peptide. Accessed 20 Jan 2022
Upendra RS, Khandelwal P, Jana K et al (2016) Bacteriocin production from indigenous strains of lactic acid bacteria isolated from selected fermented food sources. Int J Pharma Res Health Sci 4:982–990
Viedma PM, Abriouel H, Ben ON et al (2009) Effect of enterocin EJ97 against Geobacillus stearothermophilus vegetative cells and endospores in canned foods and beverages. Eur Food Res Technol 230:513–519
Wang G (2014) Human antimicrobial peptides and proteins. Pharmaceuticals 7:545–594
Wang G (2020) Bioinformatic analysis of 1000 amphibian antimicrobial peptides uncovers multiple length-dependent correlations for peptide design and prediction. Antibiotics 9:1–26
Wang XW, Xu JD, Zhao XF et al (2014) A shrimp C-type lectin inhibits proliferation of the hemolymph microbiota by maintaining the expression of antimicrobial peptides. J Biol Chem 289:11779–11790
Wang S, Zeng X, Yang Q, Qiao S (2016) Antimicrobial peptides as potential alternatives to antibiotics in food animal industry. Int J Mol Sci. https://doi.org/10.3390/ijms17050603
WC W (2010) Describing the mechanism of antimicrobial peptide action with the interfacial activity model. Acs Chem Biol 5:905–917
Welling MM, Hiemstra PS, Van Den Barselaar MT et al (1998) Antibacterial activity of human neutrophil defensins in experimental infections in mice is accompanied by increased leukocyte accumulation. J Clin Invest 102:1583–1590
Wen LF, He JG (2012) Dose–response effects of an antimicrobial peptide, a cecropin hybrid, on growth performance, nutrient utilisation, bacterial counts in the digesta and intestinal morphology in broilers. Br J Nutr 108:1756–1763
World Health Organization ROFS-EA (2018) Situational analysis on antimicrobial resistance in the south-east Asia region. New Delhi, India
Wu Q, Patočka J, Kuča K (2018) Insect antimicrobial peptides, a mini review. Toxins (Basel) 10:1–17
Wu R, Patocka J, Nepovimova E et al (2021) Marine invertebrate peptides: antimicrobial peptides. Front Microbiol 12:1–12
Xiao H, Shao F, Wu M et al (2015) The application of antimicrobial peptides as growth and health promoters for swine. J Anim Sci Biotechnol 6:1–6

Yang HT, Yang MC, Sun JJ et al (2015) Catalase eliminates reactive oxygen species and influences the intestinal microbiota of shrimp. Fish Shellfish Immunol 47:63–73
Yasir M, Dutta D, Willcox MDP (2018) Comparative mode of action of antimicrobial peptide melimine and its derivative Mel4 against Pseudomonas aeruginosa. bioRxiv 450577
Yeaman MR, Yount NY (2003) Mechanisms of antimicrobial peptide action and resistance. Pharmalogical Rev 55:27–55
Yoon JH, Ingale SL, Kim JS et al (2012) Effects of dietary supplementation of antimicrobial peptide-A3 on growth performance, nutrient digestibility, intestinal and fecal microflora and intestinal morphology in weanling pigs. Anim Feed Sci Technol 177:98–107
Yu G, Baeder DY, Regoes RR, Rolff J (2018) Predicting drug resistance evolution: insights from antimicrobial peptides and antibiotics. Proc R Soc B Biol Sci. https://doi.org/10.1098/rspb.2017.2687
Zai Y, Xi X, Ye Z et al (2021) Aggregation and its influence on the bioactivities of a novel antimicrobial peptide, temporin-pf, and its analogues. Int J Mol Sci. https://doi.org/10.3390/ijms22094509
Zheng Z, Tharmalingam N, Liu Q et al (2017) Synergistic efficacy of Aedes aegypti antimicrobial peptide cecropin A2 and tetracycline against Pseudomonas aeruginosa. Antimicrob Agents Chemother. https://doi.org/10.1128/AAC.00686-17
Zhu S, Gao B, Umetsu Y et al (2021) Adaptively evolved human oral actinomyces-sourced defensins show therapeutic potential. EMBO Mol Med. https://doi.org/10.15252/emmm.202114499
Zohri M, Shafiee Alavidjeh M, Mirdamadi SS et al (2013) Nisin-loaded chitosan/alginate nanoparticles: a hopeful hybrid biopreservative. J Food Saf 33:40–49

The manufacturer's authorised representative in the EU is Springer Nature Customer Service Centre GmbH, Europaplatz 3, 69115 Heidelberg, Germany. If you have any concerns regarding our products, please contact ProductSafety@springernature.com

Printed and bound by CPI Group (UK) Ltd, Croydon, CR0 4YY
30/09/2026
02225244-0005